Applications of the Laser

Applications of the Laser

Author:
Leon Goldman, M.D.

Laser Laboratory
University of Cincinnati
Medical Center

and

Department of Dermatology
College of Medicine
University of Cincinnati
Cincinnati, Ohio

published by:

CRC PRESS

CRC Press, Inc., 18901 Cranwood Parkway, Cleveland, Ohio 44128
Formerly The Chemical Rubber Co.

This book represents information obtained from authentic and highly regarded sources. Reprinted material is quoted with permission, and sources are indicated. A wide variety of references are listed. Every reasonable effort has been made to give reliable data and information, but the author and the publisher cannot assume responsibility for the validity of all materials or for the consequences of their use.

International Standard Book Number 0-87819-028-17

Library of Congress Card Number 73-89461

THE AUTHOR

Leon Goldman, M.D., is the director of the Laser Laboratory, University of Cincinnati Medical Center and professor and chairman of the Department of Dermatology, College of Medicine of the University of Cincinnati.

Dr. Goldman graduated from the College of Medicine of the University of Cincinnati in 1929.

Prior to his present appointment, Dr. Goldman was an intern and resident in medicine and later dermatology at the Cincinnati General Hospital. He was also a fellow in dermatology at Dermatologische Klinik, University of Zurich.

Dr. Goldman has filled appointments as co-chairman of First Conference on the Laser, New York Academy of Science, 1965, chairman of Second Conference on the Laser, New York Academy of Science, 1970, and chairman of the Gordon Conference on Lasers. He is also a member of the American National Standards Committee Z-136-Standard for the Safe Use of Lasers and is currently chairman of the Subcommittee on Lasers of the AAMI (Association for the Advancement of Medical Instrumentations).

He is the author of 3 books and has published over 75 articles on lasers.

DEDICATED TO

The John A. Hartford Foundation whose courage, foresight, and interest in the U.S. actually laid the groundwork for comprehensive laser biomedical investigations at a time when only very few seemed interested and ever fewer seemed concerned

and

the pioneers of laser technology throughout the world who have continued to work, have expanded and have shown the present and the great future of lasers making this a new and, hopefully, a better world

and

La Belle, who understood, and John, Bobbie, Steve, Carol, and Ed who gave both bench work and sympathy to the nepotist.

FOREWORD

As Director of a Laser Laboratory for more than 11 years, it is obvious that I should be aware of the impacts of the laser. This is actually literally true because I have been aware of more than a thousand purposeful impacts on my own skin as part of the long continued and pioneering studies on laser safety as regards chronic exposure. In spite of the fact that I have written some little books on laser medicine and biology, mostly inadequate, I still felt that it was necessary to try to assemble mostly superficially, what the laser has actually done, what are its current needs, and what are some of its dreams for the future. My own special fields of interest and work have been in laser safety, biology and medicine, laser biomedical engineering, law and order, social conscience of laser research, now, unfortunately, becoming but a trite phrase, and finally, most curiously, laser art and music. Unfortunately, at the last moment and without my wishes, it was not possible to be restricted to these themes. So, much to the horror of the expertise, who in spite of their protective glasses, will see red as I scan, superficially again, many fields of laser applications in order to complete this review. Courage and, of course, information have been furnished, fortunately, by associates and friends in laser technology. These have been accompanied at times by expressions of real sympathy.

In order to do this, I have been helped by many people. Chiefly, of course, is the staff at the Laser Laboratory, present and past. The Directing Physicist of the Laser Laboratory, R. J. Rockwell, Jr., M.Sc., has looked over my shoulder for a number of years and has helped to restrain the unwarranted enthusiasm and unmeasured work of biologists and physicians in the Laser Laboratory. Many studies have been done together with him. Ronald Dreffer and John Johnson, technicians in the Laser Laboratory, have labored long hours with the experiments and with tender care of many patients.

The Office Manager of the Department, Billie Wilson, and her associate, Yvonne Wing, have done countless hours of extra work in this new field of laser technology to assist me in assembling the vast amount of data. Esther Giermann, as current secretary of the Laser Laboratory, has accepted the extra burden of preparation and help for this work. Pati Kelley, recently introduced to laser technology, has gone through mountains of literature. In addition to her preparation of excellent slides, Marilyn Franzen, Director of Laser Histology and Histochemistry, has ventured into many strange fields, including laser photography in the preparation of visual educational aids to help in this monstrous task and in the exhibits from the Laser Laboratory.

Patty Smith, our photographer, has had large responsibilities in all phases of this laser work, has spent long hours assembling these pictures, and has developed some interesting techniques for laser photography as an art. Ron Dreffer is also a real pioneer in the development of laser art. I am also grateful to our summer students for teaching us, as teaching is also one of the functions of the Laser Laboratory. Summer students are young and full of enthusiasm and thrilled by the challenges of the laser, very unlike the givers of grants. They are good answers to those scientists who have fears and anxieties about working with the new technology. To those who have promised little or much and not come through, I do understand, for, again, this is a new technology. It is difficult to give up someone's things or even to loan some of one's precious darlings. However, as maturity continues in this new field, such insecurities, anxieties, and fears will no longer continue. All interested in laser technology have a certain responsibility to do and to teach.

My friends throughout the laser field in this country and abroad have given valuable services. They include Arthur Schawlow, Professor of Physics at Stanford and truly a "father" of the laser, who has stimulated all of us at the Laser Laboratory to go far beyond our antiquated instrumentation; Don Rounds in biology; Fred Brech in laser instrumentation, especially the microprobe; George Wilkening, Charles Powell, and Wordie Parr in safety; Bill Ham, Chris Zweng, and Fran L'Esperance, all in ophthalmology; and Dick Honey for offering to help in the current difficult field of attempts of standardization of laser biomedical instrumentation. First names are used purposely to show the personal interest that they have taken in our past and current studies.

In Germany, Gunther Nath, Fritz Hillenkamp, and Ernest Weidilich; in Russia, Nicholai Gamaleja;

in Japan, our former Research Associate in Laser Surgery at the Laser Laboratory, Kumio Hishimoto, and in Czechoslovakia, Zdenek Naprstek, our laser surgeon, have all helped to show the international need for laser cooperation and work.

Altogether, then, we all are trying to introduce the laser to this brave new world and to open new paths for much better, more effective, and more controlled uses of laser technology. This will be mostly for the good of man in a world made much smaller and more cooperative by the laser.

Leon Goldman, M.D.
Cincinnati, Ohio

TABLE OF CONTENTS

INTRODUCTION

A current review of application of lasers would be out of date as soon as it is written. Yet, it is necessary to show some of the directions the laser can take now and what it may take in the future. This work will present an informal, oftimes superficial view of the applications of many laser systems primarily to stimulate interest, imagination, and desire. There are many excellent volumes and journals on lasers and on specific areas of information for laser applications and for laser expertise.

The background for this is a laser laboratory with more than 11 years of detailed activity. The goals were and continue to be studies of the basic reactions of lasers in living tissues, the development of laser safety programs, and the responsibility and obligation to teach laser safety, research, and the development of the biomedical applications of the laser. Therefore, in this laboratory or laser center, one must know about applications. A staff of many disciplines helps to guide the interested but unlearned amateur through the laser field.

Perhaps, as stated previously and repeatedly, some of the material to the expertise may appear to be too superficial, inadequate, or even with the sources of information, inaccurate.

For those not familiar with lasers, there will be a review of lasers and some remarks on the fascinating new developments in such systems that do affect applications now and in the future. Wave guides to transmit lasers to make them more flexible, the important aspects of measurements to be measured and to measure, chemistry, bits of botany, and holography will be given. The vast expanse of communications, especially through the development of the new and important hybrid discipline of electrooptics will be reviewed. The military and law and order programs show their applications; all initiate the study of actual applications. Then follow metalworking, construction, pollution, and a number of miscellaneous techniques. There will be, we hope, a critical review of safety programs so necessary for the proper development of laser technology. Then follows the story of the applications in biology, medicine, dentistry, photography, art, and music. Many of the applications cross to other fields. To stimulate the youth to be interested in science, there are brief remarks about the social conscience in laser and, finally, the dreams of the future.

Details will certainly have to be added sometimes by some critics and some details may be revised, as in all fields of development. If interest is stimulated to venture into the lasers, to examine the references, or with reason to object, then its purpose is fulfilled. The reviewer will agree that this is to interest and communicate, not to develop expertise.

To the annoyance or maybe just puzzlement of some, the increased emphasis of certain phases of technology may be disturbing. The repeated detailed emphasis on laser safety, its philosophies, its goals, and its confusions should not be disturbing to anyone. Safety cannot be overemphasized and laser safety programs are to be reviewed often.

Above all, the ancient cliché of all the laser technology should be repeated again and again and again, "If you don't need the laser don't use it." If you cannot tell, then with proper controls, try it, you'll like it.

Chapter 1

AN INTRODUCTION TO LASERS AND LASER SYSTEMS

It is not well known or perhaps even appreciated that the laser was born in 1917 when Einstein in his great publication, "Zur Quantum Theorie der Strahlung," established the concepts of stimulated emission, stimulated absorption, and spontaneous emission. It was not until 1953, however, after the development of the maser, microwave amplification by stimulated emission of radiation, that Basov and Prokhorov, in Russia, and Townes and Weber, in the U.S., developed the concept of the laser, light amplification by stimulated emission of radiation. The first ruby laser was shown to the public on July 7, 1960, by Maiman of Hughes Aircraft. Since that time, tremendous strides have been made in this new darling of modern-age physics.

There is much literature available on lasers and laser physics. Only a brief review will be given here.

Laser physics includes such basic phases as population inversion, energy levels, and stimulated emission. From population inversion, the fundamental characteristics of laser light are produced:

1. Monochromatic light
2. Coherent light
3. Parallel beams of light
4. Polarized light
5. High intensity
6. Electromagnetic field strengths.

As indicated, this book will not go into the detailed background, perhaps even for what Jones[8] calls "lasers for the masses." The interested reader is referred to the books listed in the references.

It is quite proper that a brief review of laser systems be given by Rockwell, the Directing Physicist of the Laser Laboratory. This is taken from his detailed presentation to the Health Physicist Meeting with Wordie Parr as Chairman (Electronic Product Radiation and The Health Physicist BRH/DEP 70-26) in 1970.

Characteristics of Laser Systems

"*Fundamental Aspects of Laser Devices.* The list of substances which can produce laser emission is comprised of an impressive number of different solids, liquids, gases, and junction diodes. There are, however, several features which are common to the configuration of all types of lasers, namely:

1. *The Laser Media.* – This is the substance, either solid, liquid, gas, or junction between two dissimilar metals, which is capable (because of its atomic and/or molecular makeup) of sustaining stimulated emission.
2. *The Source of Excitation Energy.* – In order to generate a laser beam a redistribution is required in the number of atoms which normally exist in certain atomic energy levels of the laser media. This requires an external source of excitation energy often called the "pump" energy.
3. *A Fabry-Perot Interferometer.* – This device is a pair of mirrors which are aligned plane-parallel to one another. In the case of the laser, one mirror is placed at each end of the laser media. Usually, one mirror is a total reflector, the other a partial reflector (i.e., it allows part of the laser beam which is generated within the active media to pass outside the interferometer).

In the most general manner, the following describes how a laser operates: Excitation energy is vigorously supplied to the active media to produce the specific condition called a population inversion. In this condition, more atoms of the laser media are in a specific excited-state energy level than in the lowest "ground-state" level. This condition is contrary to the normal population of states of a system in thermal equilibrium. One manner for an atom in an excited state to release excess energy is by the spontaneous emission of light in discrete units called photons. The unique feature of a laser device lies in the fact that because of the population inversion the energy release may be accomplished by the process known as *stimulated emission.* In this case a photon released by one excited atom will cause (stimulate) an excited atom it may encounter in its path also to release a photon of excess energy. The result of this interaction is the combination of two photons with identical coherence properties (phase relationships) so that they add completely together to produce a beam of twice the intensity. As the beam progresses through the excited laser media, its amplitude will be rapidly increased while its coherence properties remain unaltered. Upon reaching the total reflection mirror the beam direction is completely reversed, thus allowing another pass through the excited laser media so that the beam may be further amplified. Upon reaching the partial reflecting mirror, a portion of the beam escapes. This escaping portion is the active emission from the laser. The process will continue for as long as sufficient pump energy is supplied to the laser media.

TYPICAL LASER CHARACTERISTICS

Laser Media	Predominant Wavelengths (nanometers)	Active Media	Common Method of Operation	Continuous Power (w)	Peak Power (megawatts)	Beam Divergence (milliradians)
Ruby	694.3	Solid	Pulsed	1.0	1–1,000	0.5–20
Neodymium-glass	1,060	Solid	Pulsed		1–500	0.5–20
Neodymium-YAG	1,060	Solid	CW	1–100	1–10	0.5–10
Helium-neon	632.8 1,150 3,390	Gas	CW	0.001–0.100		0.1–1.0
Argon ion	476.5 488.0 514.5	Gas	CW	1–20	10^{-4}	1.0–3.0
Krypton ion	647.1 568.2 520.8 476.2	Gas	CW	0.5–2.0		1.0–3.0
Helium-Cadmium	325.0 441.6	Metal Vapor	CW	0.015 0.05		0.5
Neon ion	332.4	Gas	CW	0.250		1.0–3.0
Neon-pulsed E field	504.1	Gas	Quasi-CW	.003	10^{-2}	Rectangular beam 2 x 30
Carbon dioxide	10,600	Gas	CW	10–5,000	10^{-2}	1.0–5.0
Nitrogen-pulsed E field	337.1	Gas	Quasi-CW	0.100	10^{-2}–10^{-1}	Rectangular beam 2 x 30
Gallium-arsenide	840.0	Semiconductor	Quasi-CW	1–20		Rectangular beam 1–10

The following sections will review the most important laser devices in use today. Special emphasis will be placed upon the variations possible in the outputs which these devices may produce.

Pulsed Laser Devices. Lasers which use a solid-state media are generally operated in a pulsed mode. The most common solids used are ruby crystals and neodymium-doped glass and neodymium-doped crystals such as Yttrium Aluminum Garnet (YAG). In the present state of the art, the pulse envelope is generated in one of the following manners:

1. A millisecond or multi-millisecond pulse envelope comprised of many random pulse spikes of submicrosecond duration.
2. A single multinanosecond high power pulse.
3. A single nanosecond or sub-nanosecond high power pulse.
4. A train of evenly spaced ultrashort, sub-nanosecond pulses.
5. A single ultrashort, picosecond pulse.

The distinction of the type of pulses becomes exceedingly important in the evaluation of the interaction phenomenon of pulsed laser radiation, where the rate of energy delivery is the critical factor. The following will review thé most common variations in pulse characteristics produced by commercially available laser systems:

1. *Normal Mode Operation.* – The pulse envelope from a long pulse or so-called "normal mode" solid-state laser will appear, when viewed photoelectrically, as a series of random "spikes." Although this sporadic spiking behavior is not completely understood, it is believed to be a consequence of interaction which occurs between the excited ions in the active laser media and regions of strong electromagnetic fields of the laser beam *inside* the media. This produces a sporadic de-population of the excited ions, which are observed as random pulsations during the laser burst. Each spike has a duration of 300 to 400 nanoseconds and the average spacing between spikes is in the order of a few microseconds. Recent reports have shown that each spike may consist of a large number of even shorter pulses which may last from three to ten picoseconds (10^{-12} seconds).

Ruby or neodymium laser-systems derive the excitation energy from xenon arc flashlamps that are usually placed in a highly reflecting housing so as to actually focus the lamp emission onto the laser rod. The pulse of the normal mode laser output is, in the first approximation, about equal to the length of time that the xenon flashlamp is excited; providing, of course, that the emission from the lamp is sufficient to sustain a population inversion in the laser media. The duration of the current pulse in the lamp is determined by the time-constant of the inductance-capacitance (IC) network which drives the flashlamp circuit.

There are methods to stretch the pulse-length in normal mode operation. The most common is to use additional inductance-capacitance sections which will serve to increase the pulse-length monotomically; with

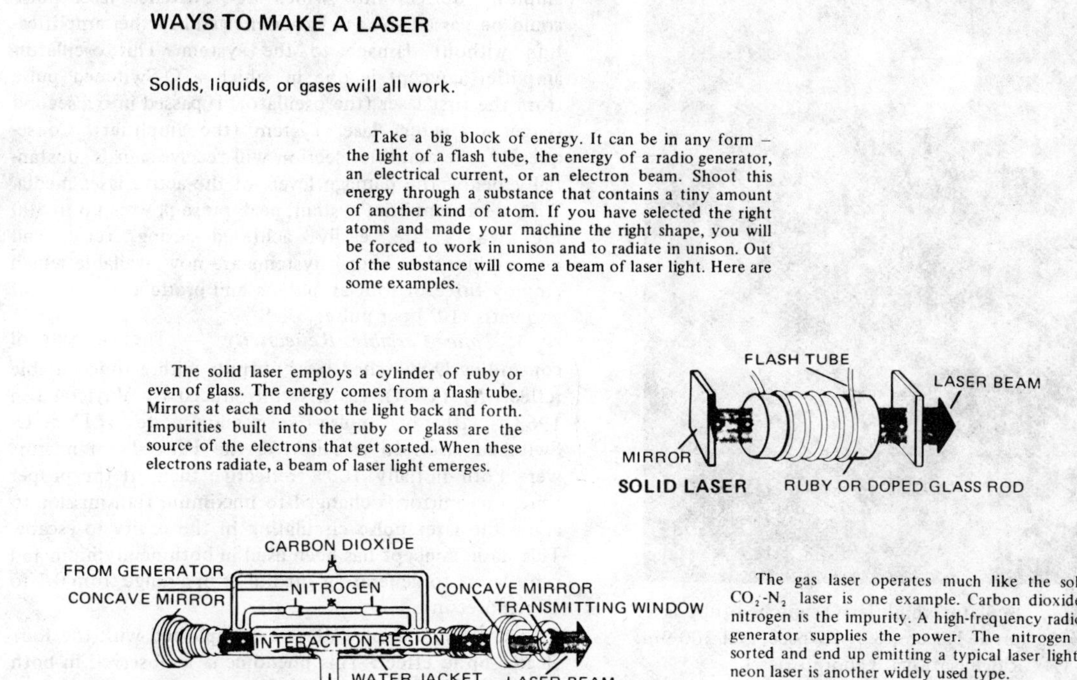

FIGURE 1-1. Ways to make a laser. (Xerox Education Publications.)

FIGURE 1-2. Oscillator amplifier Q-switched ruby laser impact showing target 1.8 cm, 8 J, 25 nsec, and 300 mμ. (Murakami, Oak Ridge National Laboratories.)

this technique, normal mode laser pulses have been stretched to 30 msec for some laser welding applications.

2. *Q-Switched Operations.* – Pulsed laser operation is also possible in a second mode of operation. In 1961 McClung and Hellwarth announced the technique known as "Q-Switching." The term "Q" is a carry over from the radio wave and microwave terminology, and it relates to the so-called "quality factor" of a resonating system. A "low Q" would refer to a system that would not easily support oscillation; thus, to switch (or spoil) from a low to a high Q (viz., Q-Switching) means to change rapidly from a condition in which the laser cannot lase to a condition in which it will. The mechanics of Q-spoiling usually incorporate some form of electro-optical or electromechanical shutter between the mirrors of the laser cavity. Recent reports, however, indicate that Q-Switching can be accomplished also by a simple misalignment of the Fabry-Perot mirrors. The effect of Q-spoiling is to force the laser to operate in a condition of maximum gain. This occurs because the overpopulation in the metastable state will be larger than in the case of normal mode laser action due to the fact that no stimulated de-population can occur in the low Q cavity. When the "switch is thrown," so to speak, de-population by stimulated emission will rapidly occur, causing the emission of an intense giant pulse, or train of two or three pulses, each lasting from ten to 50 nanoseconds. The result of such short bursts of light is the production of enormous peak-power outputs. With this method it is possible to produce instantaneous light levels exceeding 100 million watts in a single burst of light.

Because of the desire for even high peak powers, the damage threshold was soon reached for most solid state laser materials and investigations immediately began on amplifier devices into which a Q-Switched laser pulse could be passed so as to experience still further amplification without damage to the system. This oscillator-amplifier concept is one in which a Q-Switched pulse from the first laser (the oscillator) is passed into a second (larger diameter) laser system (the amplifier). Consequently, the amplifier section will receive signals substantially below the damage levels of the active laser media. With single amplifier system, peak pulse powers up to 500 megawatts are easily achieved using ruby and neodymium-glass lasers. Systems are now available which employ three or four amplifiers and produce hundreds of gigawatts (10^9) per pulse.

3. *Time Variable Reflectivity.* – The concept of controlling Q-Switched laser outputs with a time-variable reflectivity (TVR) was first introduced by Vuylsteke in 1963, with his pulse-transmission-mode (PTM) Q-Switched laser. In this device, the Fabry-Pérot mirrors were both initially 100% reflective then, at the proper time, one mirror is changed to maximum transmission to allow the laser pulse circulating in the cavity to escape. This basic concept has been used in both neodymium and ruby lasers to generate giant pulses in a range from 0.1 to 10 nanoseconds.

The technique of TVR is accomplished with the Kerr electro-optic effect. This phenomena is observed in both liquids (nitrobenzene) and solids (potassium dihydrogen phosphate: KDP) which are placed in a large electric field. Under these conditions, the material is changed into a double refracting media. Consequently, a linearly polarized light wave passing through the media will be changed into a circular, elliptical, or orthagonally linearly polarized beam depending upon the degree of retardation imposed.

The Kerr effect device most commonly used with pulsed lasers is called a Pockel's Cell; and usually employ the crystal KDP as the media. In this case, application of a large electric field (in the direction of propagation) will change a linearly polarized laser beam into a circular polarization. Upon reflection from a 100% reflector the sense of circular polarization is reversed, thus, when the beam re-enters the Pockel's Cell it is changed back into a linear polarization, but the direction of the electric field vector will be orthagonal to the original beam polarization. Consequently, insertion of a polarization sensitive prism inside the resonant cavity of the laser (viz: a calcite prism or Glan-Thompson prism) will allow the laser pulse to be reflected out of the cavity from the side.

The principal advantage of TVR switching is that the energy of the output pulse is independent of the cavity gain characteristics and the duration of the output pulse is solely dependent on the separation between the resonator mirrors. The laser pulse length is equivalent to the time for the pulse to make a round-trip circuit in the cavity. Consequently, laser pulses which last only 2 to 5 nanoseconds are easily generated (without loss of pulse energy) from standard Q-Switched ruby lasers which would normally produce pulses which are 10 nanoseconds or longer. In this manner, the peak power of the pulse is increased by at least a factor of five over standard Q- Switched systems.

4. *Mode-Locked Laser Systems.* – A laser device is a resonant oscillator of electromagnetic waves. Consequently, it can support standing wave patterns in a manner somewhat analogous to the resonance of an organ pipe. The laser "standing waves" are produced when an integral number of half-wavelengths exactly fit into the separation between the mirrors of the Fabry-Pérot cavity. In all lasers this spacing is many, many orders of magnitude larger than one laser wavelength, thus, there are many slightly different standing wave frequencies which can fit within the narrow linewidth of a laser source. This gives rise to many distinct spectral components or longitudinal "cavity modes" each with individual amplitude, frequency, and phase characteristics. In the event that the phases of these different frequency modes are synchronized in some manner (e.g.: the phases of the different longitudinal modes become "locked together") the different modes will then interfere with one another to generate a beat effect. The result will be a laser output which is observed as regularly spaced, periodic intensity pulsations.

Mode locking is usually accomplished by inserting an ultrafast intensity dependent "switch," such as a bleachable dye, inside the resonant laser cavity (i.e.: between the mirrors). The overall system operation may be explained as follows: The active laser media serves as a light amplifier, the resonance of the Fabry-Pérot Etalon and the natural line width of the laser act as a selective frequency filter. The time for the laser beam to travel the cavity length twice will serve as an optical delay line. The bleachable dye is sometimes referred to as the "expandor element," meaning that it serves as a non-linear absorber, since it provides less loss for a high intensity signal than a low level signal.

Thus, a pulse circulating intra-cavity will be amplified only if it has sufficient intensity to cause the dye to momentarily bleach to transparency. The dye also serves to produce optimum mode coupling, provided its relaxation time is shorter than the time for a pulse to make round trip in the cavity. Since many of these dyes may be "switched" in less than 10^{-13} seconds, it becomes theoretically possible to produce individual laser pulses in this time range. The individual pulses which are contained in the pulse train of a mode locked laser are "ultrashort" picosecond (10^{-12} seconds) pulses. The separation between successive pulses is that of a round trip of the beam between the Fabry-Pérot mirrors.

5. *Ultrashort Pulses.* – In 1967, Bell Telephone Laboratory scientists, Duguay, Shapiro, and Rentzepis, discovered that picosecond duration pulses may actually be present in the output from many standard solid-state Q-Switched lasers. Because these pulses are of such a brief duration, it is almost impossible (prior to the unique measurement techniques such as they devised) to individually observe these pulses; although it was suspected that they did exist. Since most solid-state lasers do have a definite width to the spectral line (or, in other terms, many cavity modes) it is possible by mode-locking techniques to generate very short duration intensity pulsations. The broader the spectral width of the laser line, the more harmonic components that will occur in the interference phenomena, and thus, the shorter will be the duration of the beats which occur. As a result, the amplitude or the peak power will increase as the width of the pulse decreases. The neodymium-glass laser is of particular interest in the generation of mode locked pulses because of its quite broad (100-200 Å) bandwidth. This would theoretically allow for the generation of subpicosecond high peak power laser pulses.

Special techniques can be used to isolate an individual ultrashort pulse. The importance of such pulse widths lies in the peak power of the pulse. For example, using a neodymium-glass oscillator-amplifier laser system, peak powers of 30-40 gigawatts (10^9 W) in a pulse width of 10-15 psec have been observed.

Scientists at the United Aircraft Research Laboratories recently observed an ultrashort laser pulse of only 0.4 picosecond duration. A neodymium-glass laser was used which was simultaneously mode locked and Q-Switched by a saturable absorber. Although it had been known for some time that the theoretical limit of the pulse width from a neodymium mode locked laser would be in the order of about one-third of a picosecond, pulse widths previously observed were usually in the order of 4 to 10 picoseconds. This discrepancy between measured and the theoretical pulse length suggested the existence of an amplitude or carrier wave modulation in the pulses themselves. The experiments confirmed this fact and showed that most of the measured spectral content of the pulse was due to an almost linear relation between the frequency of the carrier wave and the laser wavelength. The modulation was about one per cent larger at the beginning of the pulse than at the end. As a result, the pulses could be compressed to a length approaching the reciprocal of the bandwidth by passing them through a dispersing system which also had a linear relation between time and wavelength. In this manner, ultrashort pulses were compressed into a range of 0.4 X 10^{-12} seconds, near the theoretical limit.

Thus, it is seen that there is an enormous variation in pulse durations generated from the many available laser systems. The spread in time of the overall pulse envelope from the normal mode pulse to the single ultrashort pulse may be as great as 10^{10}. The specification of such pulse characteristics is extremely important, especially in any critical evaluation of the interaction phenomenon of laser energy.

6. *Wavelength Variation With Pulsed Lasers.* – The advent of Q-Switched laser systems with their associated enormous electromagnetic fields, also introduced the more common use of frequency shifting with two non-linear techniques; namely, second harmonic generation (SHG) and Stimulated Raman Scattering. The non-linear polarization field induced in non-centrosymmetric crystals such as quartz, potassium Dihydrogen phosphate (KDP), and Barium Sodium Niobate can actually generate harmonic components of the fundamental laser frequency. In this manner, high power pulses in the ultraviolet spectrum may be produced: 347 nM with the ruby laser and 265 nM using two frequency doubling crystals and a Q- Switched neodymium laser.

Commercial systems are also available which employ

Stimulated Raman Scattering to produce additional wavelengths. In one such system, (Geoscience Instruments) produces three Stokes (longer wavelength) and four Anti-Stokes (shorter wavelength) shifts. Using the ruby laser as the source, the available wavelengths are 432 nM, 492 nM, 577 nM, 694 nM, 871 nM, 1167 nM, and 1768 nM. It is evident that by combining both SHG and Raman shifting techniques together, one may achieve laser frequencies in almost any desired region of the spectrum with a single laser source.

Continuous and Quasi-Continuous Wave Laser. Simulated emission is possible when the necessary condition of population inversion is met in the active media. Although there are, in general, many competitive processes which may limit continuous laser operation (e.g., heating of the media, "self-quenching" effects, etc.), it is often possible to achieve continuous or high repetition-rate operation with many materials that have heretofore been considered as only "pulsed" laser media. The limiting factors are (1) the low efficiency of converting pump energy into laser emission, (2) retention of heat by the laser media, and (3) degradation of the components used for continuous optical "pumping." This does not mean, however, that the same levels of power as achieved instantaneously in the pulsed operation will be achieved in CW operation. The rapid de-population and subsequent high power achieved,

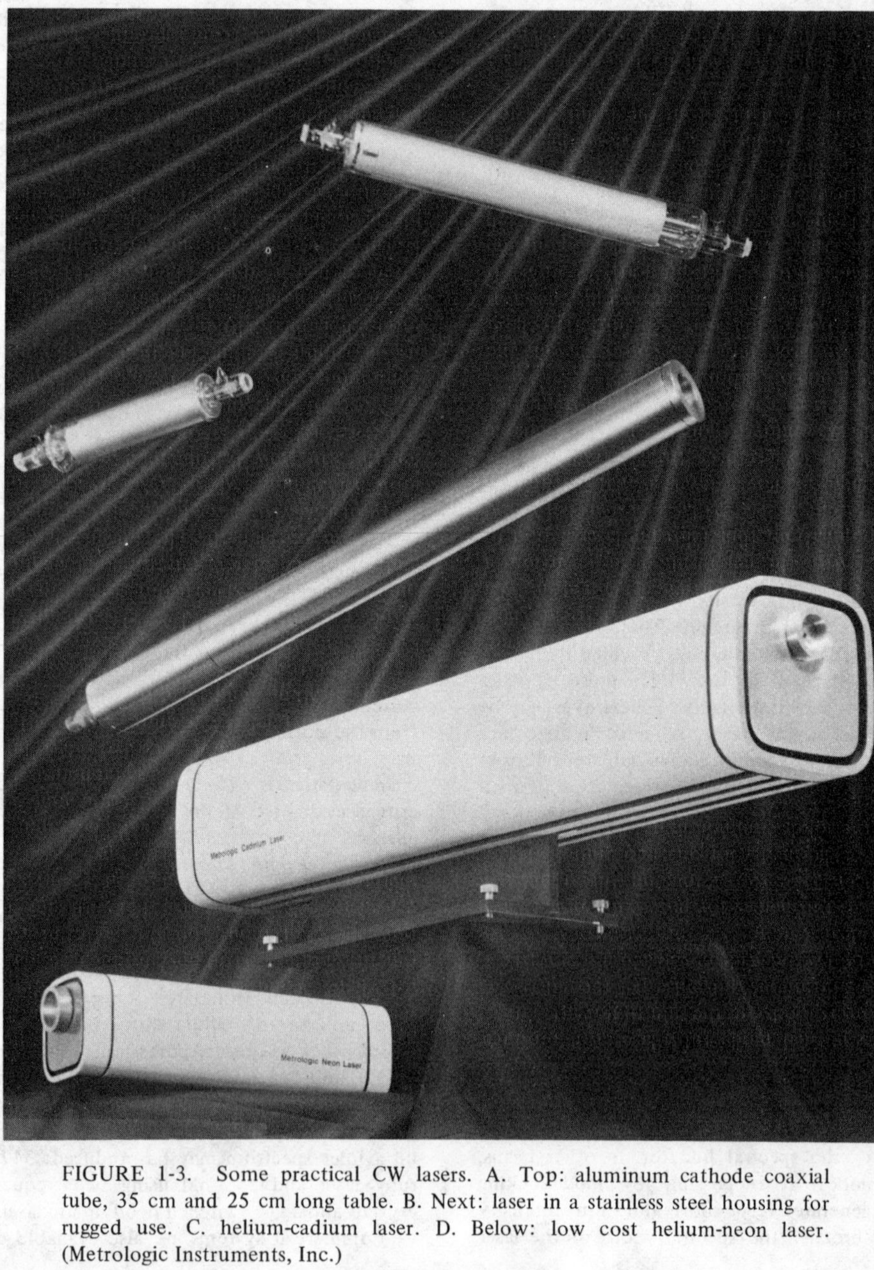

FIGURE 1-3. Some practical CW lasers. A. Top: aluminum cathode coaxial tube, 35 cm and 25 cm long table. B. Next: laser in a stainless steel housing for rugged use. C. helium-cadium laser. D. Below: low cost helium-neon laser. (Metrologic Instruments, Inc.)

for example, in a normal mode ruby laser will produce instantaneous power levels in the multikilowatt range, whereas the quasi-continuous, 60 pulse per second (pps), operation of a ruby crystal will only reach a few watts – even in the most ideal system.

The most successful medias for pure continuous wave operation have been gases (or gaseous mixtures) and the many diode lasers. The media for quasi-continuous lasers include most of the solids common in pulsed operation as well as many gases.

The first CW system was the helium-neon gas mixture. Although its first successful operation was at an infrared wavelength of 1150 nM, the helium-neon (He-Ne) laser is most well known operating at the red (632.8 nanometers) transition. The earlier He-Ne lasers were excited by radio frequency (RF) discharge. In this case, the He atoms were excited from the ground state by the RF field. This energy excess is coupled to an unexcited neon atom by a collisional process with the net result of an inversion in the neon atom population, thus allowing laser action to begin. The more recent He-Ne gas laser designs have used direct current excitation. Power levels available from the low efficiency He-Ne laser ranges from a fraction of a milliwatt to about 75 milliwatts in the largest available systems. The He-Ne laser is noted for its high-frequency stability and single-mode operation.

The family of ion lasers (argon, krypton, xenon, and neon) provides a source for over 35 different laser frequencies, ranging from the near ultraviolet (neon at 332.4 nM) to the near-infrared (krypton at 799.3 nM). It is possible to mix the gases, for example, argon and krypton in which lasing may occur simultaneously at ten different wavelengths, ranging from the violet through the red spectral region. Such an output is truly a "white light" laser. Reference to table will review the most important wavelengths available from the family of ion lasers.

There are many other gases which can generate laser emission. The carbon dioxide laser is the most efficient, and consequently, the most powerful of all CW laser devices. Continuous powers have been reported above 1,000 w at the infrared 10,600 nm wavelength. Nitrogen gas may be operated in a rapid-pulse or quasi-continuous manner. The beam is emitted in a continuous train of 8-10 nanosecond pulses of kilowatt power levels at pulse repetition rates of 100 pps, producing a maximum average power level of about 100 mw at the ultraviolet 337.1 nm wavelength. Neon gas, operated with same transverse field method, produces green pulses of 3-5 nsec duration for an average power of 3 mW at 100 pps.

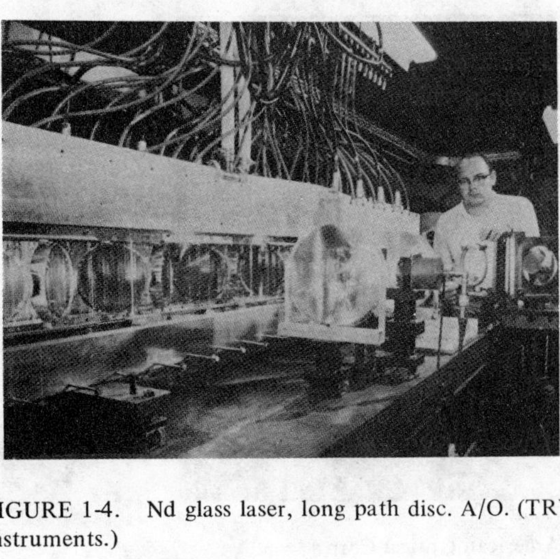

FIGURE 1-4. Nd glass laser, long path disc. A/O. (TRW Instruments.)

Recent development of a metal vapor laser has been successful using Helium and Cadmium, which is placed into vapor distribution by a cataphoresis technique. This laser produces emission at either a dark blue (441.6 nm) or an ultraviolet (325 nm) wavelength at maximum powers of 50 mW and 15 mW, respectively.

One of the most promising laser sources of the day uses a neodymium-doped crystal yttrium-aluminum-garnet (commonly called YAG). This device is optically pumped either by special tungsten or krypton flashlamps and is capable of CW outputs approaching 50 w at the 1,060 nm wavelength. The recent emergence to popularity of the YAG laser has been made possible by better growth techniques of the YAG crystals. Sufficient crystal lengths can now be obtained to provide for high CW power levels.

Most of the diode lasers are also operated on a continuous wave basis. The most common diode device uses a gallium-arsenide junction which emits a fan-shaped infrared beam at 840 nm.

Normal mode and Q-Switched solid-state lasers are often designed for a high repetition-rate operation. Usually, the specific parameters of operation are dictated by the application. For example, normal mode ruby lasers operating 1 pps at 150 joules per pulse are used in high-speed dynamic balancing metal removal applications. As the repetition rate increases, the allowable exit energy per pulse necessarily decreases. Systems are in operation, for example, which produce ten joules per pulse at a repetition rate of 4 pps. A similar ruby, operated in the Q-Switched mode, could produce a one megawatt per pulse at a rate up to ten pulses per minute.

a. *Frequency Expansion Using Non-Linear Techniques.* Methods of converting high-power CW outputs into new frequencies with harmonic generation techniques have also been most successful. Probably the most significant development has been the high power continuous frequency doubling from the neodymium-YAG laser, using the crystal: Barium Sodium Niobate ($Ba_2NaNb_5O_{15}$), commonly referred to as the "Banana Crystal." Continuous powers of 1 watt at the green frequency doubled wavelength of 530 nM have been obtained using an intracavity doubling crystal.

   A further expansion of output frequencies is available if the output of the CW frequency doubled YAG-Nd laser is directed into a parametric oscillator. This device also uses "Banana Crystals" which, when temperature-controlled in a range from 97 to 103°C, will allow a tunable output in a range from 980 to 1160 nM. Using 300 nW of 530 nM input power, such a parametric oscillator can produce up to

3 mW at selected wavelengths in the range from 980 to 1160 nM. Smith and Geusic of Bell Telephone Laboratories indicate that proper choice of oscillator mirror reflectances in a parametric oscillator could allow the tunable range to cover a range from 650 to 4,000 nM.

b. *Emergence of Tunable, Liquid and Diode Lasers.* The first multi-wavelength liquid laser was introduced in 1966 by Sorokin, and his associates at IBM. The device, using three different organic dyes, was capable of producing red, green, and yellow laser emission. The key development in this system was the excitation by an ultrafast pulse of incoherent "pumping" light. Previous liquid laser systems required excitation using very short laser pulses. Sorokin's studies had shown that lasing of organic dyes required a very fast optical pumping pulse (300 nsec. rise). Consequently, a special flashlamp system was used which delivered pulses of 400 nsec. duration at repetition rates of 1 pps. Power outputs in his early system were 1 mW/pulse with 0.2 joules/pulse.

Sorokin's early data had shown that the principal central wavelength of organic dye lasers is a function of the dye concentration. This was quantitatively shown by Schaefer, Schmidt, and Voltze, using a 5 MW 10 nsec. Q-Switched ruby laser to optically pump a DTTC – Bromide dye solution. Variation in concentration from 10^{-5} to 10^{-3} mol/liter produced a change from 805 to 865 nM in the principal laser wavelength. The output was also found to be dependent upon the cavity "Q"; with shorter wavelengths characteristic of a low "Q" cavity.

Several techniques have recently been demonstrated which provide a unique "tunable" control of the dye laser output frequency. One such system is the movable piston concept devised by Kagan (IBM). Movement of the piston actually changes the length of the lasing media, and, since the emission from a nonuniformly pumped dye laser is also dependent upon path length, a lower frequency will be produced. The longer path length serves as a passive absorber of pump energy, thus, changing the ratio of emitting to absorbing molecules. As a result, there will be an upshift in wavelength due to the decreased absorption in the organic molecules at the longer wavelengths.

Soffer and McFarland (Korad), proposed a diffraction grating resonating cavity to obtain spectral narrowing and tunability. Using dyes in the xanthene and carbocyanine families, a 40 nM shift was obtained in rhodamine 6G pumped by a frequency doubled ruby laser (347 nM in 10 nsec. @ 10 MW/cm^2). Variation in the inclination of the grating served to selectively produce high cavity gain at a given wavelength. Additionally, a significant narrowing is obtained (approximately: 0.06 nM) in contrast to the 5-6 nM spectral width of normal dye lasers. Tunable organic lasers have also been controlled simply by changing the temperature of the dye media. Schappert has shown that an

FIGURE 1-5. CW tunable laser. (Chromatix.)

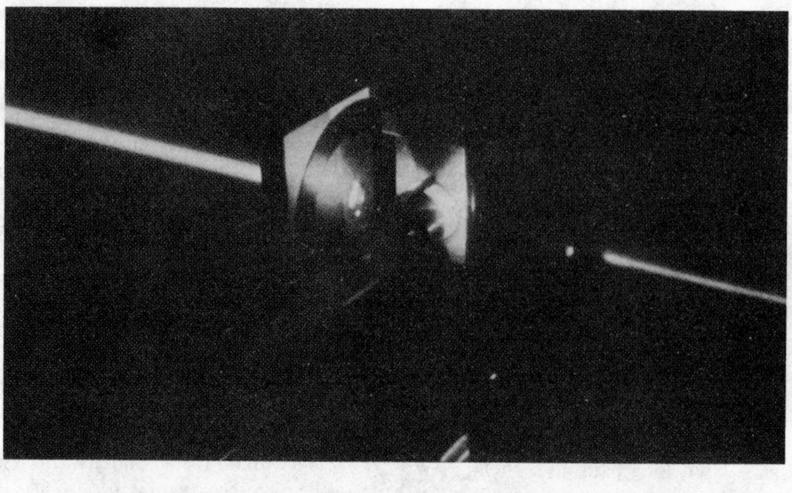

FIGURE 1-6. Ultraviolet laser. (American Optical Corp.)

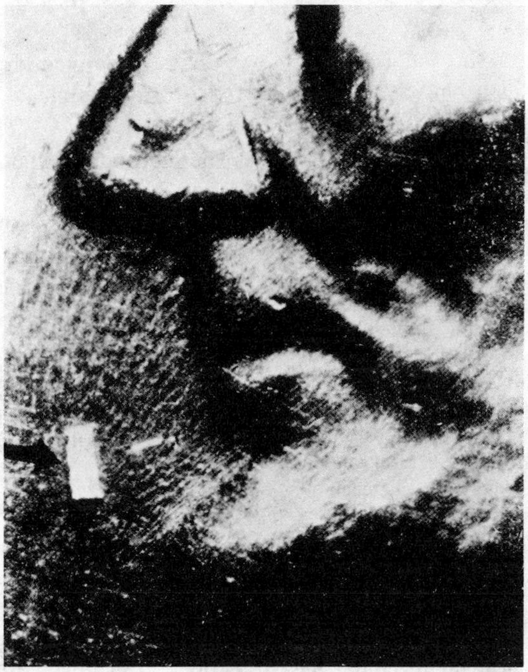

FIGURE 1-7. Semiconductor laser 0.015 x .003 in. small battery powered at room temperature. (Hayashi, Panish, Foy, Sumski, Bell Telephone Lab.)

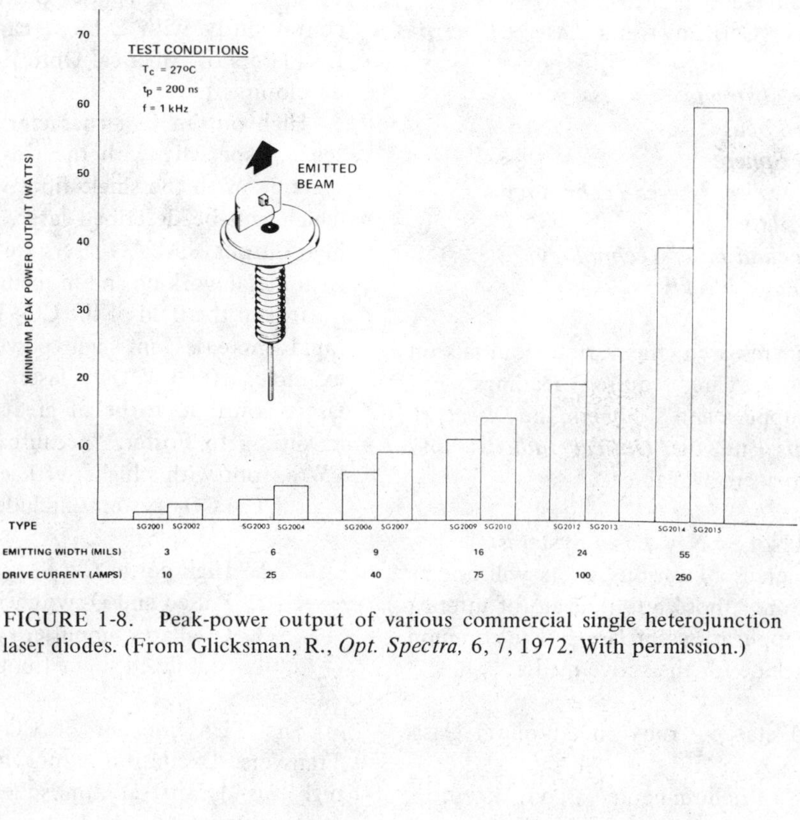

FIGURE 1-8. Peak-power output of various commercial single heterojunction laser diodes. (From Glicksman, R., *Opt. Spectra,* 6, 7, 1972. With permission.)

organic dye (DTTC in Ethanol) will display at least a 20 nM wavelength shift when the temperature is varied from −117°C to +78°C.

The first continuously tunable diode laser was announced in 1968 by Hinkley and his associates at the Massachusetts Institute of Technology. The central output frequency could be tuned continuously over a range of 50 gigahertz by simple variation in the diode current. The diode systems were fabricated from crystals of the mixed semiconductor lead, tin, and telluride; formed by a vapor deposition process from lead telluride seeding crystals. The emission could be varied from 6500 nM to 28,000 nM by simply changing the chemical composition of the diodes. The central output frequency of the diodes may be changed by simply controlling the current which passes through the diode."

To keep up with current and, as indicated, the almost daily rapid changes in laser technology, the interested reader is referred to the following current laser journals:

1. *Electro-Optical Systems Design News*
2. *Holosphere*
3. *Lasers* (Germany) and *Laser + Electro-Optik*
4. *Laser Abstracts*
5. *Laser Focus*
6. *Laser Sphere*
7. *Microwaves – Laser Technology*
8. *Optical Spectra*
9. *Optics and Laser Technology*
10. *The Laser Weekly*

For those in research, the various journals on optics and physics are required readings. For catalogues of supplies and suppliers, the *Directory of Laser Focus* and the *Optical Industry and Systems Directory* are available.

Some New and Not So New Laser Systems

Again the details of various lasers will not be reviewed here, only those bits that are of interest in the current applications of lasers. The common laser sources include for the active media:

1. Solid state – ruby, neodymium glass, Nd-YAG
2. Gas – helium-neon, argon, krypton, xenon, carbon dioxide, nitrogen, helium-cadmium[15]
3. Semiconductor – gallium arsenide[4]
4. Chemical

The methods of operation continue to be:

1. Pulsed, normal mode, Q-switched, nanoseconds, picoseconds
2. Continuous wave (CW)
3. Quasi CW – nitrogen, gallium arsenide

One recent development is the sunlight powered Nd-YAG laser of G.T.E. Sylvania. The sunlight is collected and focused with lenses and a 24-in. mirror. When the sun is not available, an auxillary LED light source may operate the laser. The economic operation is different from battery powered space devices. Focusing and cooling systems of this laser continue to need improvement. This type of laser will have a great potential application in communications, especially for satellite to satellite.

Other developments in glass lasers include the new pumping technique of American Optical with pulsing circuits to give a higher repetition rate of 19 J at 100 pps (pulses per second) or 1.9 kW continuously with a burst capacity of 140 pps. Disc lasers of Americal Optical Co. are also a new development.

High output is a characteristic of the modern scene, especially with the Nd-YAG and gas laser systems. With the single fiber wave guide of Nath, which will be described later, the applications of high output Nd-YAG lasers are increased, especially in metal working and in medicine.

It is in the field of the CO_2 lasers that there are rapid increases in high power outputs, as for example, 100 kW CO_2 laser of Avco. The CO_2 lasers continue to be of great interest and value according to Foster, "because they give a higher CW output with a higher efficiency than any other laser." The CO_2 systems include

1. High power CW
2. Pulsed and Q-switched
3. Oscillator amplifier systems
4. Stabilized single frequency

The TEA concept of CO_2 lasers stands for Transverse Excitation Atmospheric pressure and a high gas density at atmospheric pressure makes more molecules available for excitation and lasing, thereby making high peak powers. For excitation

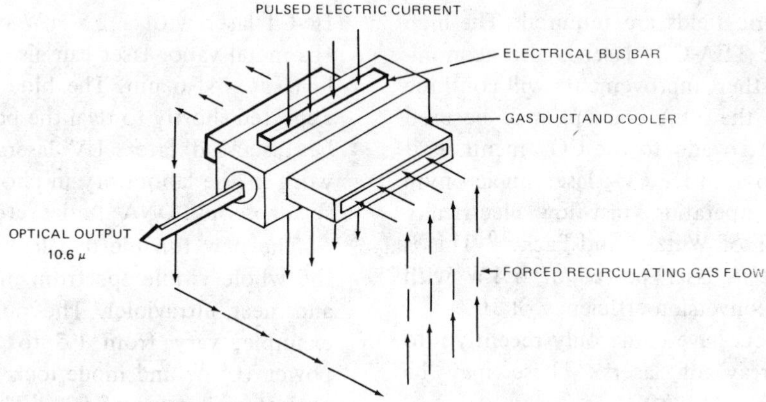

FIGURE 1-9A. Outline TEA laser. (TRW Instruments.)

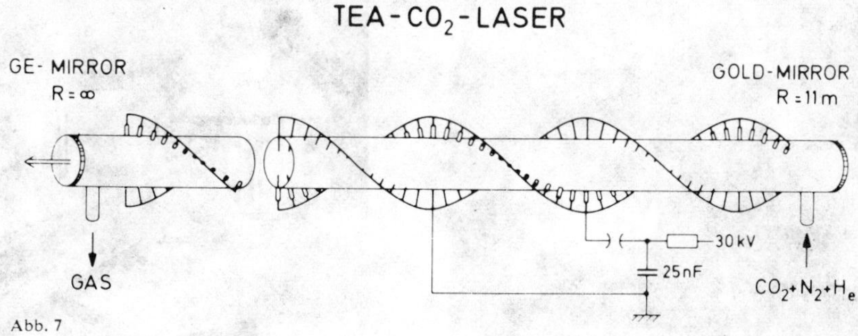

FIGURE 1-9C. TEA laser diagram – Klaus Buchle. (Laser + Electro-Optik.)

high energy electric fields are required. The high power short pulse TEA CO_2 laser is also economical. Certainly further improvements will continue in this field, for the TEA concept may be used with other gases. To add to the CO_2 menu, as it were, there is also a coffee CO_2 laser, an acronym for continuously operating, fast-flow electrically excited CO_2 laser of Wutzke and Pack.[20] This is reported to have a laser power of 6 kW with electrical-optical conversion efficiency of 10%.

Ultraviolet lasers have come only recently into the glittering array of lasers. These may be developed from:

1. Nitrogen laser
2. Fourth harmonic from Nd-YAG at 266 nm
3. Helium-cadmium (He-Cd) metal vapor

An example of a tunable system is that of Chromatix. Excitation energy for pumping a Nd-YAG laser is a krypton arc lamp. The laser emission is in 100 nsec to 400 nsec pulses. A Lit O_3 crystal internal to the laser cavity may be used for frequency doubling of the various IR wavelengths to produce red, yellow, green, or blue. These pump a parametric oscillator to provide continuous tunable radiation over portions of the visible and infrared spectrum. To obtain high power UV at 266.0 nm, 532.0 nm output of the laser is used. Chromatix reports 26 wavelengths of the frequency doubled the Nd-YAG laser.

Silfvast[15] has recently developed a 37-cm-long He-Cd laser with a 2.5 mW output at 325.0 nm. His metal vapor laser can also emit 12 mW of blue light at 441.6 nm. The blue light He-Cd laser is expected shortly to rival the popular He-Ne, as will be discussed later. UV lasers are important for work in the laboratory in photochemistry, Raman spectroscopy, DNA studies, etc.

The new tunable dye lasers can be tuned over the whole visible spectrum into the near infrared and near ultraviolet. The pulsed dye lasers, for example, vary from 1.5 to 5.0 psec with peak power 10^8 W and mode lock 10^7 giant pulse with optical efficiency of 50%. There is an increasing use of dye types that include essentially the scintillators for the ultraviolet systems, coumarins, xanthines the most practical dye group (Rhodamines), and cyanins. Water is a good solvent for the dyes. If a surfactant is added, there are no aggregates of the dyes.

CW dye lasers, 525.0 to 710.0 nm, have been developed with 2.5 W output. Much work is still needed to develop uniformly good models of tunable dye lasers. Higher output systems are also needed.

The electron beam ionizing sustaining laser, with 1 kJ at 20 μsec, produces the highest energy, at present, from a pulsed gas laser. In this new laser system, 20% of the electrical energy is changed into laser energy of high efficiency. The various electron beam lasers include the N^{371} and H^{1161}. The xenon is still under detailed investigation regarding fusion. There is much interest at present in the chemical laser. With the high output

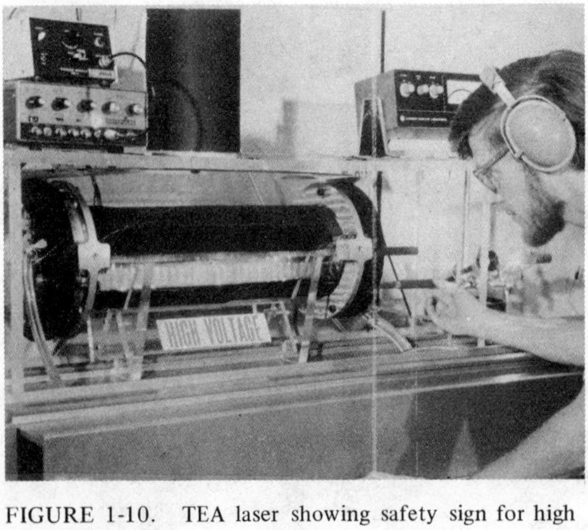

FIGURE 1-10. TEA laser showing safety sign for high voltage. (Lumonics Research, Ltd.)

FIGURE 1-11. Electron beam pumped hydrogen laser emits in the vacuum ultraviolet. (From *Ind. Res.*, May 1972. With permission.)

of the chemical lasers, some produce the kilowatt output. These lasers will become increasingly more common in the future. The efficiency of chemical lasers varies from the low efficiency of the iodine laser to the 3% efficiency of the H_2Fe_2Ne.

Chemical lasers, of course, may use combustion products of gaseous mixtures to give atoms in high-energy states. This is the new group of so-called explosive lasers. At Los Alamos Scientific Laboratory, CO_2 is used to produce 1,060.0 nm. Other gases are used to help the lasing action of CO_2.

Another chemical laser is the "flame laser" of Piloff et al.[12] of the Naval Research Laboratories. Carbon disulfide and oxygen are used in a vacuum in the laser cavity about 10 torr. For this chemical laser 1 mW emission in 3 IR wavelengths is reported.

There are increasing uses of the semiconductor[4] junction lasers in industry and in the military because of:

1. Small size
2. Low cost
3. Ease of modulation
4. Compatibility with solid-state electronics

Earlier models had relatively limited uses because of their need for cryogenics. Now their "threshold current density at room temperature has been reduced by more than two orders of magnitude and their overall quantum efficiency increased to 40% range at room temperature." According to Glicksman,[4] single diodes are available with emitting junction widths of "3 to 55 mm. which provide peak power outputs of 2 to 70 Watts at drive currents of 10 to 250 Amperes respectively." Power efficiency varies from 3 to 5%; even 12% has been reported. Heat generation is a factor which limits diode operations. In many applications to produce higher power laser diodes series are connected in linear modules and placed in parallel configuration.

The recent advance in the development of diodes has been due to large optical cavity diodes (LOC). These improved laser diodes are developed at RCA and Bell Telephone Laboratories with Ga AlAs-GaAs single heterojunction laser diodes. They are reported to have current densities of 8,000 A/cm^2 and quantum efficiencies of 40%. The homojunction laser diodes have 40,000 A/cm^2 and 20% quantum efficiency. The LOC is used for those "illuminator applications which require high average powers and in communication systems which can make use of the high repetition rate capability of this diode."

As the electromagnetic spectrum is surveyed by the dreamers and planners in laser technology, the challenge has been presented for generating coherent emission in the X-ray region. Recently, the chemists at the University of Utah believed that they had achieved this enviable goal. A gelatin containing a weak solution of copper sulfate was illuminated by a Q-switched neodymium glass laser. Kepros, Eyring, and Cagle[17] used a gelatin layer to 0.1 mm thickness copper sulfate 10^1 m*M* solution with 20 nsec pulse of up to 30 J. The detection of X-rays was from changes on the photographic plate covered with aluminum foil. The film was 110 cm from the gel and the photographic spot was no greater than at a distance of 30 cm. The frequencies of the X-rays have to be determined. Recently, Boster[1] of the Lawrence Livermore Laboratory repeated the

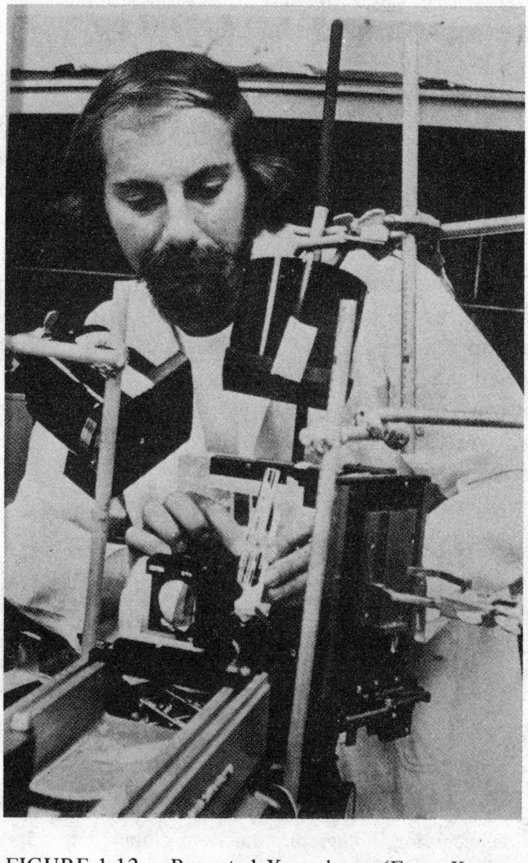

FIGURE 1-12. Reported X-ray laser. (From Kepros, J., *Ind. Res.*, 14, 23, 1972. With permission.)

experiments with film tracks of three layers of no screen X-ray. He found no spots through all three layers. Boster suggests that the spots on the front emulsion of the first film and the front emulsion of the next film were due to an electrostatic effect with the "two closely spaced emulsions active as a capacitor."

Although it is recognized that there are many great difficulties in producing stimulated emission in the X-ray region, usually less than 10.0 nm, recent studies are more encouraging. Instead of molecular stimulation, atomic ions have been stimulated. Triple ionized carbon ions have produced emission from a resonance transition. Two stimulated lines from carbon around 155.0 nm were detected. It is hoped that other elements may lase at lower wavelengths. French physicists have obtained stimulated emission in an aluminum plasma, wavelengths 11.7 nm (*New Scientist*, 31, 533, 1973).

As Brech[2] indicates for laser technology for the future, power and energy will go up, wavelengths will go up and down, and similarly prices will first increase and then decrease. The continued low cost of the helium neon lasers is an excellent example of the economics of laser instrumentation.

The Current Laser Industry Briefly

The decreasing prices of volume production of HeNe lasers introduce a brief survey of the laser industry.

The laser industry has always been optimistic. This enthusiasm was nurtured by the investigators themselves, by hysterical headlines, by science fiction writers, by the small garage shops looking for their own Polaroid® or Xerox® dreams of the future, and by serious workers. Most of the fair-weather folks have left the field to the more serious pros.

FIGURE 1-13. Acousto-optic Output Coupler to transform a CW laser into a high peak power pulsed mode for data recording, holography, micro-machining, etc. (Spectra-Physics.)

Laser research and development have been supported essentially by the military and by industry. Laser areas still include the small garage shop, the small industry, specific laser industries as Spectra Physics and Hadron coherent radiation, and the laser division of such large groups as American Optical Co., IBM, Bell Laboratories, Raytheon, G.E., R.C.A., G.T.E. Sylvania, Compagnie Générale Electrique in France, Siemen SA in West Germany, and many industries in Japan. Government laboratories include the Armed Services and also the large laboratories in the Bureau of Radiological Health and other government divisions. Lasers have a brief introduction in the high school physics laboratory, somewhat more extensive in the university physics department, and a start has been made in the areas of laser art and music in the Cincinnati area. Unfortunately, the biomedical applications, except for eye work, remain a minor phase of current medical technology.

Laser technology has been helped greatly by its association with the electronic industry. This has produced the electro-optics discipline. Its great process in 1972 was reported to be related to the following applications with lasers:

1. Control
2. Communications – transmit, display, store by holography
3. Information processing
4. Power conversion and generation
   a. Solar power (electric)
   b. Thermonuclear fusion

Industry, of course, is still optimistic. Perhaps its market, at least in the U.S., totals near 200 million. The modest (or is it immodest?) estimates increase 20% each year. Perhaps also, there is still more budgetary expenditures by the military than by industry. The recent vast expenditures for laser-induced thermonuclear fusion are considered industrial. It is predicted that those companies that have endured the early hard times can look forward to the future when lasers will do much for the "good of man."

Those of us who have tried to take, sometimes, without complete understanding, the results of the laser physicists, believe that there is much to offer in laser technology for the good of man by making the world smaller with clearer, shorter lines of communication and understanding. There have been great developments in the laser and there will be even greater developments, for irrespective of those who do not understand, the laser, like the automobile, is here to stay. It is still an infant. It is needed today and will be needed even more in the world of tomorrow.

Some difficulties slowing advances include undue concern for the basic cries, oft repeated, "if you don't need the laser, don't use it." There is then the need for large venture risk capital, usually from the government. Now there is the great concern for the consumer exposed to a new physical modality. If through hysteria and pencil, not field research, unduly restrictive and rigid safety codes are developed, then again progress is slowed until all are convinced. The often repeated plea for more interest and support for basic research will produce adequate programs for laser safety to assure the continued and inevitable development of laser technology. Yet, progress is inevitable. The limits to laser research and development are determined by dreams, facilities, expertise, and support.

Eventually there will be an X-ray laser. Boster[1] predicts one by 1978. Besides the usual concern for this laser as offensive weaponry, there is interest in this laser to show molecular structure, X-ray microscopy and holography, and to be used for cancer therapy, communications, material inspection, etc.

REFERENCES

1. **Boster, T.,** X-ray laser, *Laser Weekly,* December 25, 1972.
2. **Brech, F.,** personal communication, 1972.
3. **Foster, H.,** High-power $C0_2$ lasers. A review, *Opt. Laser Technol.,* 4, 121, 1972.
4. **Glicksman, R.,** A bright future for semi-conductor lasers, *Opt. Spectra,* 6, 17, 1972.
5. **Goldman, L. and Rockwell, R. J., Jr.,** *Lasers in Medicine,* Gordon and Breach, New York, 1971.
6. **Hagen, W. F.,** Ultraviolet lasers, *Ind. Res.,* May 1972.
7. **Heavens, O. S.,** *Lasers,* Gerald Duckworth, London, 1971.
8. **Jones, D. G. C.,** Lasers for the masses, *Nature,* 235, 232, 1972.
9. Laser: sunlight powered, *Laser Weekly,* February 7, 1972.
10. **Patek, K.,** *Glass Lasers,* The Chemical Rubber Co., Cleveland, 1970.
11. **Patek, K.,** *Lasers,* The Chemical Rubber Co., Cleveland, 1967.
12. **Piloff, H., Searles, S., and Nicholas, D. J.,** Flame laser, *Electro-Opt. Syst. Des.,* July 1971.
13. **Polanyi, T. and Tobias, I.,** Principles and Properties of the Laser, American Optical Corp., Research Division, Framingham, Mass., 1971.
14. **Pressley, R. J.,** *CRC Handbook of Lasers with Selected Data on Optical Technology,* The Chemical Rubber Co., Cleveland, 1971.
15. **Silfvast, T. W.,** He-Cd laser, *Laser Weekly,* January 17, 1972.
16. **Vanier, J.,** *The Basic Theory of Lasers and Masers. A Density Matrix Approach,* Gordon and Breach, New York, 1971.
17. X-ray laser, *Lasers Electro-Optics Ind. Res.,* September 1972.
18. **Schawlow, A. L.,** Advances in optical masers, *Sci. Am.,* 209, 34, 1963.
19. **Charschan, S. S.,** *Lasers in Industry,* Reinhold, New York, 1972.
20. **Wutzke, S. A. and Pack, J. L.,** "Coffee" laser system, *Laser Focus,* 9, 27, 1973.

Chapter 2

WAVE GUIDES AND THIN FILMS

Applications of the laser often depend on the flexibility of the system. Any aid to get the high output laser off the optical bench is important.

Fiber Optics

Because of the importance of wave guides for the transmission of the laser beam, some attention will be paid to this. Fiber optics are used as wave guides. These serve not only to conduct light images, but also they may act as magnifier converters. Kapany was one of the early pioneers in the field of fiber optics. Fiber optics may be composed of glass, crystalline or fused quartz, or plastics.

At present, plastics account for most of the fiber optics market.[11] The plastics may include acrylic and polyethylene jacketed fibers. Television sets may be operated with both audio and video transmission through a single optical fiber reported not to have interference such as occurs with the coaxial cable. There are other extensive uses of fiber optics singly and in bundles such as for display purposes for industrial systems and for medical purposes, especially for endoscopy. With this technique, light beams are conducted into cavities in the body. Such transmission can also be done with lasers so that these wave guides may also serve as operating tools.

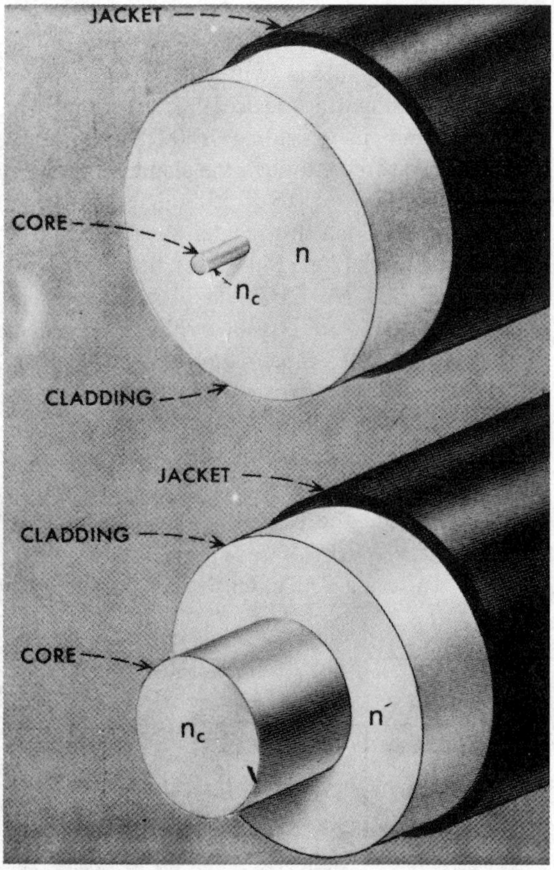

FIGURE 2-1. Optical fibers – single mode top: multimode bottom. (From Li, T. and Marcatili, E. A. J., *Bell Lab. Rec.*, 48, 331, 1971. With permission.)

Lasers require quartz fibers. Until recently, laser fiber optic systems were limited to the visible region. Now fiber optics systems are available for infrared and even UV systems.[1] The difficulty, of course, is the loss of light through transmission in the fiber optics. Some applications may not care about high losses over short distances but, for example, where communications are concerned, low loss fibers are required for transmission over long distances. Transmission losses in the usual fiber optics are about 1 dB/m or 1,000 dB/km. This is equivalent to an attenuation factor of 10^{100}. Corning[5] has indicated that its new current fiber optics wave guide looses "just four decibels per kilometer (i.e. the light level falls to a quarter

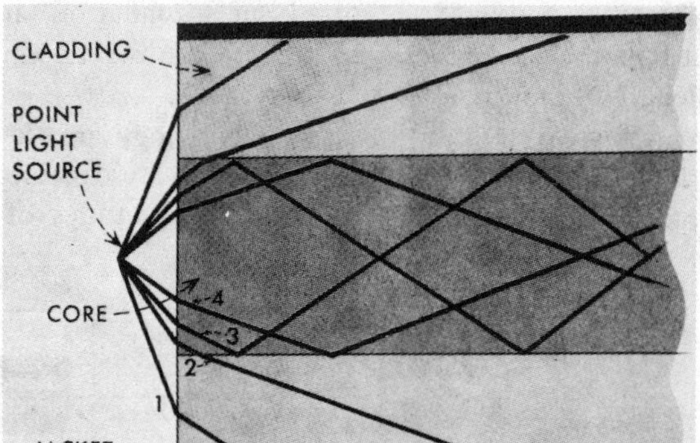

FIGURE 2-2. Multimode optical fiber showing incoherent light source as from an LED; a single mode optical fiber can be fed only by the coherent light from a laser. (From Li, T. and Marcatili, E. A. J., *Bell Lab. Rec.*, 48, 331, 1971. With permission.)

FIGURE 2-3. Light from a fiber optics bundle. (From *Bell Lab. Rec.*, December 1971. With permission.)

TABLE 2–1

Fiber Optic Wave Guide Loss

(dB/km = decibels per kilometer)

1 dB/km –	Theoretical limit fiber optic wave guide loss
20 dB/km –	Highest acceptable level for economic communication systems
13.5 dB/km –	Liquid light pipe (Bell Laboratories)
4 dB/km (0.55 km)	New Corning fiber
?	Japanese fiber
?	Single fiber of Nath

(After *New Sci.*, September 21, 1972.)

of its original level)." The theoretical limit for fiber optics wave guide is said to be 1 dB/km. At present, Corning has reached 2 dB/km in a 0.55-km fiber. Actually, for long transmission needs repeaters could be spaced at intervals. Recently Shultz[13] reported an attenuation at 1,050 nm of 2.1 ± 0.5 dB/km through a 9.12 km length of multimode optical waveguide.

Of great interest for laser technology are the single flexible fibers of Nath.[7] These are capable of transmitting CW lasers power magnitudes of 100 W, pulsed powers of 100 kW, and normal mode pulse energies of several joules. The single fiber is a double tapered single crystalline quartz fiber with an entrance and exit of 3 to 4 mm and a so-called waist diameter of 0.3 mm. This type of structure transmits the laser better than the conventional fiber light bundle. There is a shielding of flexible tubing with metal and pieces containing collimating devices. The flexibility of the fiber and its shielding is 0.5 mm rad of minimal curvature. Although the standard aperture is 4 mm, apertures of 0.5 to 5.0 mm can be made. Beam diameters that are greater than this can be focused into the fiber. Since this fiber transmits infrared Nd and Nd-YAG and ultraviolet lasers, many applications are possible with this excellent flexible instrument. Targets, not easily accessible, can be illuminated; microwelding with pulsed ruby and Nd-YAG may be done. The acceptance angle of a fiber optics is a function of length and wavelength. If a limited angle of acceptance of 1 to 2° is desired, the Nath fiber may be used to measure stray light from plasmas and from Brillouin scattering. The extensive uses of the Nath

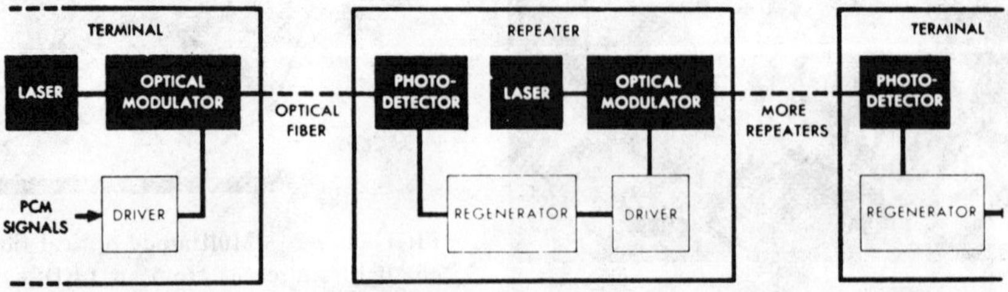

FIGURE 2-4. Optical fiber transmission system using lasers. (From *Bell Lab. Rec.*, December 1971. With permission.)

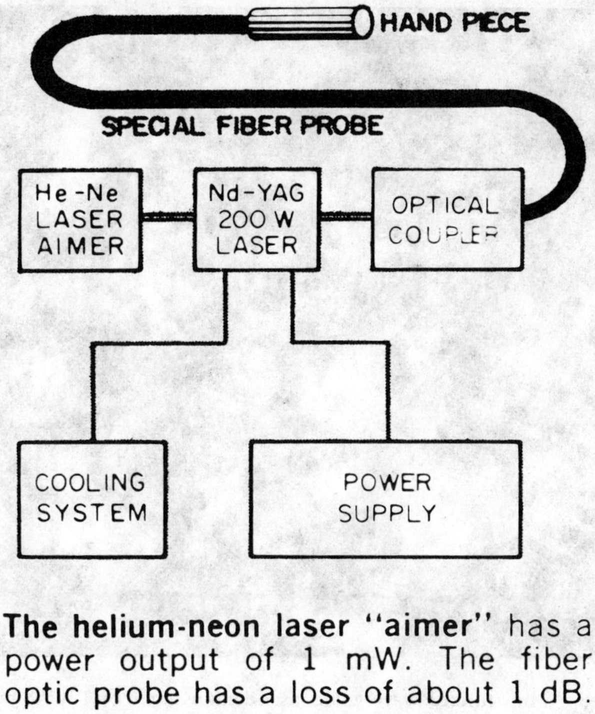

FIGURE 2-5. The single fiber optics of Nath.

fiber in biology, medicine, and dentistry will be described later.

Alexander[1] has described fiber optics into the UV range. This type of fiber optic contains a core of high purity quartz produced by Schott Optical Glass.® The cladding is a special UV transmitting "plastic material" which is not identified. A figure of 50% for 254 nm transmission is given for a 1-m bundle; for short bundles, a figure of 50% for 200 nm is given. Schott makes 70- or 100-μm individual fiber diameters. The bending radius, as indicated, is important. A 1-mm active bundle of 70-μm fibers can be bent in a radius of 20 mm. These fibers have applications in studies of fluorescence, especially in biology and medicine. Now the ultraviolet portion of the spectrum is open for applications.

An interesting wave guide, again especially for communications, is the new miniaturized wave guide gas laser described by Smith.[10] As the diameter of the discharge tube in a gas laser is decreased, the gain is increased light "guided" in this tiny tube and it can be amplified at a much higher level than usual by a gas discharge. The first experiments were done with an He-Ne laser with gains of 2.7 dB/m with a glass capillary tubing with a 430 μm bore. "Laser output powers of 1 mW were measured with a 20-cm tube. R.F. excitation was necessary to maintain a stable discharge" but now cold cathodes are used with DC excitation. With variations in the kind of gases used to fill the tubes, wave guide gas lasers are available in visible and in the near infrared.

A 400-W CO_2 laser has been used by Wenekus et al.[12] to grow single crystal fibers without a crucible of chromium-dipped sapphire with strengths up to 1.4 million lb/in^2. Titanium carbide and yttrium trioxide fibers have also been made.

Current research on fiber optics wave guides continues to attempt to make the high output laser more flexible and yet, even at a distance, maintain the brightness of the laser beam. Anything that can free the laser from the fixed optical bench will make the laser more useful. Fiber optics have shown a continued progressive development and much of its progress is due to its uses in the transmission of the laser beam. Since it is possible to transmit high output laser beams, the range of applications will be limited, as it is usually in laser technology only by the imagination and dreams of the investigator. The limitation also includes the

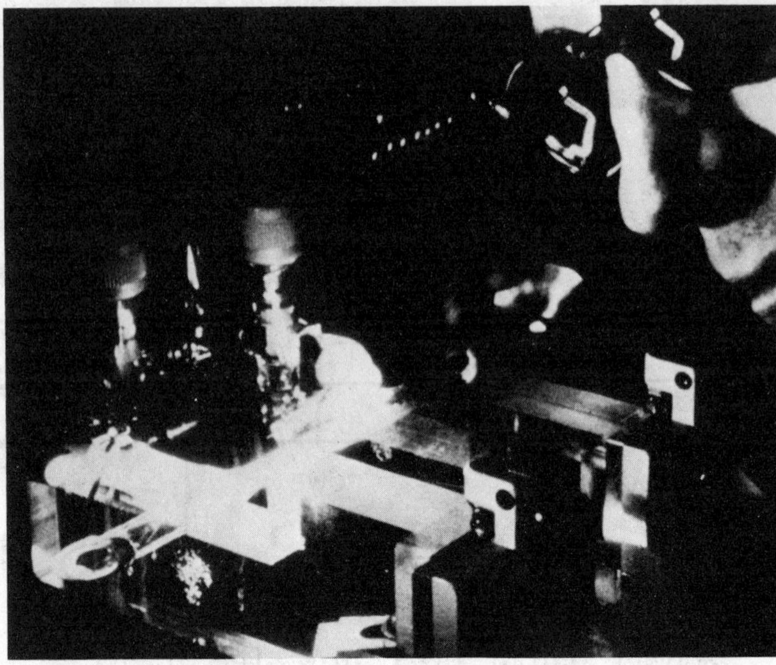

FIGURE 2-6. Miniaturized wave guide laser demonstrated by Peter Smith. (Bell Telephone Laboratories, May 1972.)

ancient excuse of "the-state-of-the-art" and the cold economics of initial pilot development even with subsequent high market volume.

Thin Films

Thin films of dielectric materials are important for many appplications of current laser systems. It is indicated that "thin films are now part of every optical system in use today."[2,9] Certain properties are required for these important coatings. Some of these are

1. Low absorption
2. Low scatter
3. Ability of the thin film to adhere to the substrate
4. Good hardness
5. Chemical purities

It is reported that the absorption is so low that it is difficult to measure values accurately. Scatter may be associated with poor evaporation techniques, dirty vacuum systems, and "improper preparation of the optical substrates." Inhomogeneity creates differences between the film thickness and the refractive index. For far UV optical coats, Mulherbe[6] has described metallic interference filters in the range of 0.12 to 0.3 μm an multidielectric coatings over 0.17 to 0.3 μm. Som multidielectric filters had passbands of 2 nm a 0.205 μm. So-called "coated" polarizers are now under investigation to attempt to replace oı supplement many of the currently used calcite polarizers. Electron beam guns can provide physically hard coatings. Durable coatings are needed especially for infrared lasers. These are coatings of metal, or metal-dielectric, or are all-dielectric. These IR lasers are CO_2, TEA CO_2, and chemical lasers.

Some new applications from advanced thin film techniques are reported by Broomer as:

TABLE 2–2

Materials
Light Metal Optics[4]

1. Aluminum
2. Aluminum alloys
3. Beryllium
4. Beryllium alloys
5. Base materials
   A. Copper
   B. Titanium
   C. Molybdenum
6. Plating – Electro-less nickel

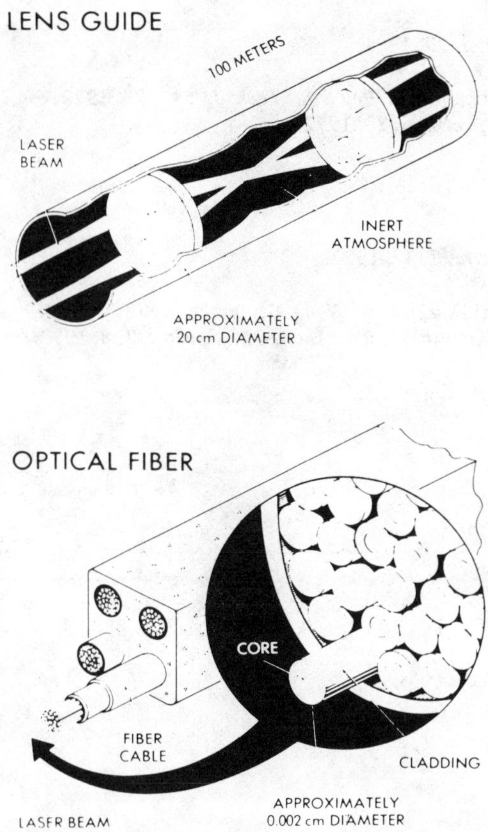

FIGURE 2-7. Laser holographic camera of Korad.

FIGURE 2-8A. Laser trap with Ronald Hill, supervisor, Sandia Laboratories Experimental Aerophysics Division. Laser beam enters from right, enters light trapping cell, strikes flat mirror left, and is reflected to and from ellipsoidal mirror right at the focal point between the two mirrors. The beam is increased by several orders of magnitude. (From *Lasersphere,* October 15, 1972. With permission.)

1. High energy lasers for controlled thermonuclear fusion
2. Navigational aids, primarily laser lighthouses and laser channel markers
3. Integrated circuits
4. Holography
5. Helicopter let-down laser aids
6. Rapid scan spectrocolor meters

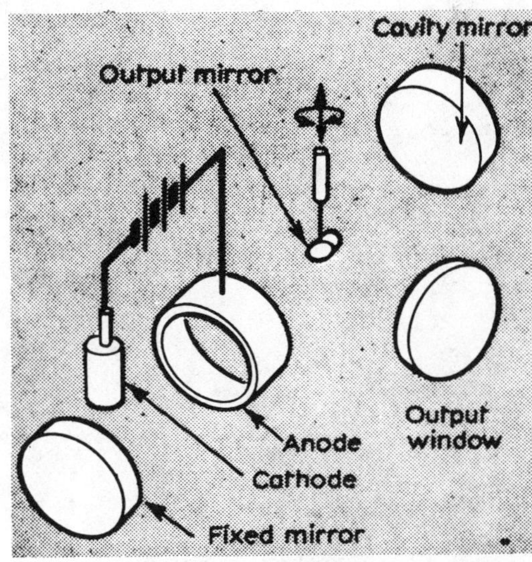

FIGURE 2-8B. The Sandia technique to extract light from an infrared laser-diagram showing the adjustable and replaceable third mirror to reflect the light through the high density polyethylene output window. (Sandia Laboratories, 1972.)

Better lines of communications are needed in this field to develop new and better coatings. Can rival research areas give freely of their own research and expect something back in exchange?

Now, electricity can be produced by pulsed impacts on thin films. This can be applied soon for thermal gradient values and for light detectors.

Light Trapping Cell

Another technique to extract light from a far infrared laser has been reported. Scientists at Sandia Laboratories[8] have used a simple economic method to do this. A small replaceable third mirror, a brass rod cut and polished at 45° to the lasers major axis, is suspended within the laser cavity by a 0.635-mm tungsten wire suspended from a length of a 6.35-mm drill rod. The rod is moved radially via an 0-ring seal. The output beam exits by an 8.26-mm-thick high density polyethylene window.

REFERENCES

1. **Alexander, W. B.,** Fiber optics moves into the U.V., *Opt. Spectra,* 5, 27, 1971.
2. **Broomer, C. J.,** Thin films and future applications, *Opt. Spectra,* 6, 34, 1972.
3. **Cohen, L. G.,** Power coupling from GaAs injection lasers into optical fibers, *Bell Syst. Tech. J.,* 51, 573, 1972.
4. **Marshall, G. F.,** Light metal optics, the new look in space, *Opt. Spectra,* 6, 23, 1972.
5. **Maurer, R.,** Corning, *New Sci.,* September 21, 1972.
6. **Mulherbe, A.,** Thin film optics, *Opt. Laser Technol.,* 6, 147, 1972.
7. **Nath, G.,** personal communication, 1972.
8. **Hill, R. A.,** Sandia laser technique, *Opt. Spectra,* 7, 14, 1972.
9. **Small, E. A., Jr.,** A door to new markets for thin films, *Opt. Spectra,* 7, 32, 1972.
10. **Smith, P.,** Bell Laboratories, May 1972.
11. **Snyderman, N.,** Fiber optics makers turning to industrial application, *Electron. News,* September 1972.
12. **Wenkus, J., Menashi, W., Haggerty, J., et al.,** New crystal fibers (Arthur D. Little, Inc.), *Opt. Spectra,* 7, 8, 1972.
13. **Schultz, P. C.,** *Laser Weekly,* July 9, 1973.

Chapter 3

MEASUREMENT OF AND BY LASERS

Obviously, it is necessary to know the output of the laser not only to evaluate and compare the laser performance, but also to monitor the output for applications and for safety.

In order to know the output of the laser, careful measurements must be taken. There are instruments available to measure optical power, energy, pulse duration, and pulse frequency. For high intensity outputs, so necessary today and to increase in the future, measurement techniques are difficult. The reader is referred to literature on measurements. For pulsed laser systems, the techniques used by the National Bureau of Standards can achieve accuracies of 3 to 10%.[4] Goldman and Rockwell[2] indicated that commercial instruments for measurement should be calibrated to an accuracy of ±15% with certain specifications for limiting aperture, response time, spectral response, and response to multiple pulse sources.

Some instruments used for measurement are

- I. Pulsed lasers
  - A. Pulsed laser energy
    - 1. Calorimeters
      - a. Cone
      - b. Disc
      - c. Liquid cell
      - d. Rats nest
    - 2. Photoelectric
    - 3. Pyroelectric
  - B. Pulse duration and peak power
    - 1. Diffuse reflection
    - 2. Picosecond
      - a. Photon absorption fluorescence
- II. CW laser power measurements
  - A. Thermopile
  - B. Disc
  - C. Photodetector
- III. Spatial distribution measurement
  - A. Photographic with densitometer traces
- IV. Beam divergence measurement
  - A. Photographic with densitometer recording
  - B. Burn spot on dark paper for pulsed lasers

Accurate measurements even of picosecond laser pulse lengths can be measured by a picosecond camera. This operates on the principle of "two photon fluorescence." The pulse length can be calculated from the photograph.

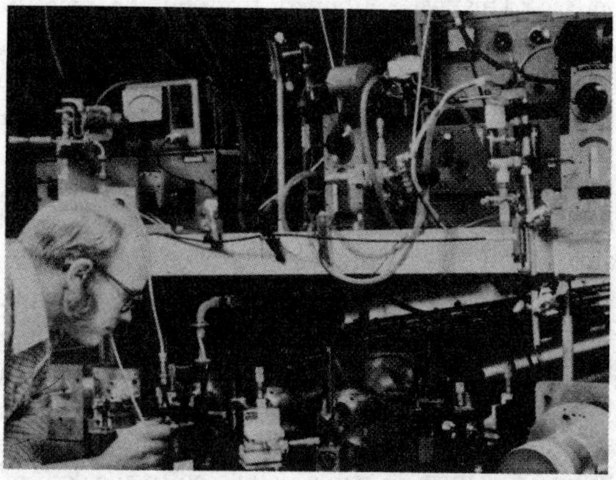

FIGURE 3-1. Dr. Ken Evenson, who directed NBS research for the new value of light. The value is 299,792.4562 km/sec ± or −1.1 m/sec or 186,282.3960 mi/sec ± or −3.6/sec. This shows his array of lasers and the cat's whisker diode to measure laser frequencies. (From NBS, *Sci. News,* 102, 356, 1972. With permission.)

It appears now that everyone is interested in accurate power level or pulse energy data. This is especially true for industrial metalworking where standardized performance especially for drilling and cutting is required. With the current concern about the low level of permissible exposure levels to lasers, accurate measurements must be made for monitoring in safety programs.

Many people are not aware that Dr. Kenneth M. Evenson received the 1972 Samuel Wesley Stratton Award for his achievement in frequency measurement. This is awarded annually by the U.S. Department of Commerce National Bureau of Standards for "outstanding scientific and engineering achievement by a staff member."[5] The laser has made it possible to obtain accurate figures for the speed of light. In 1970, a group from the Massachusetts Institute of Technology[4] made the high frequency measurement with lasers. This was a frequency measuring technique which is greater than 10,000 times more accurate than wavelength measuring techniques which are done by multiplying the frequency of the electromagnetic wave by the length of the wave.

In the work at the National Bureau of Standards Boulder Colorado Laboratories, Evenson, Day, Well, and Mullen found that the frequency emission of the helium neon laser was 88.376245 THz (trillon cycles/sec). The new figure of the speed of light is 299,792.4562 km/sec or 186,282.3960 mi/sec. This was obtained through the measurement of the frequency of an infrared laser beam.

What is the value of such accurate measurement of the speed of light? This accuracy will open up possibilities for a whole new frequency range for telecommunications. These would include, also, new areas of interplanetary communications. It also would be of value to reach their great goal. Dr. Louis M. Branscomb, Director of the National Bureau of Standards, said "ever since Albert Einstein showed that time can be considered the fourth dimension of space in which we live, scientists have looked forward to the possibility of using one gauge – one yardstick, so to speak – not only for the two dimensions of space but for the fourth dimension of time as well. To interchange clocks and rulers scientists must know the speed with which light travels, which is equal to its wavelength times its frequency. With this demonstration of both space and time dimensions of a single light source can be measured with prodigious accuracy. This goal is now within our grasp." This would mean that some day there would be a number that would be a standard used for light and time measurement.

Industrial use of this great accuracy will be shown in instrument manufacturing and precision equipment; it will help in the study of pollution by measurement of minute qualities of pollutants.

The laser itself, especially the helium neon laser, is used frequently in the physics laboratory and in industry as a standard for accurate measurement. It can be used in the machine tool industry. A dynamic yardstick for accurate measurement is now available which can be applied to any area.

There is another phase for consideration of the measuring effect of the laser, namely, the use of the laser for Doppler effects. Everyone is familiar with the increase and decrease in the pitch or noise frequency of a train's whistle as it passes an observer. This change in wave frequency when either the source of the waves or the observer moves to or away from each other is known as the Doppler effect, especially in ultrasonics. The laser can be used for flow velocity applications. This would refer to the fluid in pipes and many techniques in biology and medicine in measurement of circulatory times, obstruction of blood vessels, and so forth.

There is still another phase in laser measurement consideration, namely, the measurement of laser output itself.[8] It cannot be emphasized too strongly that accurate output data are necessary so that a standard language is available everywhere. Rockwell,[6] who has done so much work in critical evaluation of measurement of laser radiation, says, "Knowledge of the magnitude of the laser output levels is an absolute necessity if one is to attempt to quantitate the interaction effects of laser radiation, that one must consider a number of features of output parameters in order to evaluate laser radiation." Some of these are such factors as beam divergence and pulse duration. According to Rockwell, "The importance of the rate of delivery of energy density suggests that power density expressed in watts/square centimeters is a more explicit direction of laser output." The total energy is usually expressed in joules and the spot size is usually expressed in square centimeters. Since higher and higher outputs of laser radiation are being used now, the accurate measurements must be known. West[7] emphasized that because of the increased application needs and the considerable controversy regarding safety limits there is considerable pressure on laser users and manufacturers for accurate, reliable measurements of the output, power output, and energy. One is referred to the current literature on some of the difficulties and instrumentation that is required with the various types of calorimeters and photoelectric measurement, pyroelectric diffuse reflection method, and the great difficulties in the measurement of the ultrahigh output. For the continuous wave power measurements the thermal pile and the photodetector systems have been reviewed in detail by Rockwell and by West. West believes, however, that in spite of all the difficulties, the laser power and energy measurements can be performed with an error limit as low as 1%. So then, with any laser that is being used for any application there must be figures of accurate measurement available. These should be included not only in the specifications desired, but also in the high and low output figures of the manufacturer. When applications are described, the type of laser and the technique of transmission should also include the output figures and the technique of measuring the output. Only in this way will standardization be achieved so that application

FIGURE 3-2. Measurement of rims of Grand Canyon by the time it takes for laser beam to bounce from rim to rim. (Laser Systems & Electronics, Inc.)

results can be duplicated anywhere and not found only in one investigator's research and development laboratory.

Accurate standards may have other applications in the field of measurement such as the early detection and prediction of earthquakes. Laser surveying and contouring have been mentioned before. Oceanography offers another field for accurate laser measurement and techniques.

REFERENCES

1. **Cealdia, R.,** Lettere a Nuovo Cemento, *New Sci.,* 4, 821, 1972.
2. **Goldman, L. and Rockwell, R. J., Jr.,** *Laser in Medicine,* Gordon and Breach, New York, 1971.
3. National Bureau of Standards, Laser Power and Energy Measurements Technical Note 382, U.S. Department of Commerce, October 1969.
4. National Bureau of Standards, Power and energy measurement of repetitively pulsed laser, Technical Note 398; Measure of light, *Sci. News,* 101, 85, 1972.
5. National Bureau of Standards, Speed of light, *Laser Weekly,* December 4, 1972.
6. **Rockwell, R. J., Jr.,** Personal communication, 1970.
7. **West, E. D.,** A skeptic's guide to laser energy measurement, *Electric-Opt. Syst. Des.,* November 1972.
8. **Zimmer, R.,** Power and energy measurement, *Laser Focus,* 6, 39, 1970.

Chapter 4

LASER CHEMISTRY

In laser chemistry, there is much new and exciting material to appeal to chemists, in all fields, not only for those interested in photochemistry. In brief, laser chemistry may be divided roughly into the following phases:

1. Chemical lasers
2. Tunable dye lasers
3. Photochemistry
4. Laser microprobe analysis
5. Spectroscopy – to replace conventional light source
6. Strange nonlinear effects
7. Some chemistry of material processing
   A. Metal
   B. Plastics
   C. Pollution control
   D. Miscellaneous
8. Brief notes on thermonuclear fusion and plasma chemistry

Chemical Lasers

The chemical reaction can be an effective stimulus for population inversion for laser action. Chemical lasers are becoming more popular because of their relatively high efficiency, economy, and high output. In general, those chemicals that fluoresce may be capable of lasing. Much has been written about chemical lasers. An example of the high efficiency of chemical lasers is the one with helium gas plus sulfur hexafluoride through hydrogen gas; the efficiency is 10 to 20%. Selenium oxychloride deuterium and HF-DF, are among the more recent developments. Europium and terbium chelate-doped polymer rods have also been developed. Recently, Whitaker used a neodymium-containing chelate using polymethyl methacrylate (PMA) in a solvent.

In 1966, chemists at the Naval Research Laboratory[13] used the insertion of an oxygen atom from ozone photolysis into a partially halogenated methane molecule to produce halomethane. It is reported that these halogenated alcohols with halogen atom and hydroxyl group attached to the same carbon are very unstable. They decompose "almost instantaneously" into carbonyl compounds and highly excited halide molecules that lase.

In regard to high pressure chemical lasers, Russian investigators report[3] that for stimulated phototransition of electrons at the movement of contact between two nonexcited gas molecules, the elementary chemical reaction from irradiated transition of electron changes; if this reaction mixture flows through an optical resonator with mirrors, it could be a CW laser.

Many new chemical lasers are being developed, particularly for high output lasers for thermonuclear fusion.

Tunable Dye Lasers

Mention has been made of the interesting new tunable dye lasers in which various dyes have been used in a laser and new different frequent frequencies are produced.[49] The mechanism of how molecules can be excited to different vibrational levels with two separate tunable sources of photons has been reported recently in data from Chromatix. Chemicals may be used to produce second and third harmonics. One example is KDP, potassium dihydrogen phosphate.

Laser Microprobe Analysis

The laser atached to a microscope has been used as a microprobe for techniques for microanalyses. This phase of microemission spectroscopy will be considered before the other applications in spectroscopy.

For biological material, for example:

1. The assay of submicrogram quantities for elemental ingredients can be done with the laser used as its own mass balance to avoid specimen weighing and handling. No special preparation of specimens is necessary except drying or freezing of liquids.
2. Subsequently, for larger size biological specimens, multiple shot excitation techniques can be done to increase analytical sensitivity. This can be developed as a long-range automated laser analyzer.

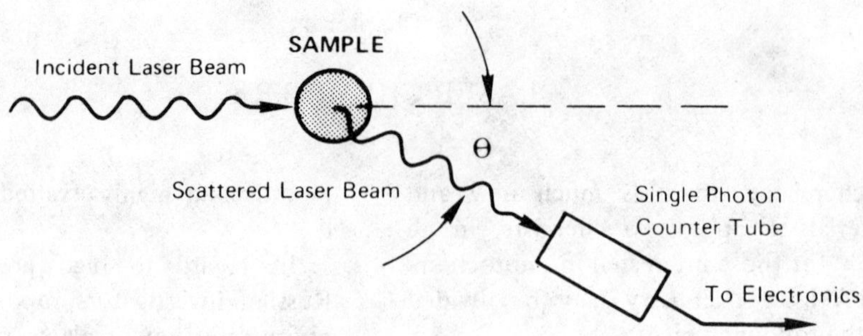

FIGURE 4-1A. Structural studies by light scattering.

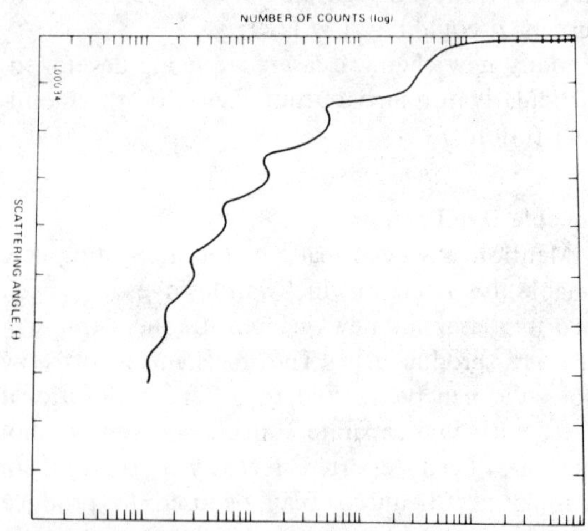

FIGURE 4-1B. Scattering angle. (Ortec, Inc., Application Note AN 35, The Single Photon Technique for Measuring Light Intensity and Decay Characteristics, 1971.)

The use of the laser microprobe as an analytical tool has been applied most frequently for qualitative analysis and only occasionally for microanalysis because the instrumentation has been predominantly of the pilot type, crude, and required fairly large samples. Preliminary studies have shown the value of the laser for analysis. Trace metals have been studied in connective tissue in muscle, kidney, and intestine. Detailed studies have been conducted by Rosan et al.[39,40] as well as Sherman[44] and Brech.[40] Recently, the most accurate studies have been done by Glick[12–15] with his much needed modified instrumentation. True standardization has been achieved, also only recently, by Glick and Rosan.[12]

Recent developments in much needed revisions of the laser microprobe have now made it possible to do submicroanalyses. The sensitivities of the method will be reviewed in detail.

Since the early days of laser microprobe of the analysis, there has always been concern about quantitative standardization. For laser microanalysis, Rosan[39,40] used standard solution materials in small disks of unfixed photographic emulsions. The silver from the photographic emulsions served as an "interval" standard.

Other attempts at standardization have been made by Katsumo with aqueous solutions of the metal salt on agar and hydrogen sulfide with precipitation of the sample with a sulfide and then desiccation. Halprin used the methacrylate base studies. The following data derived from Harding-Barling and Rosan[4] indicate that the following sensitivity limits may be possible to obtain:

Zn, Cd	10^{-16}g
Al, Ca, Cu, Fe, Nn, Mo, Pb	10^{-17}g
Ag, Cr, Mg, Sn, Sr	10^{-18}g

Early evaluation of the analytical sensitivity of the laser-microprobe method consisted of a visual observation of spectra recorded on photographic plates which gave a rough idea of detectability and elemental abundance. However, this crude method very soon was refined and attempts were made to improve the analytical precision by the use of either an internal or an external standard. More recently, calibration curves were published; some were accompanied by an estimation of their associated accuracy. Soon there will be a more critical definition of both the use and the limits of the method. With the new laser microprobes it is believed that detection limits and sensitivities could be defined on a statistical base which is the only meaningful way to describe sensitivities.

The detailed literature of the sensitivity of laser microprobe analysis will be reviewed.

Detectabilities estimated by qualitative observation of spectra recorded on photographic plates were reported by Debras-Guedon and Liodec[10] who with a 100-J ruby laser, produced craters of the order of 100 μm in diameter and without the use of cross-excitation, produced spectra of some 25 elements from pure sample or simple compounds or alloys of those elements. It is interesting to note that even with high laser energy, no spectra were obtained from Ag, B, Be, Bi, C, Cd, Co, Hg, Li, P, and Sb. Rasberry, Scribner, and Margoshes[36] signaled the usefulness of the laser microprobe for detecting Si, Fe, Ni, Co, Au, Ga, and Rh in semiconductors, wires, ceramics, PtRh alloys, and dental amalgams.

Similar qualitative detections in nonbiological samples were reported by Adams and Tong,[1] who used the laser for evaporation and performed the analysis by neutron activation and mass spectroscopy in a two-stage method, by McCormack[27] for copper inclusions in steel, and by Hellman[19] who investigated rare paintings. Tlalka[47] introduced a ceramic plate 1-mm thick, with a hole 2 mm in diameter in its center to separate the arc produced by the cross-excitation from the sample, which was a system of Al, Al_2O_3, and UO_2 in which he studied diffusions. Mode et al.[30] enhanced the plasma emission intensity with the use of an electrodeless R.F. excitation.

Qualitative detections in biological materials were reported by Rosan and his co-workers[39] on sections of brain, pancreas, and stomach tissues. Their aim was to trace certain elements like Zn; no quantitative analysis was needed.

Determinations of relative concentrations were performed by Adams et al.[1] on UO_2 samples. The sample taken was of the order of the 10^{-8} g. Detection limits were published by the Jarrell-Ash Co. for 15 elements in low alloy steel, Cu-Zn alloys, and aluminum alloys. For comparative purposes, it will be interesting to cite the following values of weight concentrations in Cu-Zn alloys which are considered to be the limit of detection when the crater is approximately 25 μm in diameter, the removed material having been estimated to be .15 μg: for Ni, 1450 ppm; for Fe, 700 ppm; and for Ag, 85 ppm.

Relative concentrations of Mg, Al, Fe, and Zn were estimated using an external standard by Rosan et al.[40] in human teeth where a crater 65 μm in diameter was produced. Other biological materials were tested for quantitative analysis by Goldman et al.[16] and Sherman and co-workers,[44] who examined supra- and subgingival calculus for their content in Mg, Al, Si, Ca, P, Fe, Cu, and Zn. Sherman et al.[44] carried out a similar study on tooth enamel, dentin, carious dentin, cementum, calculus, and bone. The concentrations of phosphorus ranged from 0.001 to 10%, of Mg from 0.001 to 0.1%, and so on: the crater sizes were of the order of 50 μm in diameter and 25 μm deep. Glick and Rosan reviewed the application of the laser microprobe method on biological materials and reported on determinations of Ca and Mg concentrations in both dog and baby stomach tissue. The need for instrumental improvements was previously pointed out in a 1966 publication. Glick et al.[15] reported on improvements of the laser microprobe and tested the instrument for its analytical capability on kidney where concentrations of Ca, Fe, Zn, Cd, Cu, Mn, Al, Pb, Mo, Sn, Sr, and Cr were estimated and where detection limits were said to range from 10^{-18} g for Cr to 10^{-14} for Mg. Laser impacts were shown on samples as small as human leukocyte and rat spermatozoon head.

There have been numerous concentration determinations based on calibration curves made from standards of the same material as the sample. Most of them are extensions of known spectroscopic methods. Typical for the early measurements of this sort is the report of Vilnot et al.[50] who, using a 100-J laser, published calibration curves for Cr and Ni in ceramics using Fe as an internal standard. Runge and his colleagues[41] superimposed four laser exposures to produce a crater 0.4 mm in diameter and 50 μm deep in stainless steel. They published calibration curves for Cr and Ni. Karyakin et al.[22] used a 10-J neodymium glass-pulsed laser to produce their calibration curves for Cr, Mn, Cu, Fe, Si, Mg, and C in feldspar, biotites, silicates, and rubies. They gave a 6% accuracy for the direct method and a 12% accuracy for a two-stage method when the vapors were first collected and then analyzed in a conventional spectroscopic way. Ryan et al.[42] reported determining the concentration of Al and Si in glass defects. The size of the used crater was 25 to 40 μm.

Microconcentration gradients of B in steel were studied by Webb and Cotterill[51] at a 50 μm spatial resolution that was sufficient to discriminate austenitic areas from "precipitate" zones where the B content was much higher.

Rather than citing many other authors in this line, one basic difficulty in establishing calibration curves will be pointed out. It was reported by Baldwin,[5] who showed that the melting step of the material removal process is thermally equilibrated. The vapor composition could be predicted from the liquid-solid equilibrium diagram in the copper-zinc system he studied. Here again, the laser was a relatively energetic ruby laser, producing craters approximately 160 μm in diameter.

A thorough study of the accuracy of the method has been carried out by Devlin and La Conti.[11] Except for the use of a ruby laser, the method itself is classical and refers the integrated intensity of the measured line to an internal standard. The crater sizes ranged from 74 to 202 μm in diameter and 28 to 118 μm in depth. The coefficients of variation were given for all the curves and ranged from 9.55 to 30%.

In garnet-bearing gneiss, Blackburn and his co-workers[9] reduced the coefficient of variation of 6 to 13% for relative concentrations by using an internal standard consisting of the added intensities of three or four of the considered elements. This was considered entirely satisfactory for most petrologic applications.

Attempts were made to measure elemental concentrations in biological materials with a known accuracy. Peppers et al.[34] published coefficients of variation for the energy of their laser (2.5 to 5.4%), for the spectral signal (2.1 to 5.9%), and their ratio (1.7 to 4.9%). These coefficients were calculated from 9 to 13 measurements of Mg in 0.1 μl samples of human blood serum which were totally evaporated.

The smallest reported laser-induced injuries to cells were of the order of 2.5 μm in the experiments conducted by Bessis et al.,[8] who reported that "craters less than 1 μ diameter could be formed." They actually show a figure of a 3-μm crater on a chromium film deposited on a rat liver section. They noted that "it has been possible to sample single cells, such as white blood cells . . . . . with production of spectral lines of sufficient intensity to be useful for analysis." This was the only mention of possible analysis of single white blood cells by the laser-microprobe method. The white cell analysis, together with elemental detections on spermatozoon head, represents to the author's knowledge today's ultimate capabilities of the laser microprobe as far as smallness of sample is concerned. The difficulty of accurate measurements, on the other hand, is well illustrated by the studies on material removal with laser beams by Ready,[37,38] Andreev et al.,[3] and Klocke.[23] The latter demonstrated the occurrence of significant reflection of a normal laser pulse from metallic surfaces. In organic materials the matrix often introduces a high level of continuous background, as was shown for borine albumine and sucrose by Glick and his co-workers.[15]

A new laser microprobe of Fisher Scientific developed by Brech and Allemand[2] has a Nd glass laser. The emission of the Nd glass laser at 1.06 μm is neither detected by the photographic plate nor by the photomultiplier, as was the red emission of the ruby laser. The immediate result is a great reduction of a background that resulted when light strayed in the spectrograph.

Both mechanical and optical stability of the laser were carefully improved and the laser output energy is adjustable in 25 discrete steps, thus allowing a fine control of the crater size. Its total energy is monitored, providing a means for correction of variations in output.

The laser produces a single light pulse and is single-moded at low energies, thus permitting the optics to operate close to its diffraction limit, and the craters to approach their theoretical limit.

Older microscopes required a different focal setting for visual observation than for firing the laser. At very small crater size the focusing becomes critical; inaccurate focusing would result in changes in the crater size. The new microscope does not require such an adjustment.

It is known from Raman spectroscopy that, especially for small light sources, it is advantageous to collect the light with a high aperture objective. High apertures always call for large aberrations that have to be compensated for. This has been taken into consideration in the new laser microprobe, which uses a collection optics of approximately f/1.6 with negligible geometrical and chromatic aberrations of the image produced on the entrance slit of the spectrograph. This fact has been found to be of primary importance in separating the continuous background produced by molten particles from the plume-emitted signal and is quite an important improvement over the simple quartz lens used for single-cell analysis. The spectrograph is a Jarrell-Ash Fastie-Ebert with a capability for both electric and photographic recording and whose aperture is f/8.6 compared to f/22 for the Wadsworth spectrograph used previously.

The stability of the system was tested by taking groups of 10 shots at 15 mJ aimed at a homogeneous steel target (NBS Standard 1168). A photographic plate was exposed without using cross-excitation and the transmission of a Fe line was measured. The coefficient of variation was 2.6%, which compares quite well with the data of Karyakin and Feld.[24]

The crater shape in steel was studied with this new instrument for craters as small as 15 μm. Brech used a replication technique followed by a profile measurement on a roughness tester to actually measure the crater profile, which is very much like those reported by Baldwin on 160 μm craters in Cu-Zn alloys. Deposited molten material even around the smaller craters could be observed. An interesting difference, however, which will be reported later, gives more information on the mechanism of the material removed.

Spectra of pure elements were taken without the use of cross-excitation to give a first idea of the sensitivity. At a 15-mJ setting, spectra of Ag, Al, Au, Be, Bi, Ca, Cd, Co, Cr, Cu, Fe, Mg, Mn, Mo, Ni, Pb, Pt, Sb, Sn, Ti, W, and Zn aluminum alloy and brass were all rich. The crater size in Fe was about 40 μm in diameter. This is quite an impressive improvement over the work of Adams and Tong,[1] who were unable to record several of the mentioned elements even with their 100-J laser. Fe lines from steel samples were detected photographically for laser energies down to 0.8 mJ producing craters of 25 μ diameter. No cross-excitation was used.

These photographically recorded spectra were compared with similar spectra taken with a Jarrell-Ash Mark II Laser Microprobe associated with a Wadsworth spectrograph without cross-excitation. About 1 J of laser energy and a 180 μm diameter crater were necessary to produce spectra of comparable intensity. If the crater shape would be comparable (which is not correct because of the multispike emission of the ruby laser as opposed to the single pulse of the new laser), it would mean that about 100 times less material needs to be displaced to produce a photographic signal in the new system. In fact, since the multispike-produced crater is deeper, the comparison is conservative.

This result was confirmed when the output of the J-A Mark II Laser Microprobe was reduced to 0.1 J and produced barely visible spectral lines of Ag and Cd but none of Cu. Comparable results are achieved with a 0.8-mJ laser beam with the new instrument, which represents 125 times less energy.

Cu lines from a brass sample were easily detected photoelectrically from craters smaller than 10 μm produced by a 50-μJ beam. The estimated displaced copper was of the order of 8 pg. It produced a full-scale signal and does not represent a detection limit. No cross-excitation was used. Data given by Blackburn et al.[9] indicate that for Ni and Ag in a Cu-Zn alloy the detection limits would be 225 and 13 pg, respectively. For a semiquantitative analysis, Adams used an estimated 100 pg of material. But it is important to note that in the two latter cases the cross-excitation was used!

The sensitivity of the detection system is such that it is capable of recording one single photoelectron-produced pulse. In fact, a Cu line from a brass sample could be detected even when the laser energy was reduced to the crater-producing threshold. The diameter of the produced crater was below 10 μm.

At that level the meaning of the detected signal becomes a question of statistical interpretation and, of course, no precision can be given unless several tens of single photoelectron pulses are recorded. Therefore, a meaningful analytical measurement at those very low light levels always should relate the result with the deviation to be expected from the statistics of the electron emission from the cathode and to the noise of the P.M. With the new instrument, the calculated probability to record one noise pulse in normal conditions is of the order of 10^{-5} to 10^{-4}, making the noise effect practically negligible. Therefore, it is evident that:

1. The electronic detection system is not limiting the sensitivity.

2. The finely tunable laser output (from which energy can be further reduced by means of simple filters) provides an adequate source for most microanalytical work and is capable of producing the smallest possible craters using 1.06 μm IR light.

3. The light collection optics optimize the signal to background ratio because of its high aperture and its good optical correction.

4. With photographic detection and no cross-excitation, the new system is more than two

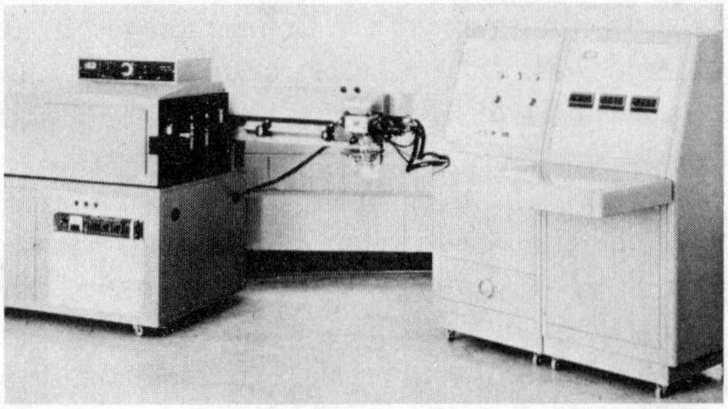

FIGURE 4-2. Advanced laser microprobed for photographic recording. (Fischer Scientific, Fred Brech and C. D. Allemand.)

orders of magnitude more sensitive than commercially available laser microprobe.

5. Currently, the actual measurements confirm the expected increase of sensitivity over older models.

6. Without the use of cross-excitation, the new system already outperforms older instruments used with cross-excitation on metallic samples.

The details of the laser microprobe are as follows: Like its predecessors, the new laser microprobe uses a focused giant pulse from a laser to evaporate a minute sample. The sample vapor is energetic enough to produce a spectrum without cross-excitation, but greater sensitivity is achieved with the use of cross-excitation. The ionized vapor short circuits an arc gap which is (or may not be) connected to a loaded capacitor bank, producing an electrical discharge through the sample vapor. The light emitted at this point is optimally collected and directed to the entrance slit of a fast spectrograph.

The new laser microprobe unit was designed as a system of well-matched components. The improvements were made along the lines suggested by Peppers et al.,[34] with some improved flexibility and the potential to be used for routine work.

One of the criteria of the new design was to ensure enough sensitivity and stability to permit the use of this unit for studies on quantitative analyses, which nowadays are more often still limited to three "bit" information like "major," "minor," and "trace," and seldom to somewhat more accurate statements in terms of relative concentration.

The instrument is divided into two parts: the excitation unit and the detection unit, which will be described separately. The excitation unit consists of the laser, the microscope, and the electrical excitation and control chassis. The laser has a neodymium glass rod emitting at 1.06 μm, a wavelength at which photographic plates are not sensitive, and is Q-switched by means of a Pockel cell. Its output beam is either not attenuated or attenuated to 75%, 50%, 25%, and 10% of its initial value by a set of optical filters mounted on a remotely controlled turret. Thereafter, the beam is focused on a small aperture that is imaged on the target as an essentially round spot 5, 10, 15, 20, or 25 μm in nominal diameter, as defined by Gaussian optics. At the wavelength of 1.06 μm used here, the diffraction-limited spot diameter is in the vicinity of 5 μm. The filters and the apertures attenuate the laser output to stepwise control its total energy between 0.13 and 15 mJ, which may be further lowered by insertion of calibrated filters. The best repeatability is achieved with repetition rates of two or less shots per minute.

The microscope is mounted in the same housing as the laser, ensuring a rigid position of the microscope relative to the laser. It is designed to be used first as a viewing microscope through which the target is positioned and then as a laser beam concentrator which aims the laser emission exactly on the chosen target which is centered by means of an illuminated reticle. The position of the reticle relative to the viewing field is adjusted to ensure an aiming precision better than 10 μm.

Magnifications up to and inclusive of those requiring oil immersion objectives are available, but for laser operation the standard magnification is 100, produced by a 10 x ocular and a 10 x objective. A scale is provided in one ocular with an apparent distance of 10 μm between two divisions when used with the mentioned optics. A 40 x high aperture reflection objective has been ordered with the hope of reducing the extension of the diffraction limit. It is expected to be delivered by fall, 1973.

Shortly before the laser is fired, a mirror is automatically swung into the light path to direct the laser beam through the objective and, at the same time, to prevent the laser emission from escaping through the oculars and damaging the observer's eyes.

The stage of the microscope may be moved both up and down for positioning purposes and finely for focusing. It may be removed completely, if necessary, thus leaving space for bulky samples.

The energy of the electrical cross-excitation is taken from a capacitor the size of which is selected between 2 and 10 μF in steps of 2 μF and which is loaded at 400, 600, 800, or 1,000 V. This energy is delivered to the sample vapor in the gap between two graphite electrodes located just above the sample; the site can be finely adjusted while the electrodes are in operating position. A multiple spring-loaded mechanism allows the electrodes to be positioned with a repeatability better than 10 μm in any direction and they can be moved together, or one relative to the other along the axis of the gap.

The detection unit comprises the relay optics, the spectrograph, and the electronic detection. The detection unit collects the light emitted by the sample and directs it through relay optics to the entrance slit of a Jarrell-Ash grating spectrograph, the output plane of which can be equipped with a photographic plate or a polychromator housing with the associated electronic detection and display. For obvious reasons, fundamentals of the design parameters are not disclosed at this time, but the chosen solution will be described.

The collection optics consist of a f/1.4 Schwarzschild mirror objective and two quartz-lithium fluoride achromats. The total magnification is 8, thus matching the aperture of the collection objective to that of the spectrograph. This relatively high magnification is particularly useful when the plume size is so small that its image entirely falls in the entrance slit. Even when this is not the case, it has been shown that a high magnification is desirable for introducing as much light as possible into the spectrograph.

The lenses are corrected for chromatic aberration in the UV at 2,400 and 3,000 Å as well as for spherical aberration, resulting in a sharp image that is needed for spatial studies of the emission as well as for the separation of the signal from the unwanted background.

The spectrograph is the Jarrell-Ash model 75-152 Fastie Grating Spectrograph that can be operated both as a photographic and as a photoelectric instrument. Its f/8.6 aperture, together with a resolution of 0.1 Å in the first order, makes it the best choice of presently available commercial grating spectrographs. With a 1,180 grooves/mm grating and a 4 x 10 in. photographic plate, it covers a wavelength range slightly over 2,000 Å at a time, the reciprocal linear dispersion being 8.3 Å/mm.

The polychromator housing was modified slightly to accommodate the shielded outputs from each photomultiplier.

The electronic detection is a fast integrating system similar to that suggested by Peppers et al., but considerably improved.

Information concerning the shape and the duration of the electric pulses collected at the anode of a low-noise photomultiplier was obtained through preliminary measurements and guided our choice. Two ways of electronic detection are practicable: the direct oscilloscopic display with or without photographic record and the fast integration that was chosen here to equip standard instruments. An integrator is switched "on" after the Pockel cell starts "opening" and later is switched "off," leaving enough time for the entire signal to be integrated on a 1,000 pF capacitor. Within the next 0.4 sec, the integrated value is read, displayed on a digital voltmeter, and held until the next laser shot command resets the integrator to ground some half second before the laser is fired again.

The small, half inch diameter R300 photomultiplier was chosen because of its size and its successful use in the field of Raman spectroscopy which also requires low-noise, fast, sensitive detectors. Its power supply circuitry is so arranged as to deliver currents to the anode in the order of 200 mA for a short time. This peak current is 10^4 times larger than the maximum allowed average current but is not expected actually to flow long enough to damage the P.M. and it is needed to ensure proper linearity; actual currents in the tens of milliamperes are common.

Three different types of interface between the P.M. and the integrator are available:

1. A most sensitive preamplifier mounted on top of the P.M. in the same cylindrical aluminum shield as the P.M. itself. Pulses produced by a single photon in the 3,500 A range will have a maximum peak value of 2.5 V when 900 to 1,000 V are applied to the P.M. dynode string. The rise time is 15 nsec and the pulse pair resolution is roughly 50 nsec.

2. A similar fast amplifier, but with a 6 times smaller gain. Both amplifiers feed into 50 Ω shielded cables.

3. A direct 50 Ω output into a cable. This

output is used when large signals are available that are fed directly into a transconductance amplifier preceding the integrator.

Specimens must be dry materials. Liquids will be evaporated to dryness by desiccators, vacuum distillation, or heating, or will be frozen in carbon dioxide or liquid nitrogen. Samples will be collected on ground glass slides covered with clean coverslips and wrapped with cigarette paper. Samples of frozen sections on slides also will be prepared.

For microanalysis and especially for trace metals the avoidance of contamination is important; the smaller the sample size the larger is the risk of contamination. Therefore, fresh tissue and cellular preparations must be used. Unlike the electron microprobe, preparation of the sample is very simple. In laser microanalysis there is no need to coat the material with metal or to work in a vacuum. For cells in biological fluids, such as white cells and red cells, a thin smear on metal-free plastic microscope slides provides a simple preparation. Cytological material also may be obtained by tissue press at the time of operation or from needle biopsies where analyses are determined of the general tissue mass.

If specific cellular localization is desired for microanalysis, such as in samples of the kidney, intestinal mucosa, or appendageal areas of the skin, thin freeze-dried microtome sections or liquid-nitrogen frozen tissues on a metal-free plastic slide and air dried can also be used. For tissue culture work done in the detailed pioneering primary studies of Glick, monolayers of cells or individual cells can be analyzed. It is important to add that fresh tissues are necessary for quantitative work. Simple analytical studies with proper controls may be done from fixed tissues. This is important especially in toxicologic studies.

Where it is necessary to localize accurately the 10 μm area for analysis, an image intensifier camera may be used. The following human materials can be investigated:

1. Selected areas of the hair shaft for Pb, As, Au, and Hg, evaporated urine, and dried blood. The patients selected can be those children suspected of clinical and subclinical lead poisoning, those suspected of arsenic or mercury poisoning, and, finally, those who have received Au in the treatment of arthritis.

2. Fragments of fingernails for Ca, P, and Fe studies. These are considered to be of importance for studies on fragilities of nails.

3. Skin scrapings for As, Au, and Pb.

4. Liver and kidney biopsies – Fe for hemochromatosis, Cu for Wilson's disease, As for arsenic poisoning, Pb for lead poisoning, and Hg for mercury poisoning. Tissue samples would be evaporated to dryness.

5. Tears from eyes, evaporated to dryness for studies of S. This is of value in air pollution with the significant sulfur oxides.

6. Scrapings from oral mucosa for analysis for Hg from amalgams and metallic dentures. In this area, microanalyses are supposed to have shown no evidence of Hg or Ni (Colstich, Department of Dentistry, University of Kentucky).

7. Microanalyses of Ca from cervical and vaginal tissues of concern in absorption studies from copper intrauterine devices for contraception. The copper in current devices has an area of 120 mm^2 and the solution rate is estimated at 10 mg/year.[54] There are no data available at present on the concentration of Cu in adjacent tissues.

Control microanalyses can also be done by atomic absorption spectroscopy, electron probe microanalyses, and also neutron activation.

According to Harding-Barling, Smetsinger, and Keil,[4,26] laser microprobe atomic absorption studies have been done already in regard to obtaining absorption spectra from refractory materials by Atwil. Hagenah was the first to study laser atomic absorption and indicated that atomic absorption begins a few tenths of microseconds after the laser pulse triggers and continues for a few tenths of microseconds after the laser pulse triggers and then continues for a few 100 microseconds. They also noted that "in a multispike mode" each laser spike caused its own atomic absorption radiation. Additional studies have been done in this area by Piepmeier, Karyakin and Kargordov.[52] In 1970, O'Brian and Anderson studied laser microprobe gas chromatography. The neodymium laser was used to fragment polymers. "The fragments obtained were less complex than those obtained by thermal pyrolysis, but their nature was dependent on the energy density of the laser beam."[52]

It is reemphasized that it is difficult and often expensive to arrange to do microsample analyses for patient diagnosis, care, and management. The currently available microanalytical techniques out-

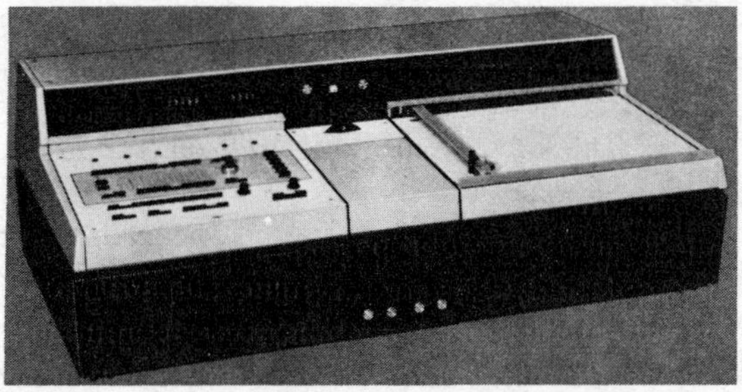

FIGURE 4-3. Laser Raman spectrometer. (Beckman.)

side of atomic absorption spectroscopy include electron probe and neutron activation. These have considerable difficulties that make their use as routine (and routine is emphasized) analytical procedures for patient care most difficult. Microanalysis techniques today are done often by the electron or X-ray microprobe. The electron microprobe has spatial revolution below 1 μm. However, it is necessary that the specimen be maintained in a high vacuum environment and the surface must be electrically conductive. These change the original condition of the specimen for analysis.

It is evident that laser microprobe analyses have great value and can be considered as a form of nondestructive sampling analysis since the test area is so small. This is possible not only for microanalysis of the cell and its components but also for such delicate art objects as Egyptian tear glass, jade, and paintings. The future will show its true standard sensitivity as compared to the results of the control instruments.

Other Techniques of Spectroscopy

In the past years, the monochromatic beam of the laser is said to have revolutionized light scattering research. Much work has also been done with Raman scattering. The scattered light, in brief, consists of the frequency of the incident light plus the light of different colors or wavelengths (sidebands). The frequency differences or frequency shifts are the vibrational frequencies of the molecules in the sample. No two molecules of material have "exactly the same number of frequencies or the same vibrational frequency, the Raman effect series as a fingerprint of the sample." In the Ortec instrument[32] a rapid single photon counter using a pulsed laser beam is used to measure the intensities of the sideband frequencies in a very short time.

Another form of excitation is the investigation of the "pure rotational spectra of gases" to provide information on the dimension of simple molecules. The laser incident beam may be scattered from both gaseous and semiconductor plasmas when laser Raman spectroscopy is combined with infrared spectroscopy. This combination "considerably adds the assignment of the symmetry of molecular species." Lasers may be used also for Brillouin spectroscopy. Raman spectra may be obtained from small quantities of powdered samples and even small crystals, Raman spectra of higher resolution now may be obtained with lasers, and even low frequency Raman shifts can be measured. Hyper-Raman lines in various systems have been obtained by high output pulsed laser and a multichannel photon detector by Maker.[28]

Mass spectrometry has presented mass spectrometric studies of laser-induced vaporization of numerous semimetallic substances. The controls were mass spectra obtained by other conventional differences in the type, relative abundance, size, and number of species present.

The Laser Weekly reports that the National Bureau of Standards of the U.S. Department of Commerce has linked a laser with a mass spectrometer to study molecules in their excited states. The new technique has been used to demonstrate that molecules that are excited by an infrared laser react faster than molecules in the ground state. This method permits vibrational excitation of a specific bond in a molecule and may allow scientists to produce novel chemical products or allow isotope enrichment.

The procedure uses a phase-sensitive, molecular-beam, quadruple mass spectrometer coupled to a mechanically modulated, continuous-wave, gas laser usable either as a CO_2 or CO laser for the *in situ* characterization of chemical reactions induced by infrared laser radiation. A molecular beam produced by the mass spectrometer sampling nozzle is modulated by microscopic changes induced by the absorption of the modulated laser radiation. Phase-sensitive detection is used to monitor the molecules responding to the laser radiation.

The system consists of a gas stream directed at the sampling cone of the mass spectrometer. The laser beam crosses the gas stream at right angles. A representative gas sample passes through a cone system into an evacuated chamber. The beam then passes through an adjustable orifice into another region containing a wheel for mechanical modulation of the beam, an ion source for partial conversion of the molecular species into positive ions, and a mass filter and particle-detection system.

The large photon flux of infrared lasers makes these devices attractive as bond- and isotope-specific excitation sources for chemical preparations on a microscopic scale. However, the large flux also means that in most reaction schemes involving infrared lasers it is necessary to dissipate anywhere from a few watts to several kilowatts of power in the reaction cell, which contains the reaction mixture at relatively high pressures. Under these conditions, vibration-to-translation (V/T) energy transfer is likely, resulting in heating of the gas mixture and subsequent pyrolysis. Identification and separation of these undesirable pyrolysis-reaction products from those due to vibrationally excited but translationally cool molecules have until now been difficult. Using the new analytical technique described here, one can simultaneously determine the V/T relaxation lifetimes and distinguish the vibrationally induced products from the thermally produced products.

The products of a chemical reaction due to interaction with the modulated laser beam also appear in the molecular beam and can be monitored by phase-sensitive detection. By comparing the phase shift of the reaction products with the phase shift of the thermalized reactants, it is possible to identify both the products of the reactants and the products of the reaction of vibrationally excited molecules separately from the products of the pyrolysis reaction. Thus, at the appropriate phase angle, the chemical reactions of only the vibrationally excited species can be monitored and studied.

To test the experimental system, flow systems in the range from 1 to 760 torr were studied. With BCl_3, reactions were observed with H_2, D_2, and NO, yielding HCl, DCl, and NOCl, respectively, together with other products. It was possible to demonstrate the reaction of vibrationally excited but translationally cool BCl_3, which would normally be hidden by the well-known pyrolysis reaction of thermalized BCl_3. The procedure was also demonstrated for systems using SF_6 as the infrared-radiation absorber. Results with this technique suggest the following uses for laser-modulated, molecular-beam, mass spectrometry in laser chemistry studies:

a. The measurement of V/T relaxation for absorbing species

b. The measurement of V/T energy transfer to a nonabsorbing gas component

c. The measurement of translation-to-translation energy transfer from an absorbing species to another component

d. The monitoring of chemical reactions of thermalized species detected as a modulation of reactant and product species

e. The monitoring of chemical reactions of vibrationally excited molecules that will also be detected as a modulation of reactant products. The presence of each of these phenomena is detectable and can be differentiated by the observed phase shift.

f. The use of molecular modulation as a test of high pressure sampling fidelity.

The technique was developed by Hastel and Sessoms of the NBS Inorganic Chemistry Section and Kaldor and Kurtz of the Infrared Spectroscopy Section.[18] Laser source mass spectroscopy by a Nd laser has also been used to analyze geological material.[43]

Now the tunable dye lasers have been introduced as tunable microscopy. There have been many disappointments, as there were with the early introduction of dye lasers. The application is still very tempting, for one can test a number of different molecules without changing lasers.

To obtain more sensitivity for laser absorption spectroscopy to study composition of gas samples, Zalewski[53] has placed the absorption cell inside an organic dye laser cavity. His system was xenon-flash lamp pumped rhodamine 6-G dye solution 50 mg/liter in absolute ethanol in a 7.5-cm dye cell.

In Russia, Kompanyets' group[24] is working on high resolution SF_6 spectroscopy with two independent CO_2 lasers at the 3-year-old separate Institute of Spectroscopy at Podolsk. The deputy director of this separate institute is V.S. Letokhov. In this same institute Tumanov is using TEA laser to study pressure shifts and broadening in NH_3. Energy transfers between isotopes of ammonia have also been studied.

The tunable infrared laser has been a tremendous advantage for infrared spectroscopy. The Heriot-Watt University device is a magneto-Raman

FIGURE 4-4. Laser Source Mass Spectroscopy to Analysis of Geological Material.

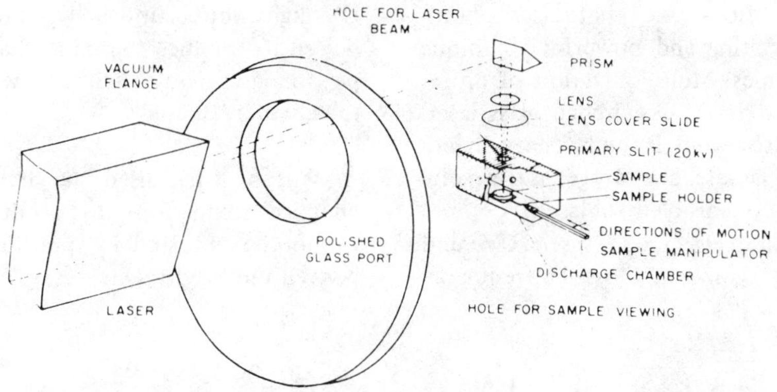

FIGURE 4-4A. Schematic diagram of modified source region of the mass spectrograph.

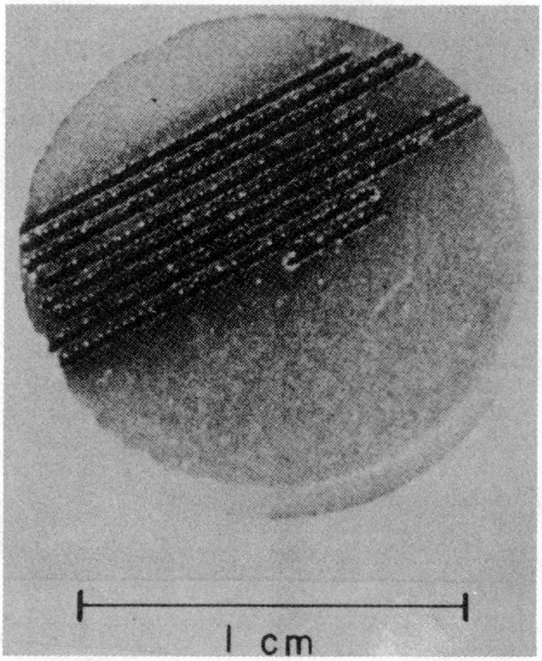

FIGURE 4-4B. Sample pellet after laser irradiation. (From Scott, F. H., Jackson, P. E. S., and Strasheim, A., *Nature,* 232, 623, 1971. With permission.)

spin flip laser with power levels reported up to 100 W and is tunable over a continuous range from 9 to 16 μm. The pumping source is a Q-switched CO_2 laser. This spin flip laser with a tellurium crystal can double the frequencies. The spin flip laser is reported to be of value:

1. To study the physics of electrons in semiconductors
2. For more precise definition of molecular spectra
3. For rapid chemical analysis
4. For chromatography
5. For spectroscopy of excited chemical states
6. For ultra short wave communications
7. For long distance monitoring of atmospheric pollution
8. In meteorology
9. In chemical warfare (?)

High resolution saturation spectroscopy has been done by Hansch et al.[17] with a tunable dye laser with a bandwidth of 7 MHz. This was an analysis of radiation only of atoms traveling perpendicular to the line of sight and consequent absence of Doppler broadening. The importance of the tunable lasers for chemistry and physics has been shown by the number of papers presented relative to spectroscopy at the American Physical Society meeting of January 29, 1973 in New York. The importance and increasing role of the laser in studies of pollution will be reviewed in the section on pollution.

Photochemistry

Laser photolysis is a new and versatile technique.[57] Sousa and Kashnow[46] have used an ultraviolet laser to study fluorescence and photochromism. They believe that the ultraviolet laser is an excellent excitation source for coloring. Nanoflash photolysis and studies on the absorption spectra of excited singlet statis in a fluorochrome have been done by Porter and Top.[35] Other investigators have also found that laser photochemistry is an exciting and powerful technique for exciting molecules. Multi-ionization of atoms has been done also by a Q-switched glass laser. Berns, Ematsu, Olson, and Rounds[7] have inactivated the enzyme lactate dehydrogenase in solution by a frequency quadrupled Nd laser.

To compare the effects of laser UV and conventional UV, Matsui et al.[29] used a frequency quadrupled, Q-switched Nd laser (2650A) to impact calf thymus DNA (deoxyribonucleic acid). Similar effects were produced by the two types of UV light, but curiously greater energy was required to produce the same changes. Other examples of laser photochemistry will be discussed in the chapter, Biology.

Wilson[52] has used the argon laser in photochemical oxidations to produce products that cannot be obtained in appreciable amounts with conventional light. One example he gave is

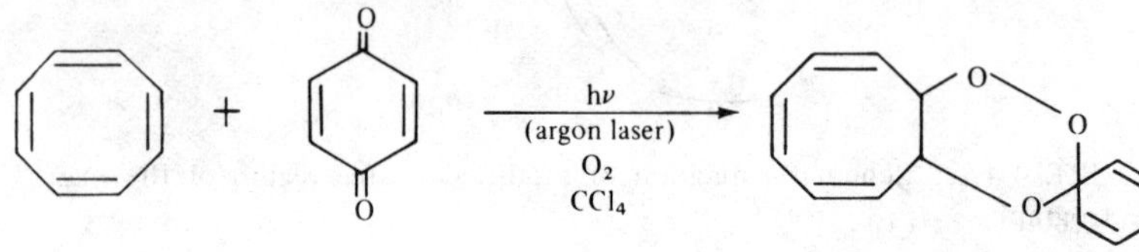

Strange Nonlinear Effects

Wilson[52] envisioned one application of the laser in chemistry to produce strange nonlinear effects. The phenomena observed here are often "highly theoretical and have no practical application yet." Some few examples of these strange nonlinear effects include:

1. Measurement of the optical Kerr effect in aqueous solutions of inorganic salts[33] and tool physical chemistry (?)
2. Short-time effects in light absorption to propose that under certain conditions, the absorption coefficient of a medium for a very short pulse of light is proportional to the duration of the pulse.[50]

It is probable that many other examples of these nonlinear effects will be observed as basic research continues, if it is allowed to continue.

Other studies of the chemistry of material processing will be shown in the chapters on Pollution and Laser Safety, especially concerning pollution. Plasma chemistry includes the important and formidable field of laser-induced thermonuclear fusion.

As indicated, progress will continue on the development of high output gaseous chemical lasers. HF and DF types of chemical lasers have outputs in the near infrared. High output visible and UV chemical lasers are desired. The safety aspects of toxic chemical lasers will be mentioned in the laser safety program.

REFERENCES

1. **Adams, M. D. and Tong, S. C.,** *Anal. Chem.,* 40, 1762, 1968.
2. **Allemand, C. D.,** *Appl. Opt.,* 9, 1304, 1970.
3. **Andreev, S. I., Dymshits, Yu. I., Kaporskii, L. N., and Musatova, G. S.,** *Sov. Phys. Tech. Phys.,* 13, 657, 1968.
4. Applications of the Laser Microprobe to Biological Samples, Harding-Barling and R. C. Rosan, 317 Laguna Ave., Palo Alto, Calif. 94306 and Pathology Department, St. Louis, Mo. 63104.
5. **Baldwin, J. M.,** *Appl. Spectrosc.,* 24, 429, 1970.
6. **Ban, V. S. and Knox, B. E.,** Mass spectrometric studies of laser-produced vapor species, *Int. J. Mass Spectrons Ion Phys.,* 3(1-2), 131, 1969; *C.A.,* 71, 1289, 1969.
7. **Berns, M., Ematsu, S., Olson, R. S., and Rounds, D. E.,** Enzyme inactivation with ultraviolet laser energy (2650 Å), *Science,* 169, 1215, 1970.
8. **Bessis, M., Gires, F., Mayer, G., and Normarski, G.,** *C.R. Acad. Sci.,* 1010, 1962.
9. **Blackburn, W. H., Pelletier, Y. J. A., and Dennen, W. H.,** *Appl. Spectrosc.,* 22, 278, 1968.
10. **Debras-Guedon, J. and Liodec, N.,** *C. R. Acad. Sci.,* 257, 3336, 1963.
11. **Devlin, J. J. and La Conti, A. B.,** Air Force Cambridge Research Laboratories, Office of Aerospace Research USAF, AFCRL 65-855(1), 1968.
12. **Glick, D. and Rosan, R. C.,** *Microchem. J.,* 10, 393, 1966.
13. **Glick, D.,** *J. Histochem. Cytochem.,* 14, 862, 1966.
14. **Glick, D.,** *Ann. N.Y. Acad. Sci.,* 157, 265, 1969.
15. **Glick, D., Marich, K. W., Carr, P. W., and Beatrice, E. S.,** *Ann. N.Y. Acad. Sci.,* 168, 507, 1970.
16. **Goldman, H. M., Ruben, M. P., and Sherman, D.,** *Oral Surg.,* 17, 102, 1964.
17. **Hanseh, S. and Schawlow, A.,** High-resolution saturation spectroscopy. American Physical Society Meeting, *New Sci.,* May 4, 1972.
18. **Hastel, J. W., Sessoms, K. A., and Kurtz, C. V.,** Mass spectrometer and laser. National Bureau of Standards, *Laser Weekly,* August 28, 1972.
19. **Hellman, H.,** U.S. Atomic Energy Commission, Div. of Tech. Inf., Library of Congress Cat. Card Number 68-62126, 1968.
20. **Pekarsi-Dokl,** High-pressure chemical lasers and light stimulated chemical reactions, *Akad. Nauk. USSR,* 1873, 5557, 1969; *C.A.,* 1, 107, 1969.
21. Jarrell-Ash, Division of Fisher Scientific Co., Product Bulletin 45-604, 1971.
22. **Karyakin, A. V., Akhmanor, M. V., and Kaigorodove, V. A.,** *Sov. Phys.,* 353, 1966.
23. **Klocke, H.,** *Spectrochim. Acta,* 24B, 263, 1969.
24. **Kompanyets, O. N. and Feld, M. S.,** Continuously tunable CO_2 laser at Institute of Spectroscopy at Podolsk, *Laser Focus,* 9, 16, 1973.
25. **Ozin,** Laser Raman spectroscopy, *Nature,* 23, 424, 1971.
26. **Harding-Barling, I., Smetsinger, K. G., and Keil, K.,** Laser microprobe instrumentation. 317 Laguna Avenue, Palo Alto, Calif. 94306, Moffett Field, California 94035, and Department of Geology and Institute of Meteoritics, University of New Mexico, Albuquerque, N.M. 87106.
27. **McCormack, J. T.,** *Metals Rev.,* 6, 1965.
28. **Maker, P. D.,** Hyper-Raman effects – Raman Spectroscopy Conference, Oxford, September 13–17, 1970; *Nature,* 228, 320, 1970.
29. **Matsui, S., Rounds, D. E., and Olson, R. S.,** The effect of laser power at 2650 A on deoxyribonucleic acid, *Life Sci.,* 104, 217, 1971.
30. **Mode, H. U., Hagenah, W. D., and Laqua, K.,** Colloquium Spectroscopicum Internationale, Madrid, 1969.
31. Naval Chemical Laser, Laser and Electro-optics.
32. Ortec, Inc., The single photon technique for measuring light intensity and decay characteristics, Oak Ridge, Tenn.
33. **Palette, M. and Orsay, F. J.,** *Chem. Phys. Physiochem.,* 65, 1629, 1968; *C.A.,* 70, 15634d, 1969.
34. **Peeper, N. A., Scribner, E. J., Alterton, L. E., Honey, R. C., Beatrice, E. S., Harding-Barling, I., Rosan, R. C., and Glick, D.,** *Anal. Chem.,* 40, 1178, 1968.
35. **Porter, G. and Top, M.,** Nanosecond flash photolysis and absorption spectra of excited singlet states, *Nature,* 168, 1228, 1970.
36. **Rasberry, S. D., Scribner, B. F., and Margoshes, M.,** *Appl. Optics,* 6, 81, 1967.
37. **Ready, J. F.,** *J. Appl. Phys.,* 36, 462, 1965.
38. **Ready, J. F.,** *Phys. Rev.,* 137, A620, 1965.
39. **Rosan, R. C., Healy, M. K., and McNary, W. F., Jr.,** *Science,* 142, 236, 1963.
40. **Rosan, R. C., Brech, F., and Glick, D.,** *Fed. Proc.,* 24 (Suppl. 14), S-126, 1965.
41. **Runge, E. F., Minck, R. W., and Bryan, F. R.,** *Spectrochim. Acta,* 20, 733, 1963.
42. **Ryan, J. R., Clark, C. B., and Ruh, E.,** 67th Annual Meeting of the American Ceramic Society, 1965.
43. **Scott, R. H., Jackson, P. F. S., and Straheim, A.,** Application of laser source mass spectroscopy to analysis of geological material, *Nature,* 232, 823, 1971.

44. **Sherman, D. B., Ruben, M. P., Goldman, H. M., and Brech, F.,** *Ann. N.Y. Acad. Sci.,* 122, 767, 1965.
45. **Smith, D.,** Tunable infrared laser, *New Sci.,* 6, 1971.
46. **Susa, J. A. and Kashnow, R. A.,** Laser photolysis; versatile new technique, *Opt. Spectra,* 1969, 4852; *C.A.,* 26977k, 1970.
47. **Tlałka, R.,** Colloquium Spectroscopicum Internationale, Madrid, 1969.
48. **Tobias, I., Rentzepis, P. M., and Balzas, N.,** Short-time effects in light absorption, *Chem. Phys. Lett.,* 1969, 205, *C.A.,* 73, 1970.
49. Chromatix, Tuning with Tunable Lasers, Mountain View, Calif.
50. **Vilnot, J., Liodec, N., and Debras-Guedon, J.,** XII Colloquium Spectroscopicum Internationale, *Exeter, Hilgh and Watts,* London, 1965, 343.
51. **Webb, M. S. W. and Cotterill, J. C.,** *Anal. Chim. Acta,* 43, 351, 1968.
52. **Wilson, M.,** Personal communication, 1971.
53. **Zalewski, E. F.,** Laser spectroscopy, *Ind. Res.,* January 1973, 25.
54. **Diczfalusy, E.,** *Symposium on the Control of Human Fertility,* Stockholm, 1970, Swedish Medical Research Council and Reproductive Endocrinology Research Unit, Karolinske, 1970.
55. **Wiley, R. H., and Veeravagu, P.,** Focused coherent radiation (laser) induced degradation of aromatic compounds, *J. Phys. Chem.,* 71, 1968.
56. **Wiley, R. H. and Reich, E.,** Focused Coherent Radiation (Laser) Degradation of Gaseous Hydrocarbons, U.S.A.E.C. Contract AT-(310)-3644, 1968.
57. **Yunis, N. G.,** Personal communication, 1968.

Chapter 5

APPLICATIONS IN ENVIRONMENTAL POLLUTION

Laser technology is becoming increasingly important in the current struggle against pollution. The role of the laser is essentially observation and detection; perhaps, in the future, it will involve prevention and treatment. Presently, the laser is the only instrument that can detect pollution within large geographical areas or remotely.[16] Lasers can be used not only in the analytical laboratories but also in the field where pollution is active or suspected.

Nitric Oxide Detection

Patel et al.[14] of Bell Laboratories measured the amount of nitric acid oxide gas in an air sample. A cylinder absorption cell received an air sample. In this cell, the gas quantity was contacted by an infrared laser and plotted on a recorder. Kreuzer, Jumar, and Patel[10] also measured the absorption strength of 10 pollutant gases at 19 laser emission wavelengths of CO and CO_2 lasers. Detection sensitivities were a few parts per billion.

Because of the importance of the pollutant nitric acid (power stations, furnaces, and automobiles), other studies have been conducted to monitor this infamous and "ubiquitous" compound (or compounds). The National Bureau of Standards uses the Zeeman shift of a nitric oxide absorption line that almost coincides with a carbon dioxide laser emission line. Since a magnetic field can split up and shift the wavelength of a spectral line emitted by a molecule, the laser is placed between the poles of a magnet. The magnetic field can shift the nitric oxide Zeeman line close to the laser line.

A small coil driven by an audio frequency amplifier generates a modulating magnetic field, sweeping the NO absorption line through the laser line. An infrared detector gives an output signal proportional to the amount of NO in the sample. The infrared light is absorbed by the NO and reemitted; thus, the amount of IR light detected is a measure of the number of NO molecules in the sample.

Measurement with the NBS apparatus on labo-

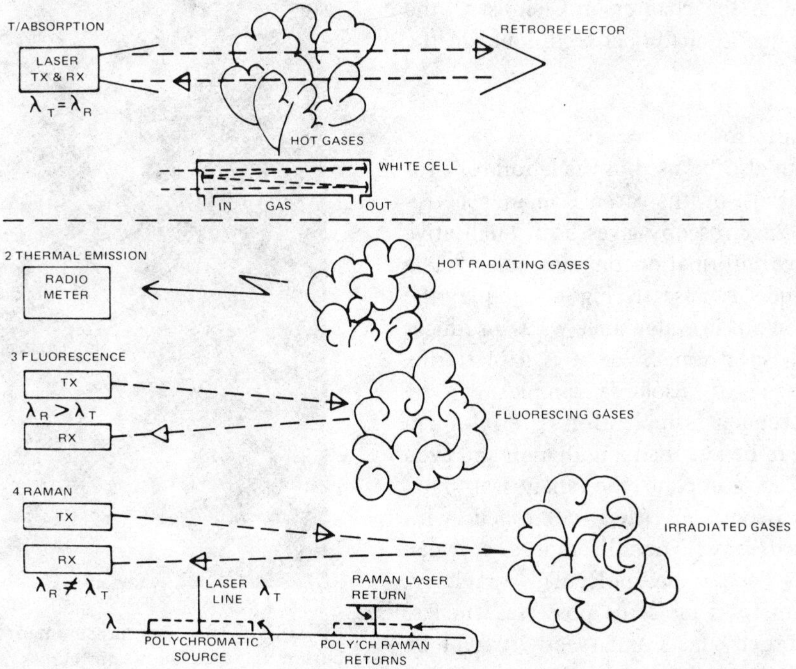

FIGURE 5-1. Various types of optical analyses. (Paul Longrigg, Electro-Optical Systems Design Conference, New York, 1972.)

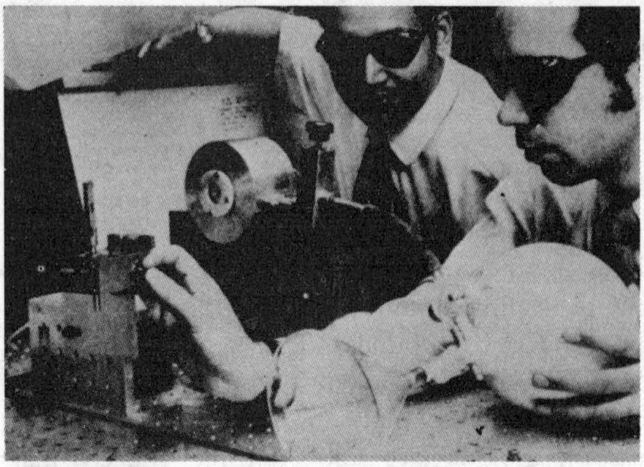

FIGURE 5-2. Infrared laser beam detection of nitric oxide in air sample. (From Kumar, C., Patel, N., and Kreuzer, L. B., Bell Telephone Laboratories; *Laser Sphere,* 3, 1, 1973. With permission.)

ratory samples of known NO content indicates that concentrations of NO as small as 60 ppm can be easily measured. By improving the detector and laser stability, it should be possible to increase the system sensitivity by at least a factor of 10. At the same time, by passing the laser beam back through the NO sample several times, thus increasing the absorption path length by another factor of 10, NO concentrations could probably be measured to better than 0.1 ppm.

As discussed in the chapter on Chemistry, the laser also has value in analytical techniques in the laboratory.

Laser Raman Spectroscopy

The laser can also be used in the laboratory for analysis by means of the laser Raman spectrometer. Raman spectroscopy gives both qualitative and quantitative information on molecules.[2] The Raman technique is a scattering technique; the specimen can be put in water since water produces a weak Raman spectrum. Since it is a scattering technique opaque or insoluble samples can be used. The specimen sample for laser Raman spectroscopy can be less than a milligram and even a nanogram. To complete the analytical data, Raman spectroscopy is often combined with infrared spectroscopy. The differences between the techniques depend on molecular symmetry. The more symmetry a molecule possesses, the less overlap of infrared and Raman activity will be observed. If the molecule has a center of symmetry, the Raman and infrared spectra will be

FIGURE 5-3. National Bureau of Standards Instrumentation.

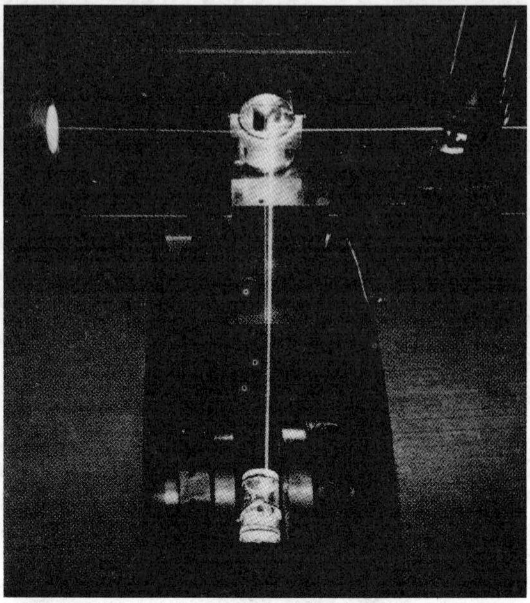

FIGURE 5-3A. National Bureau of Standards nitric oxide monitor with laser as light source at left, sample holder inside an electromagnet at bottom, beam splitter upper center and detector upper right. (*New Scientist.*)

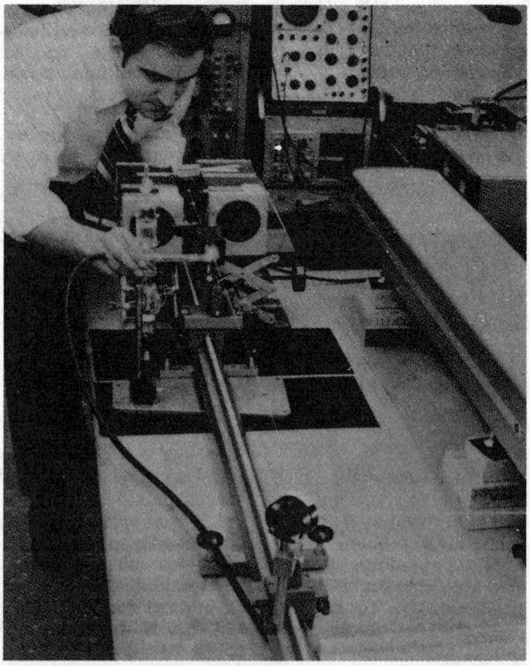

FIGURE 5-3B. For measurement of particle size distribution in dust-laden air across a laser beam. (From Gravatt, C. C., *Opt. Spectra,* 7, 35, 1973. With permission.)

mutually exclusive, i.e., no overlap. The molecule interacts differently with light in the two techniques because infrared is an absorption process and Raman is a scattering process. In practice, this usually means that weak infrared bands and strong Raman bands, and vice versa, can be detected. In some cases, this band may be so weak in one technique that it is missed; however, in general, this band will appear strong in the other, thereby insuring its identification. The only sure way of observing all of the optically active vibrational modes of a molecule and obtaining its complete vibrational picture is to use both infrared and Raman techniques. Longregg[11] states that the "advent of the laser has given a new impetus to a very specific form of molecular microscopy, using the well known Raman effect. Because of its high specificity, it is considered to be one of the most suitable forms of analysis for the combustion control task."

Raman scattering technique may be used also in field analyses of air pollution. Because of the loss by inverse squares, remote detection by Raman scattering may not be possible beyond 100 km. In daylight, visible and UV Raman scattering may be difficult to use for remote detection. To identify specific gases in the plume, high output UV laser systems may be necessary.

Various experiments have been done with distance measurements by laser Raman lidar. The laser systems act like a radar. High power laser impact on particulate matter in the atmosphere is reflected with all intensity proportional to the "concentration of particles and the magnitude of the light source." Photodiodes pick up the returning reflections. Inaba and Kobayashi[9] used laser Raman lidar for analysis of the pollutant gases SO_2, CO_2, H_2CO, NO, CO, H_2S, and CH_4. These were measured selectively in real time.

A favorite source of significant pollution in industrial areas is the chimney plume. Hamilton[5] used lidar observation of plumes in attempting to predict concentration of pollutants on the ground. Siemens[15] used a Nd laser (106.0 nm) with peak powers up to 1 mW and pulses of 20 nsec. In this system, echo detection was done with a germanium photodiode with a high power light transmitting lens as a light receiver. Recording of the echo pulse by an oscilloscope evaluates the location and density of air contamination. This instrumentation can also be used to measure atmospheric haze. The actual chemical composition of the polluted air can be measured by the laser.

A recent telescope-mounted laser of NASA, Langley Research Center at Hampton, Va. detected SO_2 as far away as 1 km. Since 1966, the Armed Services and General Electric Co. have been developing an infrared laser system for long path monitoring of air pollution for use even under varying weather conditions (ILAMS – Infrared Laser Atmospheric Monitoring System),[8] with actual identification of particles through absorption spectroscopy. The ILAMS transmits energy at multiple wavelengths (70 wavelengths are said to be available) to a remote retroreflector.

"The returned energy is gathered in a colinear transmit/receive optical system and referred to laser output energy to compensate for the power differences at each wavelength. A pattern of absorption versus wavelength is produced by energy attenuation at the selected lasing wavelengths. From this pollutants can be identified and their average concentrations over the path can be transversed by the beam.

"Pollutant effects are separated from spectral interferences through absorption patterns produced by the attenuation of laser energy at the various wavelengths. Computer programs have been developed to assist in this process. Computer programs are also employed to calculate linear weights in the system's signal processor. The processor weighs and sums returned wavelength signals and uses them to cancel interference effects and to produce direct pollution information."

In England, Thomas[17] has used a series of

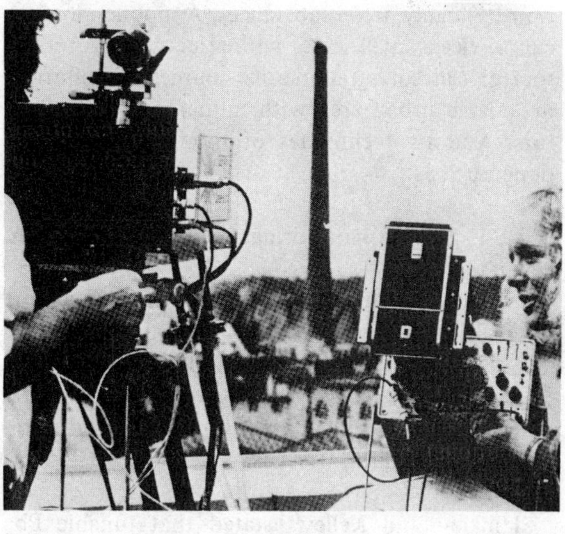

FIGURE 5-4. Siemens giant Nd pulsed laser with telescope for plume stack smoke cloud detection. (From *Laser Sphere*, 2, 1, 1972. With permission.)

FIGURE 5-5. Telescope mounted laser for detection of air pollutants as far away as 1 km. NASA, Langley Research Center. (From *Laser Sphere,* 2, 9, 1972. With permission.)

variable wavelength lasers to monitor atmospheric pollution. Stanford Research Institute's Aerophysics Laboratory has developed a so-called SRI-NAPCA-Mark VIII Lidar, in brief, a computerized narrow-beam laser transmitter and receiver, said "to be able to be reflected from invisible particles up to 5 miles away."

Mobile lidar facilities, as indicated, on trucks or other vehicles may contain detailed laboratory facilities for rapid analysis of pollution as well as the measurements of intensity and movements of pollutants. At times, this may be necessary as a flexible unit for emergency area control of pollution by highly toxic substances. A mobile monitor can service a small sector with selective retroreflectors or can survey numerous sources of pollution in a large urban area with numerous retroreflectors. Various techniques of monitoring may be done, such as:

1. Survey monitoring, as for general EPA evaluation
2. Perimeter monitoring of a high-emission, multiple source area
3. Vertical monitoring with an airborne retroreflector
4. Optical folded path for plume studies or to pinpoint pollution source.

Hinkley and Kelley[6] stated that tunable Pb, $_{-x}Sn_x$te diode lasers are suitable for detection of most of the atmospheric gases. This is so, they claim, even for long-range atmospheric propagation. However, at present, the operation of such tunable diodes must be done at cryogenic temperatures. They suggested that these diode lasers be used as tunable oscillators in the infrared heterodyne configuration.

Longregg[11] stated that a sensor that is attached near the top of the stack is less hazardous for transmission of laser beams. He believes that for an analysis of specific gases in the plume, high power UV systems will be required.

In summary, then, for stack surveillance for combustion gas emission control, the Raman technique offers the most promise, according to Longregg, for the following reasons:

1. It has a high degree of specificity; each molecule can be uniquely identified.
2. The return information is quantitative; the scattered light intensity is directly proportional to specie concentration.
3. A tunable laser is not needed, thus aiding reliability.
4. Measurements can be made rapidly, with concentrations of 100 ppm or more across typical smokestack diameters.
5. Temperature and pressure have little effect on the sample signal, an important consideration when dealing with hot stack gases.
6. Bulk volumes are interrogated for specific contaminates.
7. Water vapor absorption has no effect on Raman in the UV.
8. Cryogenic cooling is not required in gas lasers, as is the case with solid-state lasers operating in the 4 to 12 K range. This adds reliability in an industrial environment.

Longregg[11] added that considerable problems can be expected from particulate obscuration and fluorescence with short wavelength Raman. However, when cross-section enhancement by the employment of resonance Raman is used, these problems might be alleviated to some extent. In the meantime, spontaneous Raman may be employed effectively with high-grade receiver optics, at higher equipment costs.

Microemission spectroscopy is also being used in microanalyses in studies on lead pollution. In conjunction with Vahl and Brech and Allemand, many studies have been done at the Laser Laboratory of the Medical Center at the University of Cincinnati on the microanalyses of hair in children

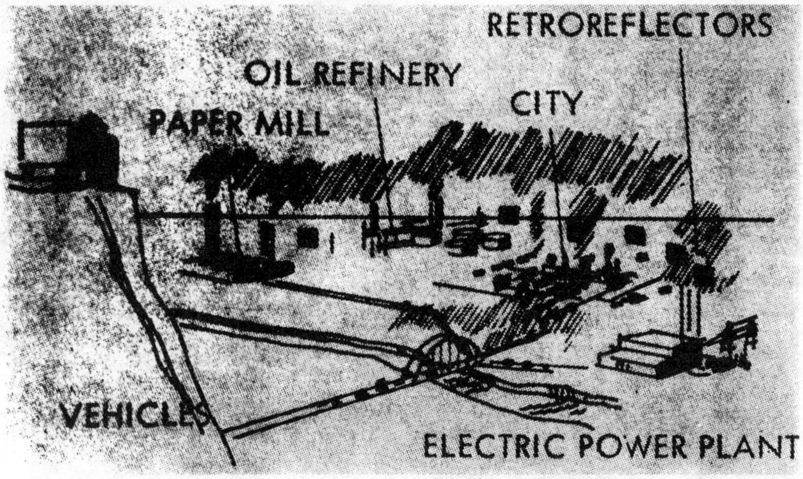

FIGURE 5-6. Schematic outline of mobile laser monitor with multiple retroreflectors for large area survey. (From Snowman, L. R. and Gillmeister, R. J., General Electric Co. and *Optical Spectra.* With permission.)

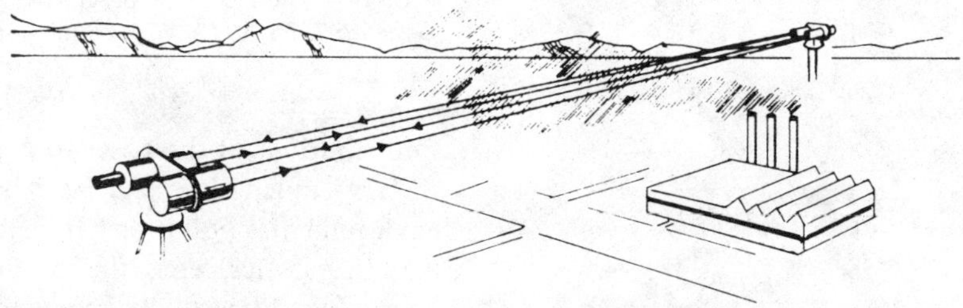

FIGURE 5-7. Diagram of optical folded path laser detector for plume studies and to localize pollution source. (From Snowman, L. R. and Gillmeister, R. J., General Electric Co. and *Optical Spectra*. With permission.)

with lead poisoning. The laser microemission technique provides quantative analyses of lead along the shaft of the hair. With impact areas of 25 to 50 μm, the hair shaft is not destroyed and, as a matter of fact, such analyses may be done rapidly and simply directly on the hair of children.

The hair also provides opportunities for detailed analyses of the laser impact area. Therefore, the absence of complete destruction of the hair and also the availability of the analytical site for detailed pictures of microstructure examination show the advantages of laser microanalysis over atomic absorption spectroscopy.

At present, studies are being done on the impact area with scanning electron microscopy, X-ray fluorescence, and X-ray diffraction. These studies will attempt to show the microscopic sources of the tissues containing lead. This may help for studies of the metabolism and excretion of lead in man.

Waste Disposal by Laser

It has been suggested repeatedly that the laser be used for the actual destruction of garbage and waste material. With high output systems now available, such an application is possible. Investigations are needed to determine the effect of the laser treatment on such organic and inorganic materials. Using material processing with the laser (solely or an adjuvant) it may be possible to develop consumer materials such as fertilizers, plastics, etc. Such laser treatments could replace the polluting garbage dumps, the garbage receptacle, and even, perhaps, the garbage man.

Laser Detection of Oil Slicks

Oil slicks are disturbing and expensive pollu-

FIGURE 5-8. Laser detection of oil slicks. (Perkin Elmer Corp.)

FIGURE 5-8A. Airborne UV laser for oil slicks detection.

FIGURE 5-8B. Outline of technique. (From *Laser Elektro-Optik,* 1972. With permission.)

tants. The laser can be used for the important period of early detection and even classification of the slick. Briefly, the laser detection and classification technique of Perkin Elmer[18] for oil slicks includes:

1. An ultraviolet airborne laser
2. Fluorescence of the oil molecules in the slick since various oils have characteristic wavelengths
3. Reflected fluorescence detected by a telescope
4. Focusing into spectrometer
5. Separation into discrete spectral bands
6. Electronic detection
7. Final comparison with laboratory standards.

DETECTION AND CLASSIFICATION OF OIL SLICK

1. U-V AIRBORNE LASER
2. FLUORESCENCE OF OIL MOLECULES
3. VARIOUS OILS - CHARACTERISTIC WAVELENGTH
4. REFLECTED FLUORESCENCE COLLECTED BY TELESCOPE
5. FOCUS INTO SPECTROMETER
6. SEPARATION INTO DISCRETE SPECTRAL BANDS
7. DETECTOR ELECTRONICS
8. COMPARISON LABORATORY STANDARDS

(PERKIN-ELMER)

FIGURE 5-9. Balloon airborne device to evaluate atmospheric propagation of laser beams and pollution patterns, Thermosonde of G.T.E. Sylvania. (From *Opt. Spectra,* 7, 35, 1973. With permission.)

Contributions of the Laser to Pollution

How much does the laser itself contribute to environmental pollution? In material processing of metals, plastics, etc., the products of laser processing would certainly be environmental contaminants. This is especially true with metalworking where the laser process is an open one. Even if the material processing is a closed system, for protection of the operator, the final exhaust may be mixed into the atmosphere outside the plant. When toxic metals, such as beryllium, are used, then the exhaust systems must be developed to avoid environmental pollution.

In the section on Safety, the toxic materials of laser metalworking and of laser plastic cutting will be detailed in the portion on respiratory hazards from laser work. It is curious that actually so little is known about these hazards in spite of the increasing application of the laser in industry. Finally, there is realization of such great needs. At present, the Laser Laboratory is doing a detailed and much needed study of the hazards to the workman of laser processing of various materials.

There is interest now in light pollution, unwanted sky light produced by man through population growth and increased outdoor illumination. Will the laser contribute to this? This will be scarcely visible lasers, unless diffuse but possibly UV and infrared.

The Laser Treatment of Pollutants

Can the laser be used to process the various pollutants instead of using them, as it were, as microretroreflectors? What is the effect of the various laser systems on the nitric oxide, the sulfur dioxides, the lead-alkyl derivatives of gasoline? To answer these questions, test model chambers could be developed with uniform exposures of selected uniform concentrations of the pollutants. Effluent air analyses could be done with atomic absorption spectroscopy and laser Raman spectroscopy. The toxicological effects could be studied with animal exposure chambers and tissue culture techniques. Such test model systems with proper controls could be used to evaluate the effects of various lasers and various wavelengths on the actual treatment of environmental pollution. Since lasers are used in studies of pollution, there should be some knowledge of what happens to the ubiquitous pollutant itself on impact of the laser, especially high output lasers.

Conclusions

As elsewhere, the laser is here to stay in many areas of the field of studies of environmental pollution. Laser Raman spectroscopy is very important and will become even more so for environmental pollution. Early detection, monitoring, and 24-hr duty surveillance can be done by the laser.

REFERENCES

1. Air pollution surveillance systems, *Science,* 170, 289, 1970.
2. Beckman Laser Raman Spectrometer Bulletin 7208, Beckman Instruments, Inc., 1970.
3. **Collis, R.,** SRI-NAPCA Mark VIII Lidar, Stanford Research Institute, Menlo Park, Calif., 1972.
4. **Goldman, L. and Rockwell, R. J.,** *Lasers in Medicine,* Gordon and Breach, New York, 1971.
5. **Hamilton, P. M.,** Application of a pulsed light range finder Lidar to the study of chemical plumes, *C. A.,* 2, 6052B, 1970.
6. **Hinkley, E. and Kelley, P. L.,** Detection of air pollutants with tunable diode lasers, *Science,* 171, 635, 1971.
7. **Hirschfeld, T. and Klainer, S.,** Remote Raman spectroscopy as a pollution radar, *Opt. Spectra,* 4, 63, 1970.
8. ILAMS, General Electric Co., *Laser Weekly,* July 10, 1972.
9. **Inaba, H. and Kobayashi, T.,** Laser Raman radar for chemical analyses of polluted air, *Nature,* 170, 1969; *C. A.,* 2, 6070, 1970.
10. **Kreuzer, L. B., Kenyon, N. B., and Patel, C. K. N.,** Air pollution: sensitive detection of ten pollutant gases by carbon monoxide and carbon dioxide lasers, *Science,* 177, 347, 1972.
11. **Longrigg, P.,** Laser stack gas sensor for monitoring and control of fossil fuel combustors, Electro Optical Systems Design Conference, New York, 1972.
12. **Morgan, G. B., Ozolins, G., and Tabor, E. C.,** Air pollution surveillance systems, *Science,* 170, 289, 1970.
13. National Bureau of Standards – Laser shifts to measure NO pollution, *New Sci.,* December 14, 1972.
14. **Patel, C. K. N. and Kreuzer, L. B.,** Nitric oxide gas analysis, Bell Telephone Laboratories, 1972.
15. **Siemens,** Nd laser pollution monitored by glass laser, *Laser Sphere,* 2, 1, 1972.
16. **Snowman, L. R. and Gillmeister, R. J.,** Laser policeman for dirty air, *Opt. Spectra,* 6, 30, 1972.
17. **Thomas, E. L.,** Laser monitoring atmosphere pollution, *Laser Weekly,* July 1972.
18. Oil Slick Detection, Perkin Elmer Corp., 1972.

Chapter 6

LASER MATERIAL PROCESSING
LASER METALWORKING

One of the important applications of the laser in industry is the use of the laser in metalworking. Laser applications in metal processing include:

1. Drilling
2. Cutting
3. Welding
4. Heat treatment
5. Inspection of metal surfaces
6. Nondestructive testing

In this conservative industry of metalworking, the laser has continued to make slow, but definite progress.

Lokerson[17] indicated that the lack of lines of communication has interfered with the continued advances of the laser in metalworking. Unfortunately, this difficutly has interfered in many fields of application of the laser. At times it may not be the lack of lines of communication, but the high cost of the laser system itself, especially to small shops for research and development, which these small shops cannot afford. It is not well known, for example, that in certain areas of metalworking the laser is perhaps the only tool that can be used.

FIGURE 6-1. Multifold 450 W CO_2 laser of Ferranti Ltd. (From Ferranti International News and *Laser Focus,* 8, 19, 1972. With permission.)

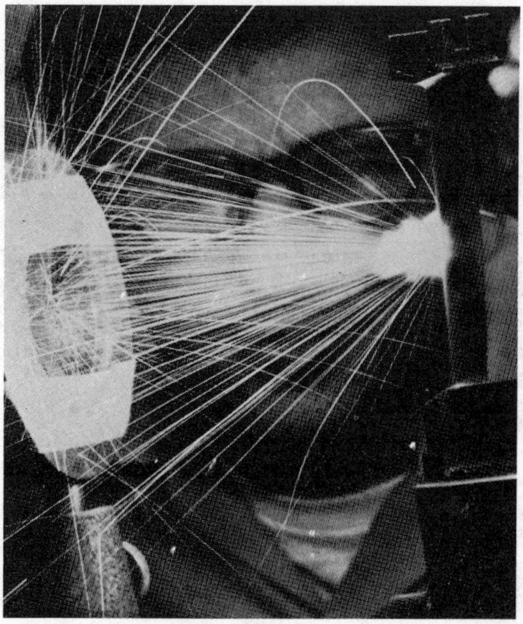

FIGURE 6-2. Molten metal streaks from laser metalworking on jet engine. (United Aircraft.)

Basic Principles for the Use of a Pulsed Solid-state Laser as a Metalworking Tool

The special features of the laser that make it of value for metalworking refer especially to the pulsed solid-state laser systems such as the ruby, the YAG, and the neodymium glass laser, according to Rockwell.[11] High output CO_2 lasers have also been used more extensively in metalworking. For the pulsed solid-state laser, as Rockwell indicates, the enormous pulsed power, the high degree of spatial and temporal coherence, and the beam's precise direction enable the light beam to be focused to an extremely small spot size. With this optical radiation density, the metal may be removed, as in drilling, hole cutting, or welding. Essentially the reaction is one in melting and vaporization because of the intense local heating. "The rapid liberation of free electrons which then interact with the ionic structure of the metal causes local violent acoustic vibrations, the

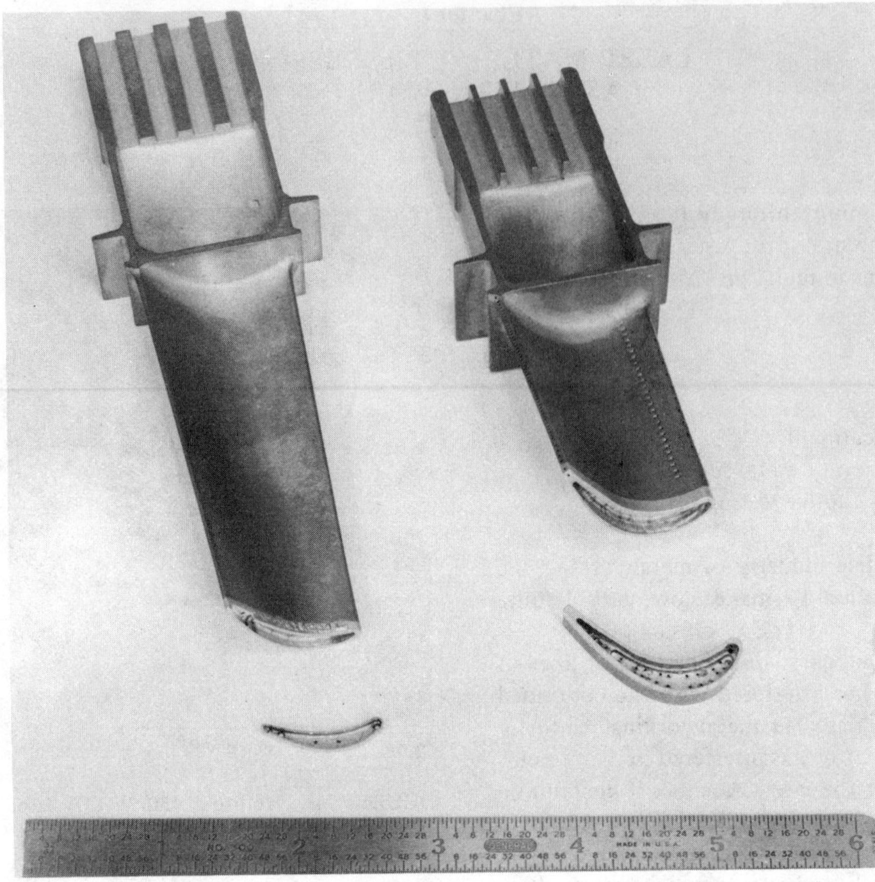

FIGURE 6-3. Ruby laser drilling of turbine blade components. (Michael Heglin, G.E., Cincinnati, Ohio.)

acoustic phonon." The result is the production of an intense thermal reaction with a broad band emission plume accompanied by a shower of metallic sparks and vaporized metal from the work piece. "The important parameter of the laser is the power per unit area of the metal surface." As indicated by Rockwell,[11] power densities in the range of 250 kW/cm^2 to 50 MW/cm^2 are required for laser metalworking. The choice depends on the type of metal on the pulse energy of the laser, the pulse duration, the beam divergence, and the focus. If the power density is not too high, the impact area will remain below the melting point. If the pulse length is shortened, a larger temperature increase will be produced due to thermal diffusion in the metal.

The following is the background of the laser-induced liquefication and vaporization of metal, as reviewed by Rockwell.[11] At higher laser powers, liquefication and vaporization processes occur.

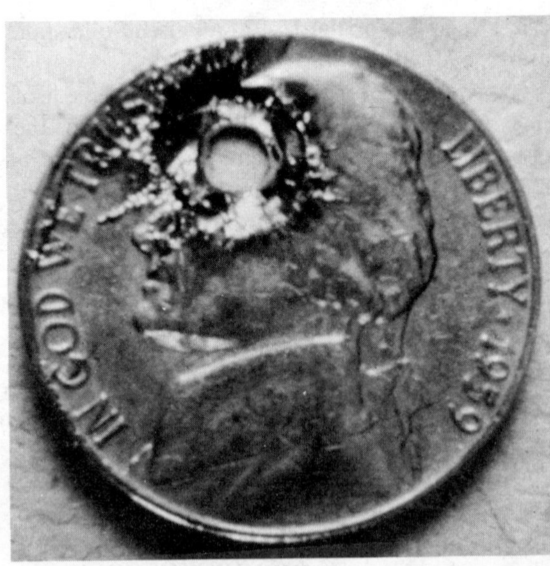

FIGURE 6-4. Pulsed laser drilling of nickel.

FIGURE 6-5. Hole drilling work station, Glen Hardway. (Space Rays Inc.)

The amount of vaporized material is a complicated function of the thermal properties of the material. In samples of low thermal conductivity, the boiling point at the surface of the metal is easily attained, but only a small amount of heat will penetrate into the material and, consequently, vaporization occurs only for a thin layer at the surface. For metals of high thermal conductivity, the depth of material that is vaporized is again small since a great deal of the heat is drained off by thermal conduction.

Large pressure waves are developed in the vaporization process. This is a result of the thermal expansion in the surface of the metal that produces pressure waves which may reach several atmospheres. The vaporized particles that are ejected from a metal reacting to an intense normal mode laser pulse will achieve velocities in a range from 2×10^4 to 5×10^4 cm/sec. These velocities are characteristic of the thermal motion of a material evaporating from a surface which has reached its boiling point.

An approximate relationship for the maximum hole depth possible can be obtained if an assumption is made that under conditions of a high power density laser beam ($\sim 10^7 \frac{\text{watts}}{\text{cm}^2}$) the speed at which the depth of the hole is formed approaches a steady-state value.[9] The basis of this assumption can be justified by the fact that, for all practical purposes, the average value of the power density of the long, normal mode laser pulse can be considered as a constant over the duration of the pulse. Thus, the speed at which the depth of the hole grows approaches an average rate that is directly related to the power density of the incident pulse.

An approximate equation relating the maximum depth (X) of the laser drilled hole may be written in terms of the incident laser power density (W) and the thermal constants of the material:

$$X = \frac{Wt}{(L + cT_V)}$$

Modern computer controlled machines operate at about 25 J/pulse. This would remove about 1.5 mg/pulse. Operating at 60 ppm, this would mean (based on 8 hr/day) the liberation of about 11,000 g of metal per year (about 25 lb). Some operations now have two or more such units operating at repetition rates of up to 4 pulses/second and indicate future requirements of up to 10 such units in the same plant area. Thus, it is conceivable that uncontrolled operation of such numbers of units could contribute hundreds of pounds per year of airborne metal fragments to the plant environment or, in enclosed systems, to be passed into some venting system, perhaps into the outside air.

At present, relatively little information is known about the composition of the laser-induced plasma produced on metals at the time of irradiation. Langer has shown that irradiation of solid targets (Be, C, W) by a Q-switched ruby or neodymium laser (30 mW at power densities up to 10^{22} W/cm^2) could produce free electrons, ion formation, and X-ray liberation. In these studies, for example, carbon ions (C^+ and C^{2+}) were measured to have kinetic energies up to 2 keV. X-rays were observed from an iron target, irradiated by a focused (40 mm lens) Q-switched neodymium laser (30 mW).

Basic Principles for the Use of a High Power CW Laser as a Material Cutting Tool

The basic reaction when continuous wave laser

energy is focused onto a highly absorbent material at a high incident power density is a process where the absorbed radiation rapidly increases the temperature of a small volume of the material and, providing sufficient energy is absorbed, vaporization and ablation occur. Since this process occurs quite rapidly, the beam can be moved at a constant rate across the material surface to produce a continuous cut. The temperature rise in the material volume described by the focal region of the laser beam will be in the order of the vaporization temperature of the material. In addition, the material bordering the cut can, by conduction, also experience a temperature rise which can cause melting or charring (the so-called kerf). The results of cutting experiments with the CO_2 laser (1,060 nm) have, however, shown that the region of damage bordering the area of a cut is usually no greater than 1 mm. When energy is delivered by prolonged exposure on the same area, the cut becomes deeper and the region of border damage is larger. The absorbed energy is expended by the processes of vaporization, reradiation away from the material surface, and conduction into material peripheral to the cut. The proper choice of laser parameters and cutting speed can, however, produce rapid cuts in many metals, plastics, and cloth base materials with minimum charring.

Estimates of penetration to a depth (z) may be obtained by assuming that the incident radiation is absorbed in the material according to Beer's exponential law of absorption: $I = I_o \exp(-az)$, where a is the absorption coefficient of the material at the laser wavelength. For example, values of the absorption coefficient in plastics at the 1,060 nm carbon dioxide laser wavelength range from 10 cm^{-1} of Lucite® (polyethylene), to 25 cm^{-1} of plexiglass (polymethylmethacrylate); thus, depth penetration for 98% absorption (az = 4) will be no greater than a 4-mm depth. To the first approximation, the latter value provides an estimate of the extent of maximum penetration of the CO_2 laser beam prior to ablation. Since scattering losses are minimal at this wavelength, the exponential absorption law can be used with some degree of confidence.

The process of cutting with a CO_2 laser to a specified depth may be considered as follows: A focused beam is held on a given point for a time (T) sufficient to ablate the material to a given depth and then moved infinitesimally to the next material volume for further ablation. If a long focal length lens is used, the variation of focused spot diameter (d) over the depth of the ablation will be small, and the speed of the cut (v) can be related as $v = \frac{d}{T}$.

Using a simple energy balance equation, the depth penetration (X) in a thermal ablation process may be expressed by Equation 1. Thus, the speed of the material cutting (v) cm/sec to a specified depth (X) may then be expressed as:

$$v = \frac{(1 - R)}{K} \frac{Wd}{X} \tag{1}$$

Thus, the speed at which materials may be cut is directly proportional to the power density of the laser and inversely proportional to the depth of the cut.

An example of the many various materials now being cut by the CO_2 laser (or with oxygen assisted cutting with the CO_2 laser) is given by Rockwell[11] as:

CO_2 Laser Cutting

Material	Laser Power (watts)	Speed (cm/sec)	Thickness (mm)
Paper	50	23	0.56
Titanium	250*	42	9.5
Carbon steel	250*	1	3.2
Stainless steel	250*	0.4	0.5
Boron (in epoxy)	250*	1.2	1.75

*Oxygen assisted cut

YAG Laser Machining of Aluminum (Ready)

Depth (mil)	Exposure time (msec)
(Surface melting)	21
25	50
50	80
75	110
100	145
128	185*

*Surface vaporization begins

Detailed data are also available (Ready) for aluminum machining using a focused 100-W CW neodymium-YAG laser (incident intensity 10^4 W/cm^2).

The lasers used at the present time may be solid-state lasers or the gas CO_2. As indicated, there are increasing uses of the high output laser systems especially for metal cutting. The solid-state laser systems may be used chiefly for drilling.

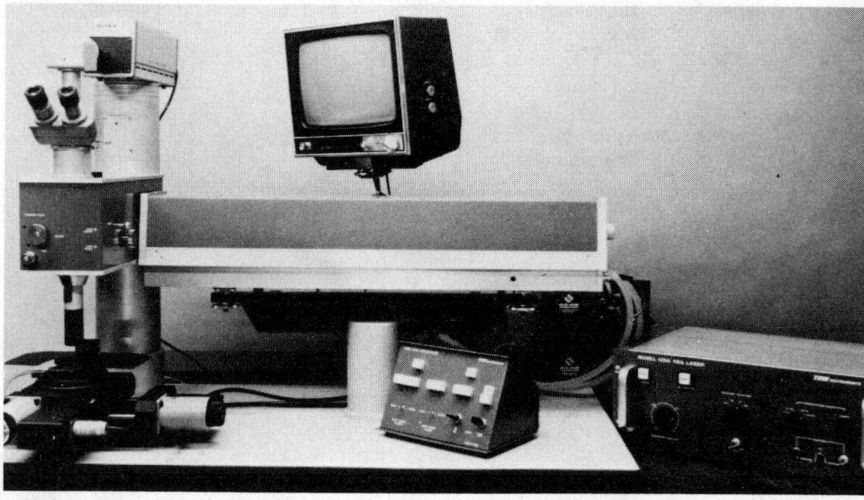

FIGURE 6-6. Nd-YAG Q-switched laser for trimmer for thick and thin film material processing. (T.R.W. Instruments.)

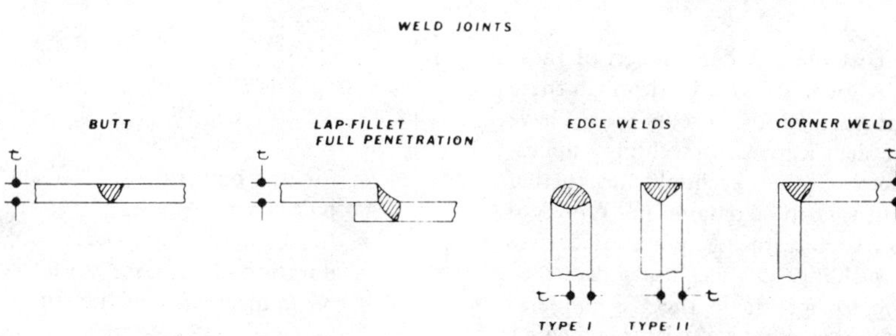

FIGURE 6-7. Weld joints, 250-W T.E.M. Mode 10.6 μm. (Coherent Radiation, Welding Engineer, 1970).

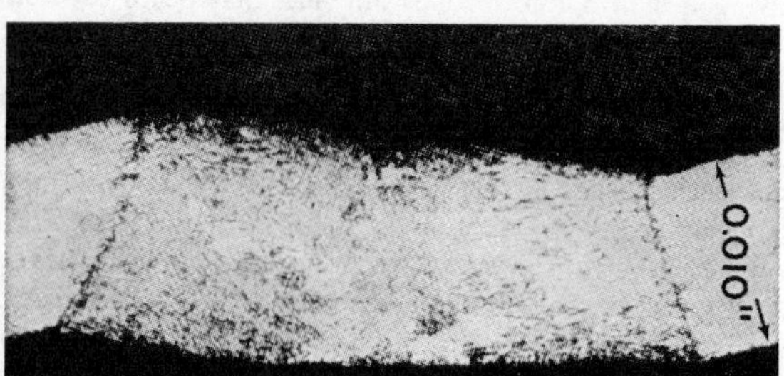

FIGURE 6-8. Photomicrograph of section through butt weld 0.010-in.-thick 302 stainless steel. (Coherent Radiation, *Welding Engineer,* 1970.)

As indicated, there are certain areas where it appears that only the laser system could be used. This may be because of the economics of hole drilling, or, for example, the hole drilling of a curved surface. At the present time, actual industry specifications are now being written that specify laser drilling for certain parts. Cutting may also be done with the laser and welding may also be done, even miniaturized welding.

Welding, for example, may be done inside various types of glass tubes so the beam can be focused on the metal part and in computerized techniques for the development of circuitry systems. Thurber[13] has reviewed the entire concept of all phases of laser welding for the Laser Laboratory of the Medical Center of the University of Cincinnati. These data are to aid the Laboratory in their study, now under way, of the occupational hazards of laser metalworking. His complete data are as follows.

Basically, lasers for welding operate either in the conventional or pulsed mode; or the CW or continuous mode. Lasers can also be operated in the so-called Q-Spoiled mode, in which very high peak powers (10^9 watts) are achieved in very short pulses (10^{-8} sec). The Q-Spoiled

Material	Thickness (In.)	Rate (IPM)	Weld Width (In.)	Remarks
Butt Welds				
321 Stainless Steel	0.005	90	0.018	Full Penetration
321 Stainless Steel	0.010	35	0.028	Full Penetration
321 Stainless Steel	0.0165	11	0.030	Partial Penetration
17-7 Stainless Steel	0.005	110	0.018	Full Penetration
302 Stainless Steel	0.005	50	0.020	Full Penetration
302 Stainless Steel	0.008	30	0.020	Full Penetration
302 Stainless Steel	0.010	10	0.040	Full Penetration
Inconel 600	0.004	150	0.010	Full Penetration
Inconel 600	0.010	40	0.018	Full Penetration
Nickel 200	0.005	35	0.018	Full Penetration
Monel 400	0.010	15	0.025	Full Penetration
Comm. Pure Titanium	0.005	140	0.015	Full Penetration
Comm. Pure Titanium	0.010	50	0.022	Full Penetration
Lap Fillet Welds				
Tin Plated Steel	0.012	20	0.025	Full Penetration
302 Stainless Steel	0.016	175	0.025	Partial Penetration
302 Stainless Steel	0.030	30	0.030	Partial Penetration
302 Stainless Steel	0.010	15	0.030	Full Penetration
Corner Weld				
321 Stainless Steel	0.010	20		
Edge Welds				
321 Stainless Steel	0.005	85		Weld Type I
321 Stainless Steel	0.010	25		Weld Type I
321 Stainless Steel	0.0165	15		Weld Type II
17-7 PH Stainless Steel	0.005	45		Weld Type I
Inconel 600	0.004	160		Weld Type I
Inconel 600	0.010	35		Weld Type I
Inconel 600	0.0165	25		Weld Type II
Nickel 200	0.005	18		Weld Type I
Monel 400	0.010	25		Weld Type I
T 6A1-4V Alloy	0.020	27		Weld Type II

FIGURE 6-9. Chart CO_2 laser welding. (Coherent Radiation, *Welding Engineer,* 1970.)

mode is not suitable for welding. A comparison of these various types of lasers is shown in Table I. Although there are many other types of lasers, only those in Table I have any practical near-term significance for welding applications. The information in Table I should be further qualified by pointing out that these data do not represent state-of-the-art lasers per se, but, rather, practical devices suitable for industrial welding. Of the lasers described here, two can be made to operate in the CW mode at reasonable output levels – a solid-state laser, in which single crystals of Yttrium-Aluminum-Garnet (YAG) are doped with Nd ions as the active medium (typically ~1% Nd); and a gas laser which employs a CO_2-N_2-He plasma (typically 6% CO_2-18% N_2-76% He) as the active medium. Continuous operation is possible with these so-called four-level systems, since those atoms which are pumped into the excited state are not required to return all the way to the ground state during stimulated emission, but, rather, only to an intermediate state. As a result, only a small fraction of the atoms need be pumped out of their normal ground state, greatly reducing the pumping requirements for laser action.

It is apparent that the laser designer has the choice of a number of parameters including output energy (or power), pulselength, wavelength, beam divergence, and pulse repetition rate, as well as certain optical controls which may be arranged external to the laser oscillator. It is appropriate at this point to ask the question, "what is the importance of these design parameters to the welding engineer?" How, for example, will selection of one wavelength of radiation, rather than another, permit him to better achieve his goals? What is the influence of pulselength on weld character:

A useful relationship in analyzing the performance of pulsed laser welders and elucidating many of these questions is:

$$F = 4E/\pi f^2 \theta^2 t \tag{1}$$

where

F = focused power density (W/cm²),
E = pulse energy (joules)
f = focal length of the system objective lens (cm),
t = duration of the laser pulse (sec), and
θ = beam divergence of the laser (radians).

If the beam divergence is defined as the cone apex angle which includes 90% of the energy flux, then F is the average power density. A ruby laser which produces a 15 joule laser pulse of 2 milliradians full angle beam divergence in 5 milliseconds and has that beam focused through a 2 cm lens, will generate a power density at the focal plane of the lens of 2.4 x 10^8 W/cm². (Refer to Table I.)

Obviously, if the beam divergence of the laser were substantially greater (i.e., 10 milliradians) then a 25 fold reduction in the focused power density would result. This emphasizes the inverse square effect of laser beam divergence in characterizing the output of a pulsed laser welder.

An excellent analytical approach to pulsed laser welding has been published by Fairbanks and Adams[2] using the transient heat transfer equations for a semi-infinite slab to which heat is transmitted uniformly over a circular spot. It must be realized that the actual laser beam is not uniform over its cross-section, but this simplifying assumption is warranted in most preliminary analyses. In this case, the area not directly impinged by the laser beam is treated as an adiabatic region. The solution to the heat transfer equation for the point at the center of the spot on the surface (r = O, z = O) is:

$$T_S - T_O = \frac{2H\sqrt{\alpha t}}{K} \quad \frac{1}{\sqrt{\pi}} \text{ -ierfc } \frac{a}{2\sqrt{\alpha t}} \tag{2}$$

Table I. Typical Operating Parameters of Lasers Suitable for Welding

LASING MEDIUM	OPERATIONAL MODE	WAVELENGTH (μm)	MAX. PULSE REPETITION RATE (Hz)	PULSE DURATION (millisec)	PEAK * OUTPUT POWER (watts)	PULSE ENERGY (joules)
Ruby	Pulsed	0.694	2	0.8–10	1500–19,000	15
Nd:Glass	Pulsed	1.06	0.5	0.8–10	1000–75,000	60 in 0.8 ms 20 in 10 ms
Nd:YAG	Pulsed	1.06	60	0.3–1.0	5000–17,000	5
	CW	1.06	–	–	300	–
CO_2-N_2-He	Pulsed	10.6	5000	0.2–1000	3000 @ 50 Hz and 0.2 ms	0.6 in 0.2 ms
	CW	10.6	–	–	1000	–

*Note: Calculated from quoted pulse derivation and energy.

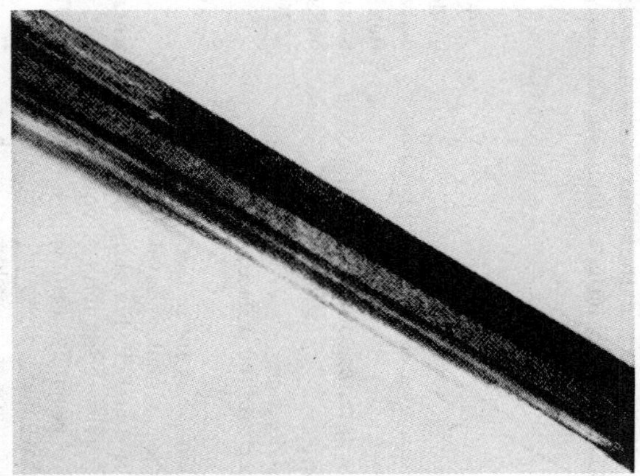

FIGURE 6-10. CO_2 welding stainless steel tube, 10 mm diameter 0.6 mm wall thickness seam welded. (Karl Gürs, Batelle, Frankfurt.)

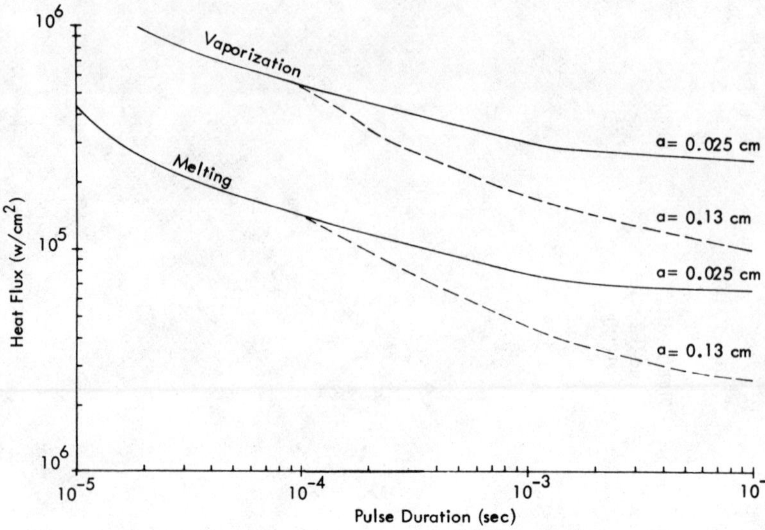

FIGURE 3. Plot of Heat Flux versus Laser Pulse Duration to Induce Melting and Boiling in Semi-Infinite Aluminum Slab for Various Laser Spot Sizes.

where

T_s = surface temperature of a circular spot of radius a, at a time t,
T_O = initial uniform temperature of the material,
α = thermal diffusivity,
K = thermal conductivity, and
H = heat flux absorbed into the surface.

H is related to F in equation (1) by the reflectivity (R) of the metal surface at the particular wavelength and temperature extant, i.e., H = (1-R)F. For short times or large spot diameters the error function term of equation (2) can be neglected. It is generally assumed by most workers in the field that there is some maximum temperature to which the surface of the weldment can be heated without excessive vaporization or metal expulsion. For simplicity, this may be assumed to be the normal boiling point of the metal. If, for a given set of welding conditions, this maximum allowable temperature is obtained at a heat flux H_1 and a welding pulsewidth of 1 millisecond, then to a first approximation the same temperature will be realized with a 4 millisecond pulse if the heat flux $H_2 = H_1/2$.

Using Equation (2) and appropriate materials parameters, it is possible to define an envelope of appropriate welding conditions between the melting and the boiling points of the base metal. Such an envelope is presented for pure aluminum in Figure 3. Graphs are presented for two different spot sizes (0.025 and 0.13 cm). It is apparent from Figure 2 that for pulse durations of less than 10^{-4} sec the allowable heat flux for welding is independent of spot size over the range considered. It is also obvious that increasing the laser beam spot size will reduce the heat flux required to produce a phase change.

A further iteration on this semi-infinite solid heat transfer model has been made by Cohen and Epperson[3], who derived analytical solutions for the case of a constant heat flux using an analog computer. They report that for most metals, the maximum depth of penetration into the body before initiation of surface vaporization is reasonably approximated by:

$$S_{max} = \frac{0.16}{\rho LH} (H^2 t_e - H^2 t_m) \tag{3}$$

where

S_{max} = depth of penetration
ρ = density
L = latent heat of fusion
H = heat flux
t_e = time at which the surface begins to vaporize
t_m = time at which surface begins to melt

The quantity $(H^2 t_m)$ can be determined from material properties as follows:

$$H^2 t_m = \frac{\pi T^2{}_m K^2}{4\alpha} \tag{4}$$

where

T_m = melting point
K = thermal conductivity, and
α = thermal diffusivity

$H^2 t_m$ can be calculated using an equivalent formula. A simplifying assumption made in deriving Equation (3) is that the ratio $\frac{L}{T_m C} = 0.45$ (C is the specific heat) which is reasonable to ± 10% for many metals.

Table II. Comparison of Heat Sources for Welding[4]

HEAT SOURCE	POWER INTENSITY, W/CM²
Laser	up to 10^9
Electron beam	up to 10^9
Electric arc (argon – 200 amp)[a]	1.5×10^4
Oxyacetylene flame[a]	10^3
Oxyhydrogen jet burner[a]	3×10^3
Black body radiation, °K[b]	
6,500	10^4
11,500	10^5
20,500	10^6
36,500	10^7
65,000	10^8
115,000	10^9

[a]Experimentally determined for heat transferred to flat surface.
[b]Calculated.

From these relations, it is apparent that the maximum fusion front penetration can be simply represented by the equation:

$$S_{max} = \frac{Km}{H} \quad (5)$$

where K_m is a constant specific for each material. Values of K_m have been calculated as follows:

METAL	K_m*	K_m**
Cr	0.13×10^2	4.49×10^2
Cu	1.76×10^2	60.9×10^2
Au	1.89×10^2	65.4×10^2
Fe	0.31×10^2	10.7×10^2
Mo	1.51×10^2	52.2×10^2
Ni	0.34×10^2	11.8×10^2
Pt	0.61×10^2	21.1×10^2
Ag	1.60×10^2	55.3×10^2
Ta	0.42×10^2	14.5×10^2
W	1.14×10^2	39.4×10^2

*H in Btu/ft² –sec and S_{max} in ft.

**H in W/cm² and S_{max} in cm.

Obviously this equation has real meaning only for those values of H sufficiently high to exceed the melting point of the material. H in turn can be obtained from Equation (2) as discussed previously.

While H must be sufficiently high to induce melting, it must not be so high that penetration as calculated from equation (5) is unacceptably low. As we have previously seen, melting can be achieved at lower heat fluxes by broadening the laser pulse. It is for this reason that all good pulsed laser welding equipment contains special networks to lengthen the normal 0.5 to 1 millisec pulse to values of 5 to 10 milliseconds.

When applying this heat transfer theory to practical problems, it is of considerable interest to compare the focused power density (H) of laser energy sources with other more conventional welding energy sources. Typical values are presented in Table II.

Anderson and Jackson[4] have made a number of instructive calculations comparing the efficiency of pulsed laser and arc welding. For example, a model was developed in which it was assumed that 0.001 inch of iron on the surface of a semi-infinite slab was melted with a welding arc of 10^4 watts/cm² intensity and that a similar thickness of metal was melted using a pulsed laser sufficiently intense to achieve boiling at the slab surface. These calculations indicate that, with the laser, the heating time is reduced 100-fold, and the input energy by a factor of 6.5. The net result is that the extent of the weld heat-affected zone is drastically reduced. The marked differences in temperatures within the unmelted material which result from these two different energy sources are graphically shown in Figure 4, which is taken from Reference 4.

The discussion thus far has treated only pulsed energy sources. As indicated in Table I, continuous wave lasers are also useful for welding. The applicable equation for the surface temperature at the center of the spot generated from a continuous source is:

$$T_s = \frac{W}{\pi Ka} \quad (6)$$

where W is the absorbed power and the other terms are as previously defined[5]. This solution is exact only for situations where the material thickness is large compared to "a". Firth[5] points out that equation (2) approaches equation (6) when the pulse duration is greater than $\frac{5a^2}{\alpha}$. For a laser beam focused to a 10 mil spot, this condition is realized at pulse durations of 0.9 millisec for aluminum and 20 milliseconds for type 304 stainless steel. Further, the temperatures

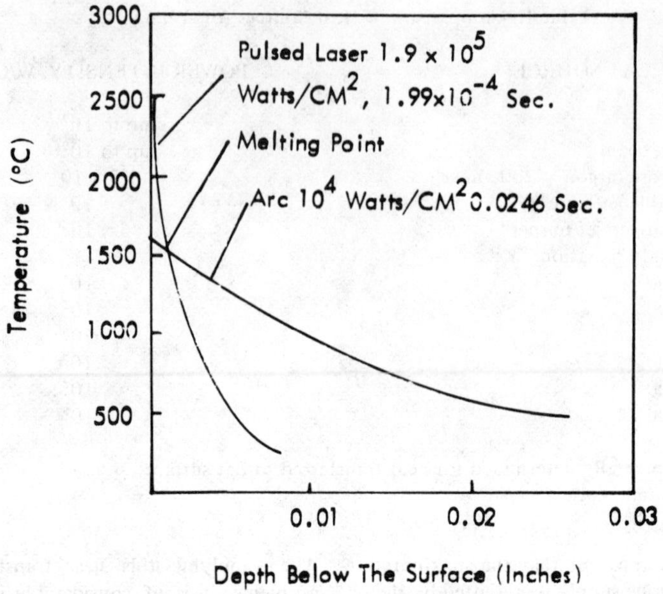

FIGURE 4. Maximum Temperature Reached as a Function of Depth for Iron for the Pulsed Laser and the Arc.

(T_R) at the hemispherical surfaces of radius R from the center of the beam are given by:

$$T_R = \frac{W}{2\pi KR} \quad (7)$$

EQUIPMENT

A point which will be subsequently developed is that most of the work which has been done to date using lasers for welding applications has involved the pulsed ruby or glass lasers described in Table I. These solid-state lasers were developed first and, therefore, more time has been available to convert them from laboratory devices into pilot plant or production welding equipment. It is only recently that the continuous wave lasers, using either CO_2 or YAG, have been available with adequate power outputs to affect useful welds.

The general situation at the present time is that pulsed ruby and glass lasers have been packaged into industrial devices specifically designed for welding; whereas, continuous wave lasers have been used for welding only under laboratory conditions where the basic laser has been integrated with rudimentary optics to produce experimental weldments. Figure 5 shows a typical pulsed laser welder. This particular unit consists of a work station, a power supply, and a closed-cycle refrigerated water cooler. Viewing optics are available at the work station to accurately locate the welding spot in microscope cross-hairs.

Closed-circuit TV can also be used for positioning the work under the welding beam. When pulsed lasers of this type are used with ruby crystals, pulse repetition rates of up to about 60 pulses per minute are generally obtained. It is, of course, possible to replace the ruby crystal with a neodymium-doped glass rod and obtain higher energy outputs, but the repetition rate for welding is reduced because the thermal conductivity of the neodymium-doped glass is only about 1/17 that of the single-crystal ruby. As was pointed out earlier, a YAG crystal could also be used in a similar arrangement for pulsed welding.

Figure 6 shows a typical CO_2 laser system. The components include the laser head, the power supply for excitation of the CO_2-bearing gas, and a service console which contains the gas handling equipment and the cooling equipment. One potential advantage of this CO_2 laser is that they are extremely efficient in comparison to other laser devices. Typically, efficiencies in excess of 10 percent are achievable with CO_2 lasers, while efficiencies with pulsed ruby lasers are on the order of one-half percent. Appropriate optics and work handling equipment must, of course, be added to the CO_2 laser system shown in Figure 6, in order to make it a useful welding tool. Development programs are in progress at several companies on ultra-high powered CO_2 lasers capable of CW output in excess of 10 kw. Such devices as they become available may provide welding capabilities equivalent to electron beam devices.

Figure 7 is a YAG laser which might also be used for laser welding. The particular laser shown here has the capability of generating 200 watts of CW power from a power input of approximately 12 kw into the Krypton arc lamps. Included in the photograph are the laser head, the power supply to drive the pump lamps, and the water cooling system required to remove the waste heat from

FIGURE 5. Typical Pulsed Ruby/Glass Laser Welder with Closed-Circuit TV for Remote Viewing. (Korad Model KWD)

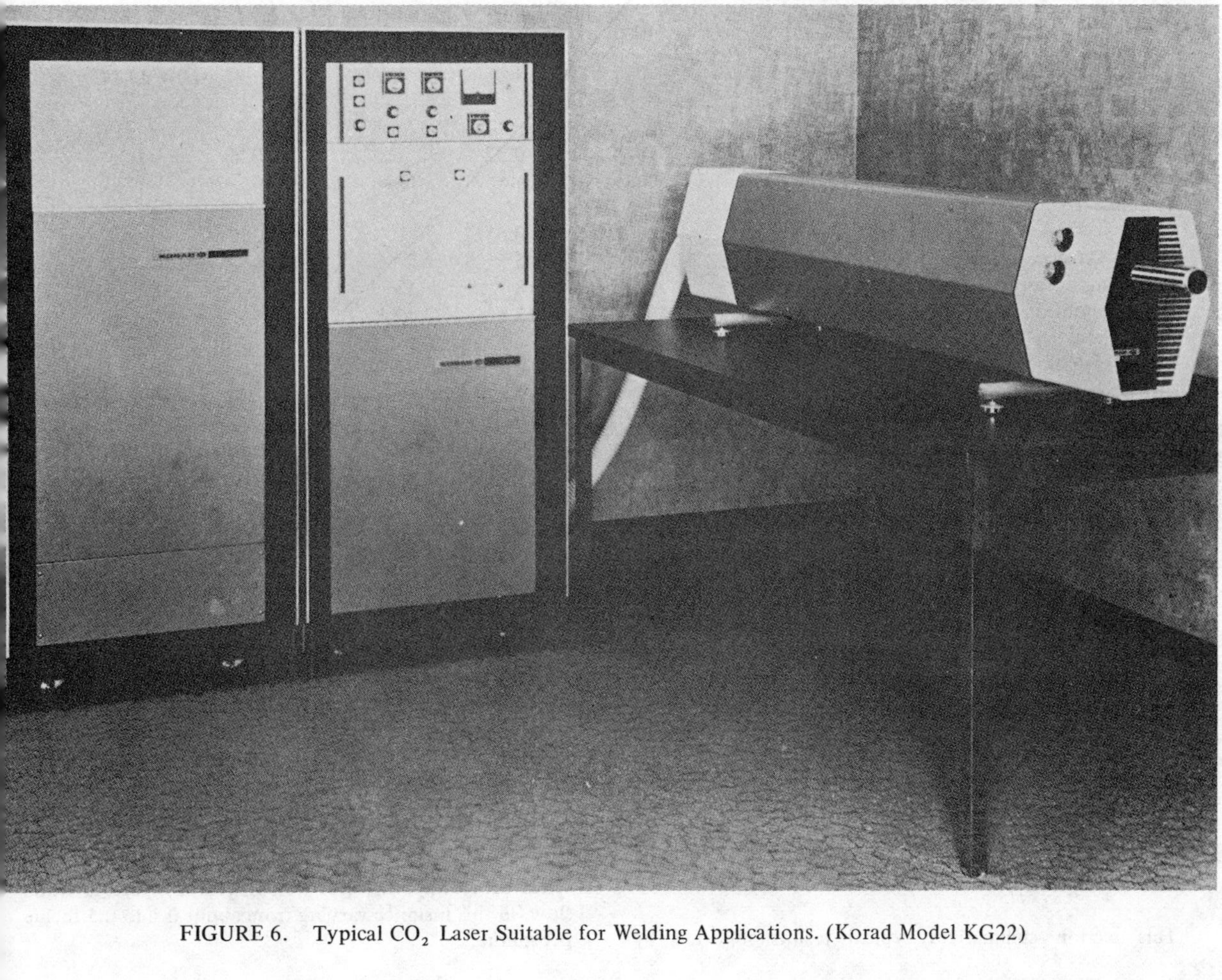

FIGURE 6. Typical CO_2 Laser Suitable for Welding Applications. (Korad Model KG22)

FIGURE 7. Typical High Power CW YAG Laser System with Power Supply and Refrigerated Water Cooler. (Korad Model KY5)

Table III. Laser Welding Conditions for Producing Butt Welds with Pulsed Ruby Laser[6]

LASER WELDING CONDITION	Ti-8A1-1Mo-1V ALLOY		321 STAINLESS STEEL		RENE 41	Ta-10W ALLOY	
Sheet Thickness, inches	0.023	0.062	0.005	0.027	0.025	0.025	0.025
Number of Passes	One	One	One	One	One	One	One
Spot Size Diameter, Inches	0.062	0.063	0.030	0.056	0.060	0.050	0.047
Energy Output, joules	23.5	36.0	*	25.5	23.0	*	51.5
Pulse length, msec.	18	18	9	18	18	18	18
Pulse Repetition Rate, pulse/min.	17.8	10.5	39.2	17.8	17.8	10.5	17.8
Weld Travel Speed, Inches/min.	0.46	0.14	0.39	0.39	0.30	0.21	0.32 (top) 0.50 (bottom)
Spot Spacing, inches	0.026	0.013	0.010	0.022	0.017	0.020	0.012 (top) 0.028 (bottom)

*Not Measured.

the relatively inefficient energy conversion process. It is interesting to compare Figure 6 with Figure 7 and note that the YAG laser head is substantially smaller than the CO_2 laser head for equivalent power outputs. As with CO_2, appropriate optics and work handling equipment must be added to the basic YAG laser system to make it a useful welding tool. For any of the laser systems discussed here the user has the choice of manual positioning of his work under the laser beam or using automated work handling equipment, which is actuated either with numerical or computer controlled drive circuitry.

WELDING WITH PULSED LASERS

This section summarizes typical results reported by various investigators on the application of pulsed laser welders to a wide variety of practical problems.

Results of a study on the welding of aerospace structural alloys in sheet geometry with a pulsed ruby laser have been reported by Witt, Kaehler, and Kloepper[6]. Ease of welding was reported for Ti-8A1-1Mo-1V, Type 321 stainless steel, René 41, and Ta-10%W alloy. However, problems were encountered with cracking in 6061-T4 aluminum alloy. Parameters employed to obtain high quality butt welds are summarized in Table III. Since, with a pulsed laser, the butt welds are created from a series of overlapping spot welds, welding speed is rather slow, in this instance varying from about 0.1 to 0.5 inches per minute.

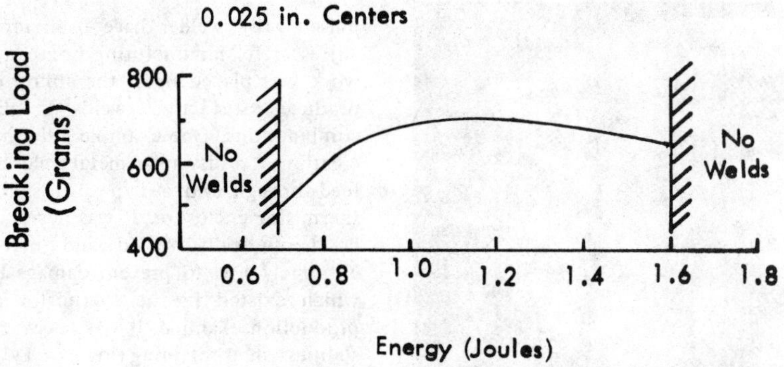

FIGURE 8. Effect of Energy on Strength of Welds. 60 Degree Weld, Edge Incidence, 4.9 ms.

One optical procedure which has been used to increase the welding speed attainable with pulsed lasers required a cylindrical, rather than a spherical, lens as the objective in the laser welder optical train. With the cylindrical lens the beam is elongated in the direction of welding, and, thus, the number of overlapping spots required to create a linear weld is reduced. Length to width ratios obtainable with the cylindrical lens are on the order of 10 to 1 or greater.

In the microelectronics field Cohen, et al[7] have reported on the feasibility of using a pulsed ruby laser for welding of an experimental plated-wire magnetic memory device. For this special application it was necessary to interconnect many 5 mil beryllium-copper wires, which were plated with 1.5μ of Cu and 0.6μ of the magnetic material Permalloy. The problem was made more difficult by the need to maintain a very close spacing (i.e., 25 mils) between adjacent pairs of welded wires in the device. Because of these constraints, laser welding was selected. Since it was felt that the laser energy could be focused to a small spot, only simple fixturing would be required. The size of the heat affected zone and attendant changes in the magnetic properties of the wire would be minimized. The authors concluded that the pulsed ruby laser, was indeed, suitable for the interconnection of a multitude of fine wires and that the probability of producing short circuits even in closely spaced areas was minimal. As can be seen in Figure 8, optimum parameters for this particular application involved energy outputs of about 1 joule with a pulse length of 5 milliseconds from the ruby laser. Energy inputs of less than 0.6 joules were inadequate to create a weld nugget and energy inputs greater than 1.6 joules were equally ineffective, presumably resulting in vaporization rather than weld bead formation. This illustrates a fundamental problem in laser welding; namely, defining the narrow zone between too little and too much energy.

Moorhead and Turner[8] have discussed the use of a pulsed ruby laser for attachment of a thermocouple vacuum gauge to the Apollo Lunar Sample Return Containers. These containers were designed to convey samples of the moon's surface back to the earth for analysis. For this particular application, brazing, mechanical assembly, and

FIGURE 9. Kovar Thermocouple Gauge Header, Coextruded Transition Cylinder, and Laser-Welded Subassembly.

soldering were ruled out on the basis of weight, contamination sources, or possible thermal damage to the container. Electron beam welding was initially chosen as the preferred joining technique and the specific problem was to effect a structurally-sound, helium leak-tight joint between a Kovar header and a Type 304 stainless steel transition cylinder (See Figure 9). Sound welds could not be produced due to unpredictable deflection of the electron beam by the magnetic Kovar pins in the header. Consequently, laser welding, which is not affected by magnetic field, was chosen and twenty-five subassemblies were produced using an appropriate rotary fixture. Welding parameters employed were: 6.6 joules/pulse, 5.2 milliseconds pulse duration per pulse, 1/32-in spot size, 30 pulses per minute, and 50% spot overlap. Of the 25 subassemblies, two were found to leak initially. However, after repair using the same parameters, all welds successfully passed the helium leak test at $<1 \times 10^{-9}$ atm-cc/sec. The cross-section of a Kovar-to-stainless steel weld is shown in the photomicrograph of Figure 10.

In another weldability study using a pulsed ruby laser, results obtained on Type 302 stainless steel and an 18% Ni maraging steel are described by a research team at the Pennsylvania State University[9]. With an energy input of 5 joules (other laser parameters unspecified) the welding of the stainless steel was possible in sheets of 0.005-in.

FIGURE 10. Laser Weld of Kovar Thermocouple Gauge Header to the Transition Piece.

thickness, and marginal in sheets of 0.010-in. thickness, while no penetration was obtained in the 0.015-in. thick sheets. However, suitable welds were made in each of the same three sheet thicknesses of the 18% Ni maraging steel and welding conditions appropriate to each thickness in terms of pulse energy and focused power density were determined. For those conditions where the input energy was inadequate or where the beam was sufficiently defocused, no penetration was observed. At the other extreme, excessive energy or a sharply-focused beam will produce metal expulsion and vaporization closely akin to hole drilling. Between these extremes, various combinations of parameters were defined where reasonable welds could be effected. The difference in weldability of the Type 302 stainless steel and the 18% Ni maraging steel is probably associated with a higher thermal diffusivity in the latter alloy (4×10^{-6} m²/sec versus 7×10^{-6} m²/sec).

A significant production bottleneck in the fabrication of klystron tubes was solved with pulsed laser welding at Varian Associates of Canada, Ltd[10]. In this application it was necessary to weld 0.002-in thick tantalum sleeves to molybdenum collars. Resistance welding was initially attempted, but the welds produced were mechanically weak and the finished components were often contaminated with microscopic droplets of copper condensed from the resistance welding electrodes. Rejects ran about four in five. Laser welding was explored as an alternative and it was found to be completely successful for this particular application. The rejection rate dropped to practically zero. The weld between these two refractory metals was produced under a glass bell jar which was filled with argon to eliminate contamination and maintain the integrity of the joint. This, incidentally, indicates another virtue of laser welding; specifically the ability to produce joints while welding through glass. Varian also demonstrated, with this equipment, the ability to repair defective klystron tubes using this same property of the pulsed ruby welder; namely, welding through the glass envelope of the vacuum tube.

A number of successful job-shop applications using a pulsed ruby welder have been reported by workers at AiResearch Manufacturing Company[11]. Emphasis in this work was placed upon the ability of the laser welder to produce satisfactory welds in dissimilar metal alloy combinations. One example cited involved the successful creation of a dissimilar metal weld between a B-1113 steel feed-through and a Type 321 stainless steel casing in a thermistor enclosure. It was necessary that the weldment be hermetically sealed and that the heat input be extremely low to prevent damage to a glass ceramic seal which existed in the thermistor assembly. In another production example, it was necessary to weld a Type 321 stainless steel retaining ring to a Type 440C high-temperature bearing race. Here again, it was essential to limit heat input during the welding process to prevent thermal damage to the part (i.e., softening of the martensitic stainless steel). A number of bearings were welded and tested and it was found that the laser weld between the retainer ring and the housing in no way jeopardized the performance of the component. These results substantiate in a qualitative way the laser welding advantages presented previously in Figure 4 regarding the extent of the heat affected zone.

Many investigators have demonstrated the practicability and frequently the practicality of making sound wire-to-wire or wire-to-sheet welds using the pulsed laser. A typical example is the precision attachment of thermocouples to an object whose temperature must be accurately measured. In the case of thermocouple welding, it is possible to simultaneously create the junction, attach the thermocouple to the substrate, and trim the rough cut wires to exact length. Laser-produced thermocouple welds have been used to attach measuring probes to transistors, turbine blades, nuclear reactor fuel elements, and the like.

Important considerations in these applications are the ability to produce dissimilar metal welds and very low energy input requirement in order to minimize thermal damage to adjacent materials.

Anderson and Jackson[4] have discussed optimum procedures for welding integrated circuit flat packs to printed circuit boards in microelectronics fabrication. Figure 11 shows a typical P.C. board to which two flat packs have been welded, while Figure 12 shows a close-up of two of the laser-welded interconnections between the leads and the pads of the P.C. board. Successful welds have been made to boards fabricated with nickel, copper, kovar, and multilayer claddings. These authors have also noted the laser's ability to produce atachments of insulated wires to lugs without stripping. In one example, connections of polyurethane-coated wires of a relay coil were produced by simultaneously vaporizing the organic coating and fusing the coil wire.

The Korad Department of Union Carbide Corporation[12] has conducted many applications studies with pulsed laser welders. In several instances, the ability to produce (without insulation stripping) sound welds between enameled wire and terminals has been demonstrated. Figure 13 shows an example of this. Here the 0.040-in.

FIGURE 11. Two Integrated Circuits Welded to a Printed Circuit Board.

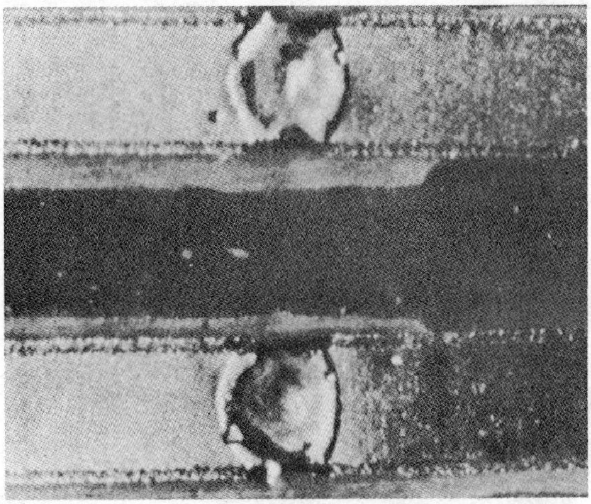

FIGURE 12. Close-up of Welds Shown in Figure 11.

FIGURE 13. Weldments of Coil Wires to Commutator Segments in Electric Motor. Welds were Produced with Pulsed YAG Laser.

enameled-copper wire from the armature coil of a motor was successfully welded to a commutator segment. In this particular instance, the energy available from the 4-by 3/8-in. diameter ruby crystal ordinarily used with the laser welder did not provide adequate energy to create a sound joint. However, when the ruby crystal was replaced with a YAG crystal of the same size, the high quality welds illustrated in Figure 13 were easily produced. This situation stems from the fact that the conversion of pump lamp energy to laser energy in YAG is considerably more efficient than in ruby.

Most of the pulsed welding discussion presented thus far has involved the use of ruby lasers. Ruby is generally preferred to neodymium-doped glass because the substantially higher thermal conductivity of the crystalline material permits the ruby laser to be pulsed at a higher repetition rate without exceeding thermal stress and beam divergence limitations. Hence, it is possible to achieve higher welding speeds. In addition, the higher thermal conductivity insures better shot-to-shot energy reproducibility and, weld-to-weld consistency when random welding cycles are employed. Nevertheless, fusions have been made in aluminum, carbon steel, stainless steel, molybdenum, niobium, tungsten, tantalum, titanium, Inconel, nickel, copper, and brass with glass laser welders. Unfortunately, details on welding conditions are not discussed[13].

WELDING WITH CONTINUOUSLY PUMPED LASERS

Virtually all work published as available literature deals with pulsed lasers (generally ruby lasers). Very little information exists on the use of continuously pumped lasers for welding. The reason for this situation, of course, has been the fact that YAG or CO_2 lasers of adequate power for useful welding operations have not been available as long as the venerable ruby devices have been available.

An interesting comparison between the CW welding with YAG and CO_2 has been made by Jackson of Union Carbide Corporation[14]. In his developments, Jackson has defined a threshold power, which is the laser power required to sustain a molten puddle (i.e., to weld at zero speed). For power levels above the threshold power, he defines a critical welding speed, which is the maximum speed at which full penetration welds are obtained.

When welding in excess of the critical speed, less than full penetration welds are produced. Test welds were produced in 5, 10 and 18 mil thick Type 304 stainless steel using a 200 watt YAG laser and an empirical correlation was developed between critical welding speed, S (in./min), laser power, P (Watts), and sheet thickness, t(mils), as follows:

$$S = 3.5e^{-0.146t}(P-22e^{0.0835t}) \quad (8)$$

Experimental results are presented in Figure 14. Similar studies were conducted using a CO_2 laser and it was

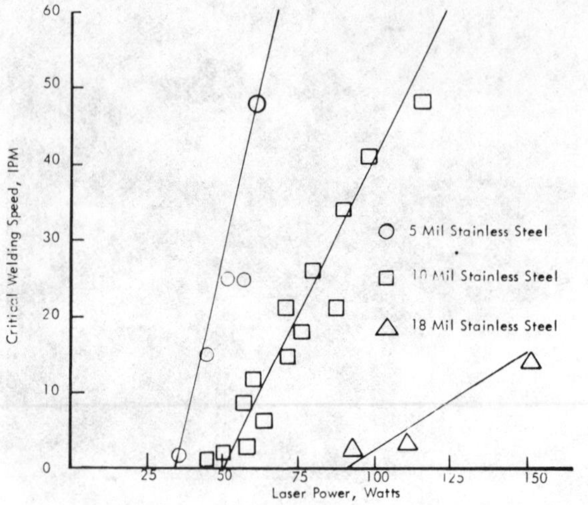

FIGURE 14. Effect of Laser Power on Welding Speed for Stainless Steel with the YAG Laser (CW-Pumped).

determined that power required to achieve a given welding speed was three times as great with CO_2 as with YAG. For example, to weld a 10 mil sheet at a welding speed of 30 ipm required an output of 85 watts from the YAG laser and 260 watts from the CO_2 laser. This trend is not surprising, since there are major differences in the absorptivity of metals to the 1.06μ radiation emitted by YAG and the 10.6μ radiation emitted by CO_2, the absorption being less at the longer wavelength. From the absorptivity values presented in Table IV, it is obvious that the stainless steel absorbs radiation of 1.06μ wavelength about two times as effectively as at 10.6μ, which supports the experimental welding observations described above. Of course, as the temperature of the workpiece increases, materials become more absorbing and the effect of these wavelength differences is less pronounced.

Further, in the context of comparison between YAG and CO_2, it should be noted that if the laser operates in the fundamental transverse mode, the beam divergence (θ) and, thus, the spot size (S) is proportional to the wavelength. Factoring this information into Equation (1) reveals that the focused power density from a diffraction-limited YAG laser is 100 times greater than for a CO_2 laser with the same power output.

Preliminary studies on the welding of super alloys using a CO_2 laser have been reported by Alwang, et al[18]. Samples of AMS 5525 (A-286) and AMS 5544 (Waspalloy) in thicknesses ranging from 18 to 31 mils were continuously butt welded. The AMS 5544 was joined using 195 watts of multimode output power focused to a 0.050-in. spot size equivalent to a power density of 15,400 w/cm^2. The welding speed was about 0.2-in./second, and an argon cover gas was employed. This speed is comparable to what might be achieved by the gas tungsten arc method. Laser welds in 18 mil thick AMS 5525 were compared with electron beam welds produced under similar conditions. Although minor differences in the chemistry and hardness of the comparison welds were noted, the two processes yielded similar results.

More recently, Webster[19] has summarized welding studies using a 250 watt CO_2 laser. Details of the thin gage welding studies are presented in Table V. Webster points out that metals with high thermal diffusivity, and reflectivity such as aluminum and copper, cannot be welded with present commercially available CO_2 lasers. Incidentally, operating costs for the 250 watt CO_2 laser are quoted at $1/hr. for gas and electric power consumption. In speculating on future developments, the author alludes to the fact that an experimental CO_2 laser operating at 4.6 kw has been used to weld 1/8-in. thick stainless steel at 14 in./min.

In certain welding applications, the problem of high reflectivity of metal to the 10.6μ radiation from a CO_2 laser can be overcome through the use of pulsing. When the plasma in the laser tube is electronically pulsed the peak power in the pulse is higher than the average power and the metal surface can be heated to a sufficiently high temperature that the reflectivity is reduced. Melting can then be achieved and a reasonable fusion zone sustained. Figure 15 shows an example of a microelectronics fabrication produced with a CO_2 welder. In this particular instance, a TO-5 transistor can was hermetically sealed with minimal heat input, thereby reducing the possibility of thermal damage to the circuit element.

Conti[20] has reported welding studies with a 100 watt (CW) CO_2 laser. His results agree with those of Webster in that he was also not successful in welding materials such as copper, beryllium-copper, and aluminum which are highly reflective at 10.6μ. Successful welding of thin sheets of stainless steels and certain copperbase alloys was accomplished, however. Results are summarized in Table VI. It will be noted that these welds were generated by operating the laser in the pulsed mode, rather than the continuous. The maximum depth of penetration obtainable from the 100 watt laser operating multimode and pulsed was about 0.009 inches.

Engineers at Holobeam, Inc.[21] have reported very limited welding studies with a 760 watt CW Nd:YAG laser. Welds were made in bulk 1/4-in. thick Type 304 stainless steel and AISI 1030 carbon steel at 15 in./min. Penetration was only 20 mils, which is scarcely better than that quoted in Table V for a 250 watt CO_2 laser. This rather surprising result may be due to the fact that the YAG laser spot size was too large in these preliminary tests.

LASER WELDING CHARACTERISTICS

It is apparent from the foregoing discussion that a certain range of applications exists where the unique properties of the laser beam may be used to advantage for metal joining. It is equally apparent that many jobs exist for which the laser is not suited. Basically, the laser must be regarded as a specialty tool. Some of the specific

Table IV. Spectral Absorptance of Various Materials

MATERIAL	TEMPERATURE (°F)	SPECTRAL ABSORPTANCE* 1μ	10μ	REFERENCE NUMBER
Titanium	RT	0.4–0.55	0.09–0.12	15
Titanium	930	0.53–0.6	0.12–0.15	
Titanium	1380	0.5–0.75	0.2–0.25	
Type 301 SS	86	–	0.23	15
	230	–	0.3	
Type 321 SS				
(Electropolished)	1200	0.3	0.13	15
(Sandblasted)	1200	0.33	0.41	15
AM–350	480	~0.7	0.42–0.49	15
	930	0.58–0.8	0.41–0.51	
	1380	0.7–0.75	0.53–0.61	
17–7 PH SS	480	~0.6–0.7	0.21–0.25	15
	930	0.56–0.65	0.21–0.25	
	1380	0.52–0.73	0.24–0.29	
Inconel X	480	~0.5–0.7	0.27	15
	930	0.42–0.52	0.23–0.28	
	1380	0.41–0.5	0.23–0.31	
René 41	480	~0.7	0.42–0.48	15
	930	0.56–0.6	0.48	
	1380	0.53	0.44–0.5	
Hastelloy X	480	~0.4	0.2–0.28	15
	930	0.53–0.87	0.2	
	1380	0.65–0.7	0.23–0.47	
Haynes 25	480	–	0.29–0.32	15
	930	0.57	0.28–0.36	
	1380	0.6–0.7	0.48–0.56	
Molybdenum	1800–2200	0.27–0.35	0.09	15
Tantalum	2200	0.3	0.1	15
Tungsten	1700	0.38	0.06	15
Al_2O_3	1886	0.42	0.99	15
Aluminum	Room Temp.	0.07	0.02	17
Silver	Room Temp.	0.0064	0.0047	16
Gold	Room Temp.	0.014	0.0061	16

*Spectral Absorptance = Spectral Emittance

advantages which have been alluded to previously are summarized as follows:

1. Welds can be made in areas impossible to reach by more conventional processes, such as the bottom of blind holes. This may be paraphrased into the saw, "If you can see it, you can weld it with a laser."

2. Because of the high energy density achievable with a laser, the extent of the heat-affected zone is minimal. This reduces undesirable metallurgical changes or permits welds to be made close to heat sensitive parts.

3. Welds can be produced in a magnetic field without deflection of the laser beam.

4. The weld can be precisely located (e.g. to within ± 0.0005 in.) and the weld bead made very small (~0.001 in.), thus providing a process suitable for microwelding applications.

5. Welds can be made through glass and other transparent media.

6. No physical contact is necessary between the joining media and the workpiece. Consequently, only simple work-holding fixtures are required and the strains associated with hold-down fixtures are eliminated.

7. Temperatures achievable with laser welding are sufficiently high to fuse all metals.

8. Welds can be made in air, vacuum, or inert atmosphere.

9. Dissimilar metal welds are relatively easy to make,

Table V. Results of Welding Various Materials with CO_2 Laser

MATERIAL	THICKNESS (t), IN.	RATE IN./MIN.	WELD WIDTH, IN.	REMARKS
BUTT WELDS				
Type 321	0.005	90	0.018	Full penetration
Type 321	0.010	35	0.028	Full penetration
Type 321	0.0165	11	0.030	Partial penetration
17–7 PH	0.005	110	0.018	Full penetration
Type 302	0.005	50	0.020	Full penetration
Type 302	0.008	30	0.020	Full penetration
Type 302	0.010	10	0.040	Full penetration
Inconel 600	0.004	150	0.010	Full penetration
Inconel 600	0.010	40	0.018	Full penetration
Nickel 200	0.005	35	0.018	Full penetration
Monel 400	0.010	15	0.025	Full penetration
CP titanium	0.005	140	0.015	Full penetration
CP titanium	0.010	50	0.022	Full penetration
EDGE WELDS				
Type 321	0.005	85	–	Weld Type I
Type 321	0.010	25	–	Weld Type I
Type 321	0.0165	15	–	Weld Type II
17–7 PH	0.005	45	–	Weld Type I
Inconel 600	0.004	160	–	Weld Type I
Inconel 600	0.010	35	–	Weld Type I
Inconel 600	0.0165	25	–	Weld Type II
Nickel 200	0.005	18	–	Weld Type I
Monel 400	0.010	25	–	Weld Type I
Ti-6A1-4V	0.020	27	–	Weld Type II
Ti-6A1-4V	0.010	300	–	Weld Type I
LAP FILLET WELDS				
Tin-plated steel	0.012	20	0.025	Full penetration
Type 302	0.016	175	0.030	Partial penetration
Type 302	0.030	30	0.030	Partial penetration
Type 302	0.010	15	0.025	Full penetration
CORNER WELD				
Type 321	0.010	20	–	–

NOTE: Welds made with Coherent Radiation's Model 41 250 w CO_2 laser equipped with a Model 141 pulsing accessory; 0.100 in. nozzle at a stand-off distance of 0.100 in.; shielding and backing gas, argon at 15 cu. ft. per hr. All materials welded in as-received annealed condition.

providing the interface between the dissimilar metal does not vitiate sound metallurgical principles (e.g., with regard to formation of brittle intermetallics, etc.).

On the debit side of the ledger, the following factors must be assessed:

1. The capital cost of the equipment is high.
2. Deep penetration welds cannot be created with present commercially available lasers.
3. Good fit-up between parts to be welded is required.
4. Pulsed laser welding is considerably slower than conventional processes.

Thurber believes in the basic credo of laser applications, "If you don't need the laser don't use it," not as a technique to supplant the more well-established welding process, but rather to do jobs not otherwise possible.

As will be shown later in the discussion of laser safety in metalworking, there is a remarkable lack of data of the various hazards from laser metalworking, including the eye hazards and the respiratory hazards from air pollution induced by the laser operation. Obviously, with closed systems such laser hazards are not significant except where the exhaust may contribute significantly to environmental pollution.

Table VI. Summary of Edge-Lap Welds on Thin Metal Sheet Obtained with 100 Watt CO_2 Laser

SAMPLES WELDED	LASER POWER (% OF FULL OUTPUT)	PULSE WIDTH (SEC)	REPETITION RATE (SEC/PULSE)	TRAVEL SPEED (IN./MIN)	PARENT MATERIAL ULTIMATE STRENGTH (LBS/IN.^{2})	PEEL STRENGTH (1% OF ULTIMATE)
0.0045 Inch Phosphor Bronze "A" (95% Cu, 5% Sn)	95	0.30	0.35	5	49,000	65%
0.004 Inch 304 Stainless Steel Foil	95	0.0002	0.001	3	86,000	45%
0.004 Inch 304 Stainless Steel Foil	95	0.22	0.28	6.5		
0.004 Inch Silver 752" (65% Cu, 18% Ni, 17% Zn)	95	0.065	0.075	3	55,000	19%
0.004 Inch Monel (70% Ni, 30% Cu)	95	0.07	0.2	3	66,000	23%
0.004 Inch Cupro-Nickel (70% Cu, 30% Ni)	95	0.30	0.35	3	75,000	12%

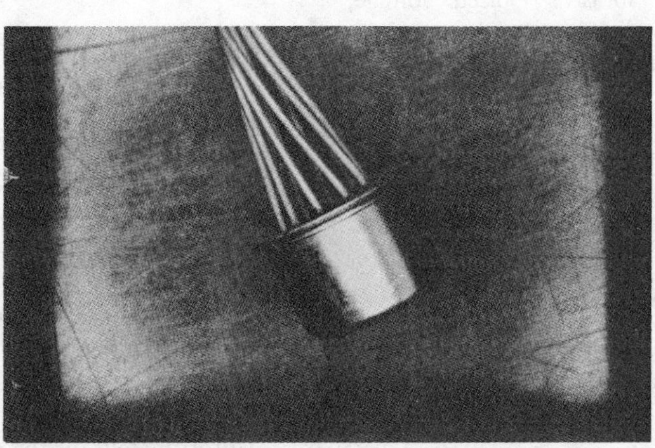

FIGURE 15. TO-5 Transistor-Can Weld Produced with Pulsed CO_2 Laser. Welding Conditions: 8.5 rpm (2 Revolutions), 200 Pulses per Second, 25 Millisec Pulse Width, 80 Watts Average Power.

Regarding laser welding and drilling, Moorehead[10] concluded:

1. The laser beam can weld or drill after passing through a window (utilized to obtain a gas sample from a sealed reaction vessel).

2. The process has a low energy input (an aluminum temperature-indicating capsule was welded without melting the low melting point ring inside).

3. The laser is unaffected by stray magnetic fields (a ferromagnetic thermocouple vacuum gage header was welded to a bimetallic cylinder). thermocouple vacuum gage header was welded to a bimetallic cylinder).

4. The laser has an off-axis welding capability (used to weld a thermocouple to the inner wall of a tube).

5. It is possible to locate precisely the laser

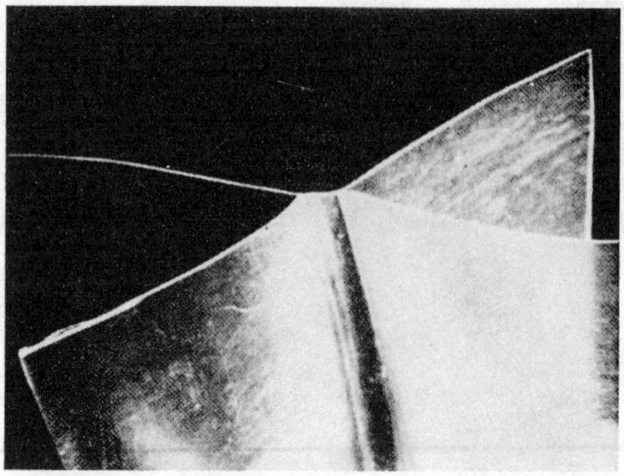

FIGURE 6-11. Laser welded polyethylene sheet. (Karl Gürs, Batelle, Frankfurt.)

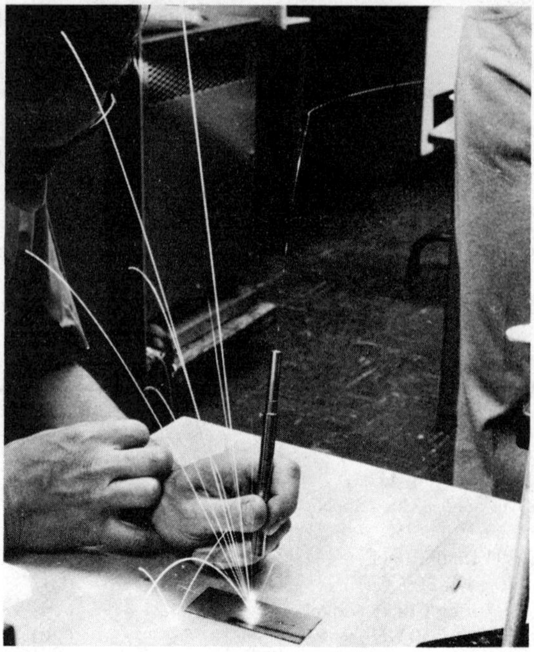

FIGURE 6-12. Q-switched Nd YAG welding through single fiber of Nath. (Gunther Nath.)

spot (used to attach different types of fine-wire thermocouples to various parts and materials).

Resistor trimming is another important phase of laser material processing. The techniques for resistor trimming are abrasive, pulsed voltage, mechanical scribing, anodization, radio frequency, and laser. The photons act as an abrasive trimming. Walters and Weiner[14] of Korad have reviewed resistor trimming with Nd-YAG laser. These details are taken from their report.

The difference between thick and thin films is the process by which they are made, not their thickness. Actually a thin film could be thicker than a thick film.

Thin films are vacuum deposited in a "bell jar" to thicknesses from several hundred angstroms to several microns. Materials used for resistor compositions are nichrome, tin oxide, tantalum, and cermet conductor.

In contrast, the thick film process is characterized by screen printing followed by firing in a furnace. Resistor compositions encompass systems based on palladium, ruthenium, thallium oxide, and precious metals, and may be several mil thick.

Both the thick and thin films are generally placed on alumina (Al_2O_3) substrates. A glass underglaze may be put on the substrate to present a smooth surface for the thick or thin film. Also, an overglaze may be applied to protect the resistor. General advantages of all laser trimming equipment include:

1. No mechanical contact with the work, and thus no wearing of parts
2. No electrical contact resulting in ability to trim electronically active circuits
3. Capability of automation and computer control
4. Operation in ordinary manufacturing environments or in clean rooms
5. Fast shut off resulting in high accuracy trims
6. Trimming so as to maximize power dissipation characteristics by optimizing the shape of the "L" cut.

Unfortunately for the microelectronics industry (and fortunately for the laser manufacturers), both thin and thick film resistors are generally fabricated to within 5 to 25% of the desired value. Consequently, trimming or adjusting to a more precise resistance value is frequently required.

Particularly with the high production rates made possible by thick film processes, trimming currently is, or promises to be, a major bottleneck in the manufacturing of hybrid microelectronics.

Trimming by laser vaporization in the Q-switched mode is a process by which resistor material is removed, thus increasing the resistance. Laser annealing by heating is also possible with the CW mode where several watts of continuous power are available.

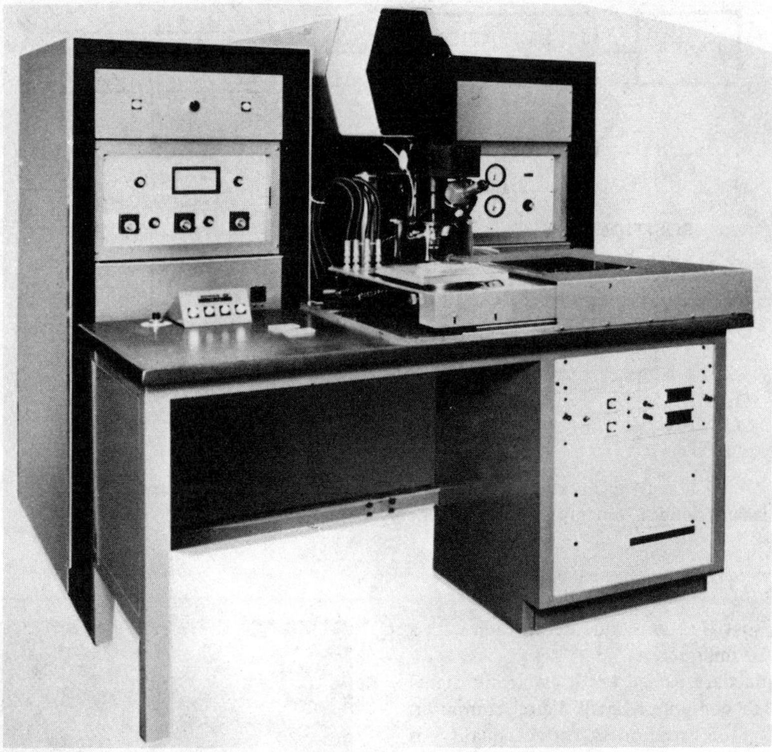

FIGURE 6-13. Surface recognition equipment for surface scanning for microfinish and defects. (Bendix Recognition Equipment Systems, W. G. Wolber.)

The laser's low average Q-switched power and wavelength in the near infrared result in negligible resistor heating and substrate damage. This means that with power density properly chosen for trimming, the laser beam can be held continuously on one spot with no discernable effect on the substrate, coupled with no change in the trimmed value. Also, the trimming or cutting rate may be set anywhere between zero and maximum, even though the laser is pulsing at a fixed high rate. In operation, laser power must be adjusted just above the vaporization level of the film and below the discoloration threshold of the substrate.

Howard and Allen[5] have examined laser trimmed, thick film resistors by scanning electron microscopy (S.E.M.). This technique is used more and more in many phases of laser technology. The lasers used included CO_2 and YAG types. Their conclusions are reported in detail. Results of parametric studies of laser-trimmed resistors are

a. Laser trimming affords high accuracy and precision in "as trimmed" parameters. With high performance resistor compositions, standard deviations less than 0.2% are routinely attainable.

b. A strong interaction between the laser technology and ink composition profoundly affects trimming accuracy and precision. Resistor ink suppliers may wish to provide compositions especially adapted to laser trimming.

c. Attaining precision and accuracy in laser trimming requires careful selection of measuring and control equipment. This equipment must be selected to use the full capability of the laser technology employed.

Inspection of Metal Surfaces

Sjolin[12] reported the use of a laser beam to inspect surfaces of rolled or sheet metal. This type of automated system is much more accurate and much more rapid than the "skill and mood" of the inspector. The Sjolin technique of surface inspection is based on an intensity Fourier transformation. It measures 1,600 scans/sec.

"The beam is modulated by grooves and possible surface defects, which cause different components of the Wiener spectrum. The spectrum is spatially filtered and projected onto a photomultiplier, which integrates the transmitted radiation. The system consists of one or more identical measuring heads in modular form each contain-

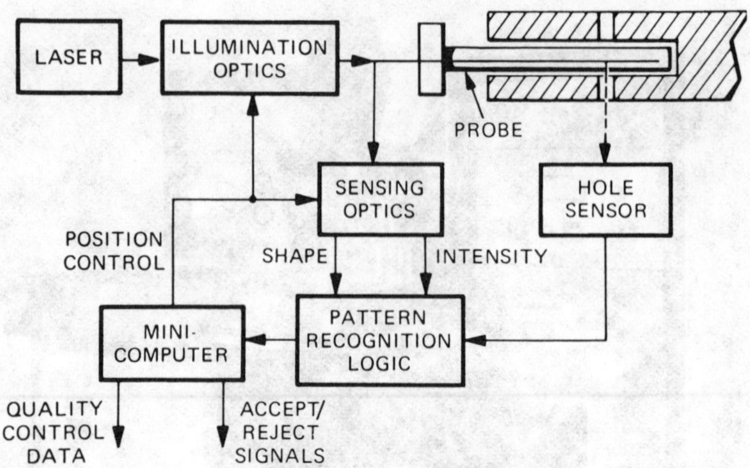

FIGURE 6-14. Brake cylinder inspection system with low output helium-neon laser. (Bendix Recognition Equipment Systems, W. G. Wolber.)

ing a laser, scanner, spatial filter and detector, each with a scanning width of 250 millimeters.

The detector signals are passed to an electronic signal processing unit which contains a small digital computer. The result, a measure of variation in surface quality, is presented as a function of the length or width (or both) of the metal strip."

Other industrial applications of metal inspections beside rolled and sheet material have been with cartridge cases. Brake linings may also be inspected by laser beam. As indicated, scanning electron microscopy (S.E.M.) can provide records of surface of metal and other materials.

Many new models are being developed for laser machine tools. One machine of the Burroughs Electronics Components Division makes tiny holes in thin stainless steel parts. This machine of Lasermation[8] weighs about 2,000 lb. Assembly lines in aircraft and automobile industries include laser machines, especially for drilling and resistor trimming.

Miscellaneous Material Processing Laser in Ceramics

Machining ceramics may be done with the laser especially in the flat alumina stock of ceramic material. The CO_2 laser at present is an accurate and controllable source of energy for this work. Neodymium-YAG lasers may also be used. The ceramic material is located at the focal point and there are melting of the ceramic and vaporization from the absorbed laser beam. Process control will assure the proper processing without any effect on adjacent material.

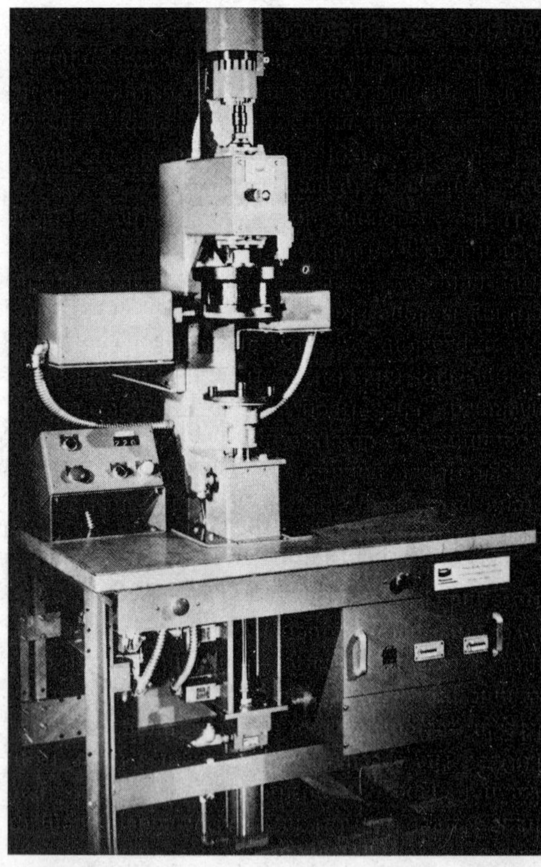

FIGURE 6-15. Laser ceramic scriber. (Union Carbide – Korad.)

Some examples that are given by Lasermation[8] are drillings in the 0.018 in.-thick ceramic, hole diameters up to 0.03 in. ± 0.0005, 0.012 in. diameter, 0.015-in. diameter, and with holes and shapes larger than 0.1 in. ± 0.002 independent of size and shape. The position tolerance is less than 0.0025. The speeds in fired ceramics range from 0.15 in./sec cutting to 4 in./sec scribing. Speed for ceramic processing is related to material thickness; unfired ceramic is easier to machine because of lower thermal conductivity and lower thermal stress and easily removed binders.

Weick et al.[15] of Bell Laboratories describe a laser machining process to form tiny electronic circuit patterns directly into ceramic substrates in a single step. Computer storage to assist the laser is programmed to describe the type of circuits to be machined. The effective use of laser drilling in ceramic substrates for hybrid circuiting will make for expansion of this application. Substrate scribing is done also on silicon. Many companies offer industrial laser equipment suitable for the particular type of job requirement.

If the machining done is open, the operator must wear glasses to prevent burns.

In the chapter on Laser Art it will be shown how laser design and sculpture can be done on unfired and fired clays in an interesting fashion. Here too, with these procedures, glasses must be worn because of the intense light.

Rubber Drilling

The CO_2 laser can be used for rubber drilling. It is especially valuable for nursing bottles to assure proper, accurate hole drilling in the nipple.

In addition to infant nursing bottles, laser applications in rubber drilling refer to valves, gaskets, seals, and insulation. The hole diameters may vary from 0.05 to 0.5 in. Holes 0.004 in. can be made, depending on the material and thickness. Coherent Radiation uses a 50-W CO_2 laser as well as a 250-W laser. The CO_2 laser drilling in rubber by Coherent Radiation[2] is in the pulsed mode. Unlike mechanical punching or cutting, there are no burrs or residual drill tailings. Some materials are said to have "sealed finished quality." Since there is no contact with the operating CO_2 laser, unique product configurations can result according to Coherent Radiation.

Plastics

The CO_2 laser can be used in plastics for cutting and drilling. The reactions in plastics depend, of course, on the laser used, duration of the thermal decomposition process, and the type of plastic used. The safety program concerning eye protection and toxic products polluting the air will be discussed in the chapter, Laser Safety.

FIGURE 6-16. CO_2 laser computer controlled cutting of cloth for Genesco. (From *Lasersphere,* 1, 1, 1971. With permission.)

The use of the ruby laser to produce decorative patterns inside of a block of plexiglass will be listed in the chapter, Laser Art. With color materials incorporated inside plastics such as plexiglass, argon and Nd-YAG lasers may be used and designs can be inscribed through the clear plastic onto the embedded materials.

Applications for Diamonds and Semiprecious Stones

The laser drilling of holes in diamonds for diamond dies has a very important industrial use. The holes may vary from 0.01 to 0.060 in. with the neodymium-YAG laser. The drilling process may be observed safely by the operator with closed circuit television. The laser has been used also with laser drilling of diamonds to remove imperfections such as may occur with gaps or variations in crystal growth patterns inside the diamond from various types of inclusions. The vaporization of these imperfections by the YAG

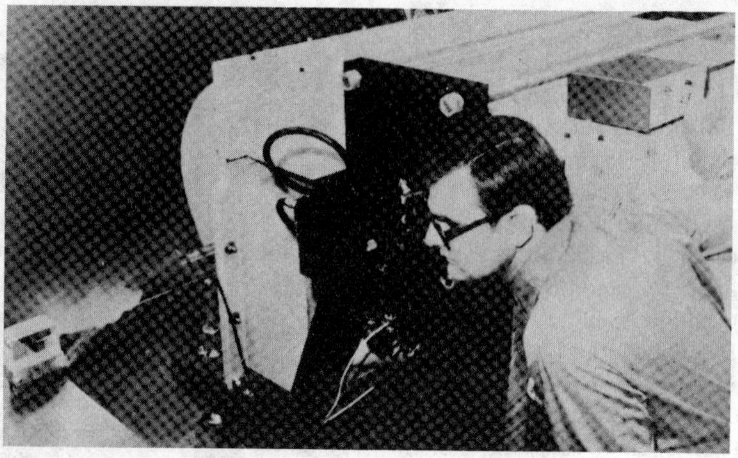

FIGURE 6-17. Laser impact of plastic block – eye protection necessary because of intense light. (From *Lasersphere,* October 15, 1972. With permission.)

laser as done by Raytheon increases the value of the diamond.

In the Swiss watch industry, laser microdrilling of synthetic sapphire and similar hard synthetics has produced accurate holes in a rapid economic fashion. The laser drilling of jade with bleaching of green jade will be listed in the chapter on art.

Laser Textile Industry

Computer controlled textile cutting has been done by laser systems. These include single pattern cutting as well as multiple pattern cutting. Enthusiasm varies as one hears such phrases as "the greatest advance in textiles since the cotton gin" and "the greatest advance since the sewing machine." An example of this application is by Hughes Aircraft Corp. with the CO_2 garment cutting machine for Genesco. The basic arrangement of this application is

1. A computer which programs pattern and cutting instructions
2. A positioning device utilizing a linear motor that lies above the flat surface
3. The laser
4. The fabric conveyor

The laser is a CO_2 laser with 250-W output. The beam movements are controlled by silicon mirrors into a pentaprism arrangement. Focusing is done by gold-plated aluminum mirrors. A cooling system controls the temperature around the area of cutting. It is said that one suit can be produced every 3 min.

Multiple pattern cutting also has been done by lasers. This is being compared with single pattern cutting as regards speed, accuracy, and distortion of the fabric. With laser beam cutting, ends of the cloth are not frayed, but are welded by the laser beam. The safety problems are minimal because of the enclosed nature of the laser operation. Similar techniques are now in use in the shoe-making industry.

Woodworking

Precise clean wood dies have been made with the CO_2 laser of Coherent Radiation. In this instrument, an air jet coaxial with the focused laser beam is said to modify the thermal cutting process and consequently allows deep penetration cuts at high speed. The high velocity of the gas controls the vaporization of the wood and no ignition of the impact area develops. An automated die making system eliminates the "variability of human skills" for the making of accurate dies.

Lasers can be used also to decorate wood materials for design shaping. Perhaps the "clean cutting" of such materials may reduce the pollution potential of this process.

The new needle-sharp water jet cutting tool may be a rival. Very few control studies have been done with the lasers for such materials as corru-

TYPICAL APPLICATIONS AND CHARACTERISTICS
OF LASERS USED IN INDUSTRIAL PROCESSES

Application	Typical Energy or Power Required	Pulse Width	Wavelength	Laser Type
1. Drill Metal, Diamond	1-100J	10^{-3} sec	0.693μ	Pulsed Ruby
2. Drill Non-Metal	10^{-1} J	10^{-4} to	10.6μ	Pulsed CO_2
3. Spot Weld Metal	35J	10^{-2} sec	1.06μ	Pulsed YAG
4. Trim Resistors	10^{-4} J	10^{-7} sec	1.06μ	Q-switched YAG
5. Scribe Silicon Wafers	10^{-3} J	10^{-7} sec	1.06μ	Q-switched YAG
6. Seam Weld Metal	100W	cw	1.06μ	cwYAG
7. Seam Weld Heat-Sensitive Devices (e.g., Thermocuples)	10J	10^{-2} sec (overlapping spots)	1.06μ	Pulsed YAG
8. Scribe Ceramic Substrates	10^{-1} J	10^{-4} to 10^{-2} sec	10.6μ	Pulsed CO_2
9. Cut Titanium	100W	cw	1.06μ	cw YAG
10. Trepan Holes in Metals and Opaque Non-Metals	10^{-3} J	10^{-7} sec	1.06μ	Q-switched YAG

FIGURE 6-18. Chart of Precision Laser Machining from Korad Division of Hadron Inc. (From *Electro-Optics Systems Designs.* With permission.)

gated board, plastic board, chipboard, foam rubber, and plywood.

It is evident that industrial laser systems have just entered many new fields of laser material processing. If these systems are better, faster, more economical, and safer, they will be used and most likely helped by computers.

REFERENCES

1. **Carstens, J. and Brown, C.,** Rock cutting by laser, Society of Petroleum Engineers, October 1971, personal communication.
2. Coherent Radiation, Inc., CO_2 laser welding joins the parade, *Welding Eng.*, 1970.
3. **Gürs, K.,** High-power CO_2 Lasers, Battelle, Frankfurt Information, February 12, 1972.
4. **Heglin, M. L.,** Laser drilling, applications and equipment, 20th Annual IEEE Machine Tools, Industry Conference, October 20, 1971.
5. **Howard, R. T. and Allen, R. V.,** Characterization of laser – trimmed thick film resistors by scanning electron microscopy Space Systems, Electronic Systems Center, Huntsville, Ala., 1971.
6. **Jackson, J. E.,** Packaging with Laser Welding, Korad P/B-010, 1970.
7. Korad, Union Carbide Corp., Ceramic scriber.
8. Lasermation, Inc., Now, the laser machine tool, *Am. Mach.*, June 14, 1971.
9. **Molines, J. L.,** The laser – new tools for machining ceramics, *Ceramic Age,* August 1971.
10. **Moorehead, A. J.,** Welding and drilling applications, *Welding J.,* February 1971.
11. **Rockwell, R. J., Jr.,** Personal communication, 1972.
12. **Sjolin, U.,** Optical inspection of metal surfaces, *Laser Focus,* 7, 42, 1972.
13. **Thurber, W. C.,** Laser welding, theory, status and prospects, Korad, Union Carbide Corp., Santa Monica, Calif., 1969.
14. **Walters, R. L. and Weiner, M. J.,** Resistor trimming and other micro-machining with a YAG laser, Korad, Union Carbide Corp., Presented at the Pennsylvania State University Engineering Seminar on New Industrial Technology, July 7-9, 1969.
15. **Weick, W. W., Cohen, M. I., and West, J. W.,** Laser machining in ceramic substrates, *Laser Weekly,* May 22, 1972.
16. **Langer, P. et al.,** Laser induced emission of electrons, ions, and trays from solid targets, *I.E.E.E. J. Quan. Elect.,* Q. E. 2-2, 499, 1966.
17. **Lokerson, J. W.,** Personal communication.

Chapter 7

IN THE CONSTRUCTION FIELD

The laser industry is interested in the definite expansion of the construction industry because of the rising importance of laser construction, alignment, and ranging in this field.

Perhaps the history of this interesting application of the laser began in 1965 when a laser of Spectra Physics was used in the San Francisco Bay area for construction of a rapid transit system. Since that time there has been considerable expansion of the laser in this particular field. Actually the use of the laser can start in the planning stage. Here laser holography or three-dimensional imagery, as will be explained later in the chapter on Holography, may be used to develop three-dimensional images of the plans and construction. Whether these are more detailed, more informative, or creative and more colorful than the construction models developed by architects is questionable, but at least this interest in holography in architectural design indicates that some research is going on in this particular phase.

Lasers may be used for surface inspection and nondestructive testing of materials for construction. In the actual construction work, the laser is used in a number of different ways. The laser theodolite is the surveying instrument for measuring horizontal and vertical angles. The theodolite instrument may be combined with the transit line. In this technique the laser may use a narrow vertical line with proper heading. For example, in a dredge range control, the beam is reflected by a retroreflector. To assure proper alignment, the retroreflector has to reflect the beam back exactly in the same direction. The atmospheric transmission does interfere in atmospheric propagation technique. However, according to experts, the laser can be used in a light haze, but not in a heavy fog. The beam dispersion given by Spectra Physics is 4 in./mi in the narrow direction and 850 ft/mi in the long direction. Spectra Physics believes that if the laser being used in dredging, which is one of the applications used in the San Francisco Bay area, the dredging accuracy better than 3 in. is routine.

The laser may also be used in tunneling operations to assure proper alignment. Recently,

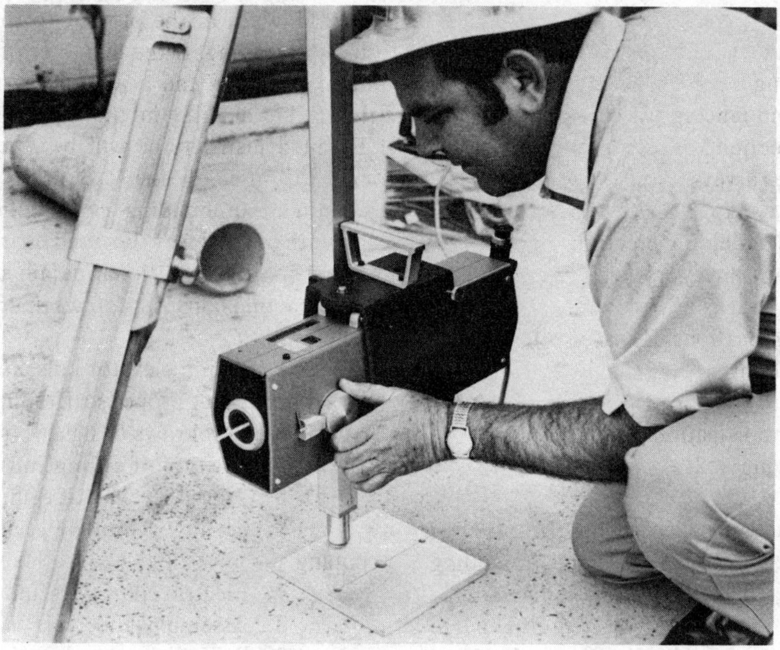

FIGURE 7-1. Dialgrade laser. (Spectra Physics.)

FIGURE 7-2. Dialgrade laser for line and grade for pipe laying. (Spectra Physics.)

the laser can be combined, in automatic fashion, with the drilling instrument. Applications for theodolite engineering include the following:

1. Mining
2. Dredging
3. Trenching
4. Piling and fencing
5. Steel erection
6. Baseline surveys
7. Layout work
8. Industrial building erection
9. Overhead conveyors
10. Aligning tracks
11. Marking locations in space
12. Shipbuilding
13. Night surveying
14. Canal and irrigation
15. Bricklaying

For specific details in pile driving, the piles can be driven accurately with the laser beam guiding the actual drilling. In dredging, as indicated for work in the San Francisco Bay area, the barge operator can check the alignment of the laser beam even at night, except in heavy fog. For position and profiling, the laser projects a spot on to a distant surface or target and the second instrument can focus on the laser spot and get common sighting. In pipe laying the laser can be used directly in the pipe or to show the direction for pipe laying. If the diameter of the pipe is large, the laser beam can be used directly inside the pipe by fixing a point in the center line of the pipe. Detailed instructions will be given in the manual for the use of the lasers. In securing proper alignment in building construction, one can see how the laser may be used in setting levels and plumbs. Some additional details that may be given in the actual construction work are

1. Plumb elevator rails and shaft
2. Erect proper scaffolding
3. Level rows of light fixtures
4. Position of ceiling and floor outlets
5. Actual leveling of ceilings
6. Setting swimming pool bond beams and coping
7. Electric conveyors and escalators
8. Assemble tank cars and mobile homes
9. Premark anchor locations
10. Install sewer lateral

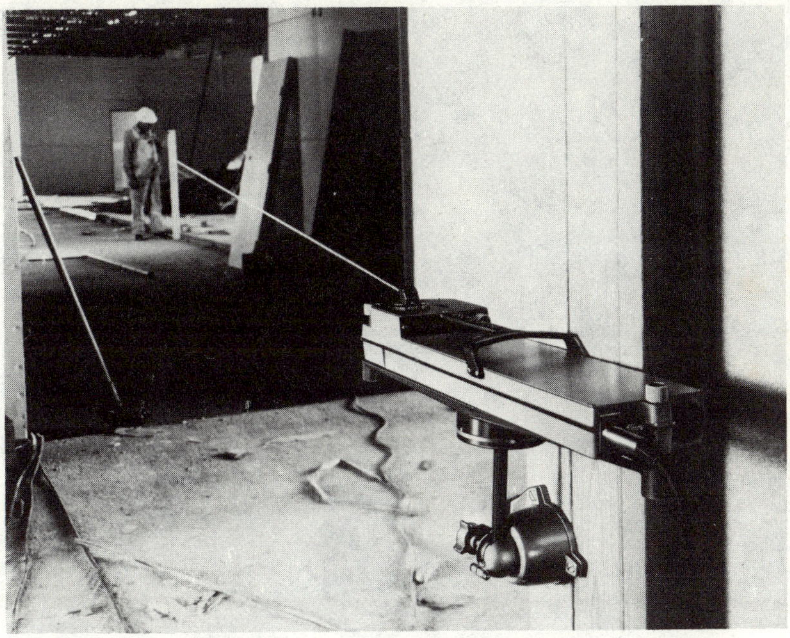

FIGURE 7-3. Building laser. (Spectra Physics.)

11. Install piping industrial plants
12. Check columns and forms
13. Highway grade

Therefore, it is evident that there are many significant uses of the laser in these operations.

Most of the lasers used in this area of industry are ordinarily helium-neon lasers of milliwatt outputs, varying to the type and need. Lasers will be marked with appropriate warning signals for safety. There is, of course, great concern in the laser industry because of the vast numbers of low output laser systems used throughout the country. If a number of lasers are used on a construction job, it would be necessary to have one individual be considered the laser safety officer. His responsibility would involve checking the laser systems, understanding what they are, and protecting personnel in their use, operation, and maintenance. If there is extensive use of lasers, it is recommended that he have some type of formal training in laser safety. Otherwise, his instruction would be the responsibility of the manufacturer. The Laser Safety Officer would instruct the personnel that the laser, even He-Ne, has definite safety precautions. No one should look directly into the beam or align it directly with his eye. Since all the jobs in the construction industry will be so-called open jobs, the danger of reflection especially from bright surfaces, bright tools, or retroreflection is real. The alignment of the reflected beam, if necessary, directly from any part of it on the construction material should be done carefully to avoid impact on the eye. The beam can give significant reflections from bright surfaces even if the He-Ne laser of only 1 mW is used. With diffuse reflection there is usually not much danger. Since little is known of chronic exposure, precautions must be taken at all times with the use of the laser. As more data accumulate on atmospheric propagation of laser systems, more definite information will be available. The workman should be taught that the laser is a hazardous instrument of great value and should not be flashed around or played with. This is important because the alignment possibilities in many phases of industry and agriculture are endless. The laser should be treated as a very valuable instrument to be handled with care and to be used with knowledge and understanding.

Further uses of the laser for construction essentially relate to architectural designs. High output laser systems may be used to create decorative designs on building materials; whether these will be done on the job or by the supplier is not definite as yet.

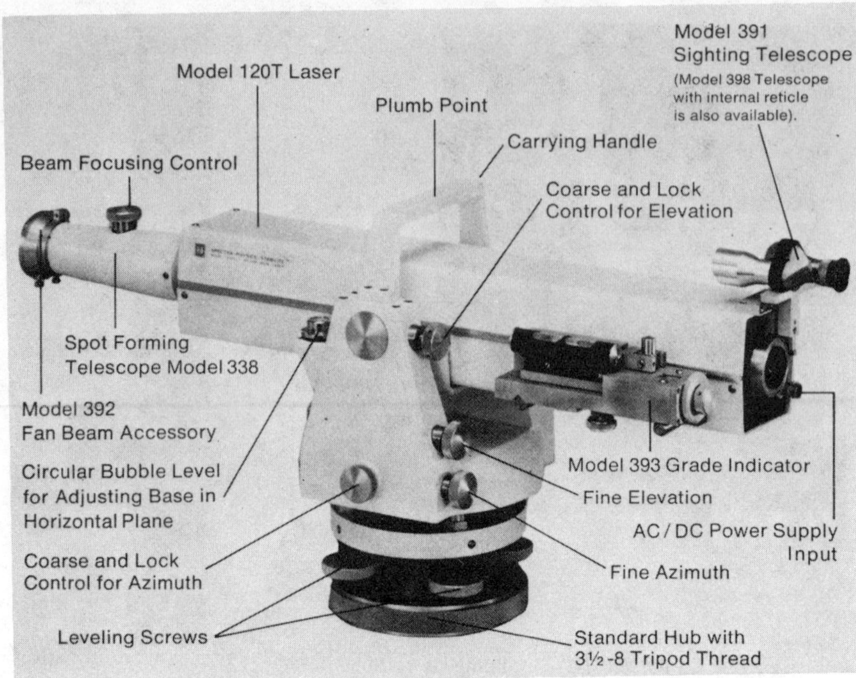

FIGURE 7-4. Transit-Lite Model LT-3 for alignment. (Spectra Physics.)

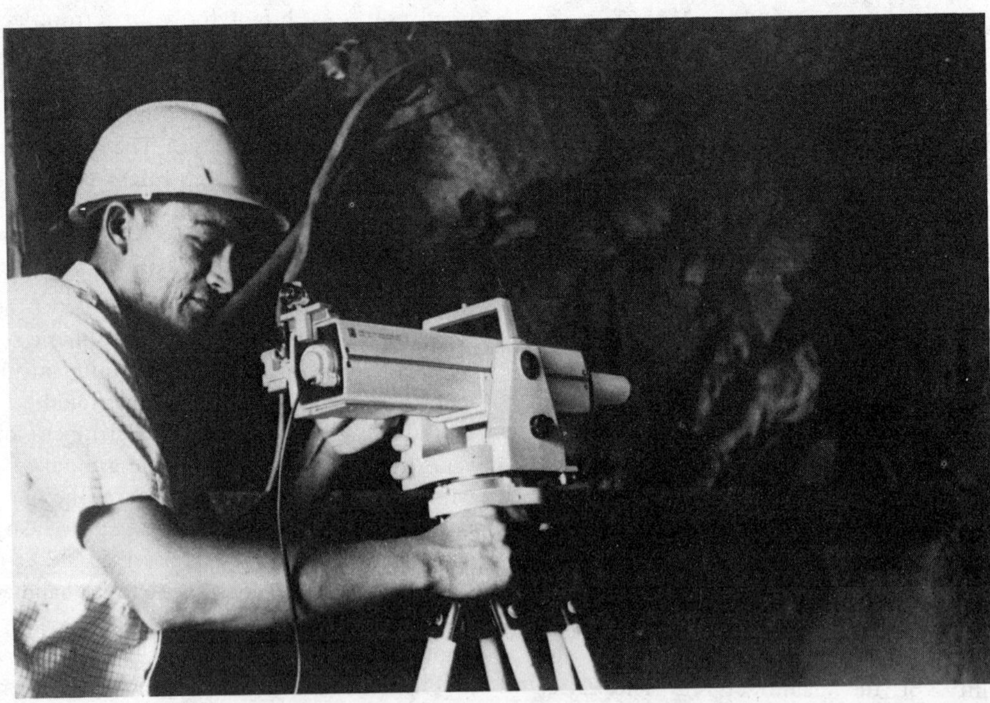

FIGURE 7-5. Transit-Lite Laser LT-3. (Spectra Physics.)

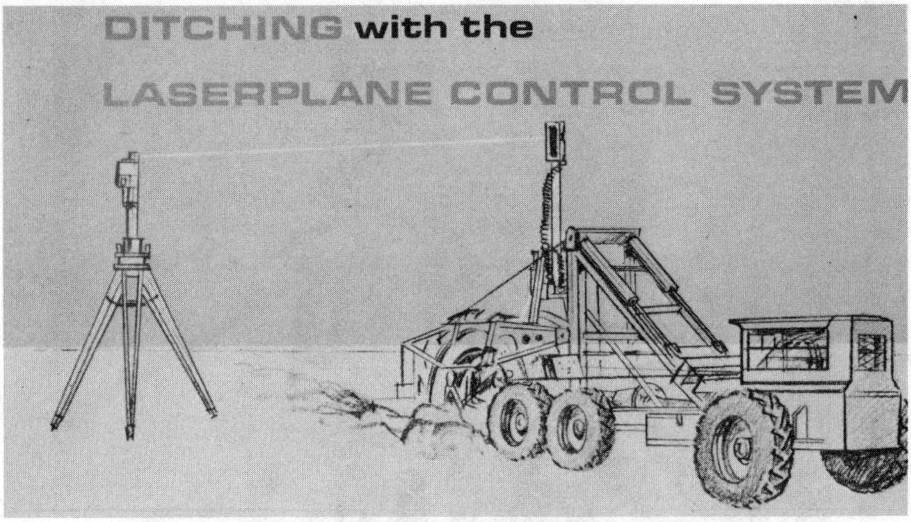

FIGURE 7-6. Laserplane control system. (Spectra Physics.)

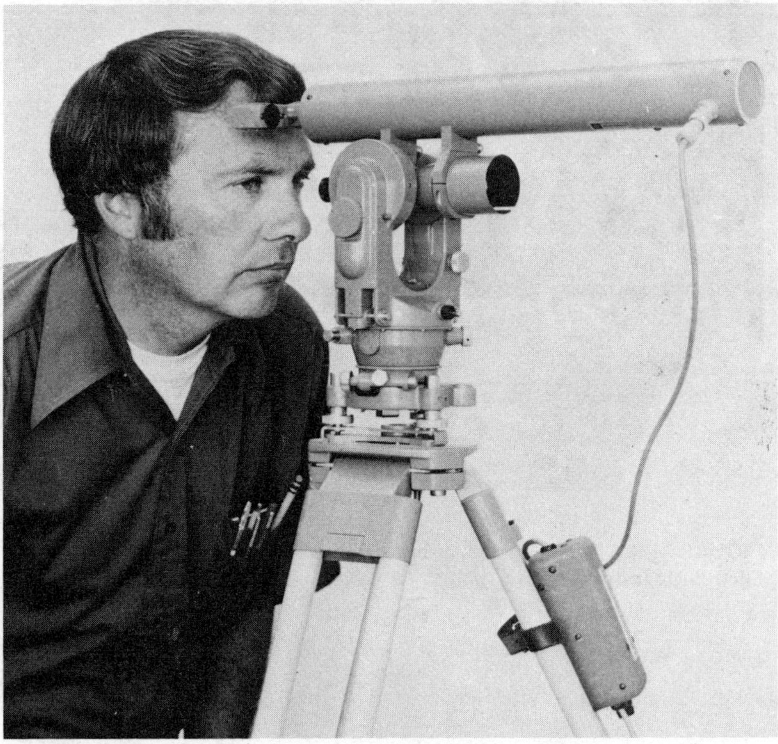

FIGURE 7-7. Transit and theodolite laser. (Spectra Physics.)

For material processing, especially of building materials, high output laser systems will have to be used, such as the CO_2 lasers and Nd-YAG laser. Detailed and elaborate safety programs are necessary for the use of such high output laser systems. In fact, even wood, and especially iron, steel, and aluminum can create hazardous situations. It is more proper that these high output lasers be used by experienced operators under controlled conditions.

Because of the fascinating designs (that will be discussed in Laser Art), it is possible that laser

FIGURE 7-8. Laser surveying instrument with laser cylinder at bottom. (Metrologic Instruments, Inc.)

beams may be used for decorative purposes in designs of a completed building. These would include probably low output He-Ne lasers and low output argon lasers. These can be developed, as will be explained in the chapter on Laser Art, to constitute fascinating designs without hazards to the viewing public.

Laser surveillance, usually with infrared systems, may be used not only during the construction period, but also in the finished product. Some techniques of surveillance by lasers will be given in the chapter, Law and Order.

Conclusion

This chapter reviews only briefly some of the beginnings of the applications of the laser in the construction industry. As the technology develops, the laser will be used more and more where it is needed.

REFERENCES

1. Florida Contracting and Engineering, June 1970.
2. Laser beam for potato cutting, Eastern Idaho Farmer XLII, June 3, 1971.
3. Laser Weekly, Alignment Costs for Construction Projects, August 21, 1972.
4. Spectra Physics – Dialgrade.
5. Spectra Physics, Laser Aligns 9 Mi. Submarine Pipeline.
6. *World Dredging & Marine Construction,* February 1968.

Chapter 8

SOME ASPECTS OF INFORMATION HANDLING

Pierce[13] has said about communications, "It is not only the essence of being human, but also a vital property of life. Technological advances in communication shape society and make its members more interdependent." The development of the new coupled discipline of electrooptics has been of considerable interest and value to the communication engineer.

Electrooptics is a rapidly expanding new field of laser technology with many excellent lines of communications between electronics and optics. In fact, it is becoming increasingly difficult to work in many new applications of the laser without the cooperation of the expertise on electronics. For the important field of communications only a few scattered features will be considered superficially to show possibilities. Those interested, and especially those in communications, will be referred to current literature and current journals such as *Electro–Optical Systems Design, Optical Spectra, Optics and Laser Technology, Lasersphere,* etc.

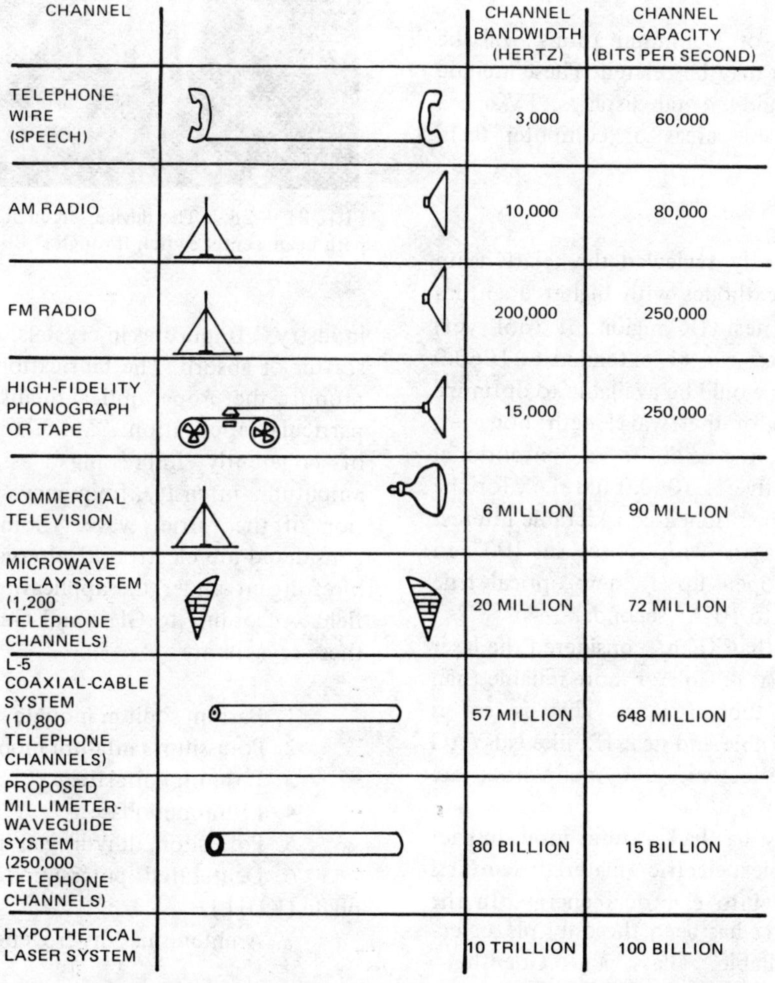

FIGURE 8-1. Capacity of various communication channels measured by Shannon's theory in bits per second. (From Communications and *Sci. Am.*, 227, 33, 1972. With permission.)

FIGURE 8-2. Thin film switch (Bell Telephone Laboratories) for aid in laser communications. Magnetic field produced by a minute electrical current passing through circuit causes light beam to change polarization and hence the direction in which light is refracted as it is coupled out of film by a prism. Information is coded on light beam by switching the beam in or out of its original path in a controlled pattern of light pulses. (From *Laser Sphere,* December 15, 1972, 5. With permission.)

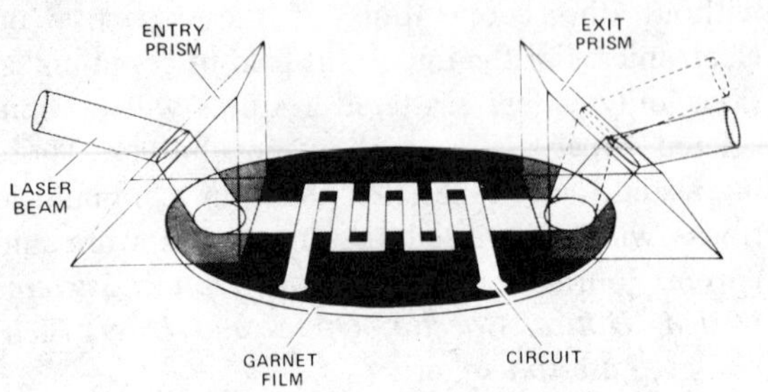

FIGURE 8-2A. Path of laser beam.

FIGURE 8-2B. The device – pencil shows approximate path beams enter switch through a prism.

Various phases of communications will be considered;[29] some may be related. These include crystals, fibers, liquid crystal displays, TV space technology, and some areas of computer technology.

Photocathodes

Eberhardt[6] recently reviewed the relationship of the new photocathodes with higher quantum efficiency in the near IR region; if (not yet) possible the response can be extended to 1060.0 nM. The YAG laser would be available to optimize the photocathodes for that wavelength; however, he did not believe it possible for a conventional photocathode sensitive at 1060.0 nM, i.e., for the CO_2 laser. He also mentioned considerable interest in mode locked lasers with pulses in 10^{-9} to 10^{-12} region to open up a "new optical true domain from 10^{-9} to 10^{-12} seconds."

It is of interest that Kahn[9] considered the laser system potentially as reliable or more reliable than the cathode ray tube (CRT). This is so as solid-state lasers, visible and near IR, like Nd-YAG and solid-state deflectors, and modulators are perfected.

Laser technology has had a tremendous impact on crystals. A piezoelectric material converts mechanical energy into electrical energy. In the past 30 years quartz has been the only piezoelectric material available. Glass and Guenther[7] claimed "that the requirements on crystal properties for laser use are now as stringent as those placed on semi-conductors by the electronics industry." Impurities in crystals, for example, may scatter or absorb. The fabrication of crystals must produce the proper optical qualities needed by the particular application. "Modulation is the process of coherently impressing a signal up on the amplitude, intensity, frequency, phase of polarization of the carrier wave." Some of the crystals considered are electrooptical crystals that become birefringent under the application of an electrical field. According to Glass and Guenther,[7] some of these crystals are

1. Barium sodium niobate
2. Potassium tantalum niobate
3. Lithium tantalate
4. Lithium niobate
5. Potassium dihydrogen phosphate (KDP)
6. Deuterated potassium dihydrogen phosphate (KD+P)
7. Ammonium dihydrogen phosphate (ADP)

With high output lasers, nonlinear effects on crystals, such as frequency doubling, may occur.

This means that two photons of one frequency can be converted to a frequency of twice the incident radiation. There is an increasing line of high output CO_2 lasers today. It is obvious that if there were optical absorption there would be "catastrophic failure." Liquid crystals displays will be discussed in subsequent sections.

Transmission Fibers

For the field of communications, there has already been emphasis on the single mode optical fiber which can be fed only by the coherent beam of the laser. Incoherent light sources such as light emitting diodes (LED) can be used for multitude fiber packs. For the laser driven fiber, a low optical loss fiber is important in this field. Some of the characteristics of these fibers have been covered previously. Also, miniature, but high output solid-state lasers will also be an ideal accessory. The usual headaches of economical considerations and service needs follow the initial research and development.

Basignies[3] claimed that "communications technology will skip much of the low-millimeter and sub-millimeter regions (of the spectrum) and jump directly to optical wave lengths which are readily generated by lasers and light emitting diodes and carried by light pipes." In brief, the laser beam is considered as a high capacity channel to carry voice, video, digital, and other signals.

Laser and TV

The often repeated enthusiastic statement is that a single coherent laser beam can carry in principle the equivalent of several thousand television channels. Therefore, lasers make the world smaller and smaller.

At the Expo-70 Exhibit in Japan, He-Ne and argon lasers aided in producing excellent color images on a very large television screen. RCA has developed an LBIR (Laser Beam Imagery Reproducer)[5] to reproduce TV pictures with more than 20,000 line resolutions. This is of interest considering that the average television broadcast picture consists only of 525 lines. The RCA system has been developed primarily for data on earth resources and satellite photo-taking missions. This device is reported not only to have high resolution quality but also enables long distance television transmission "to be recorded on film and in four (4) seconds – which is about 20 times faster than electronic recorder currently on the market." The LBIR system can take its high resolution pictures from tape recorded video signals or even from film transparencies. The basic system includes a He-Ne laser, a light modulator, a variety of optical components, and continuous film transport mechanisms. The apparatus works, according to RCA, with the intensity of the laser beam regulated by electronic signal fed through an electrooptic light module system.

> "The diameter of the beam is enlarged, focused and then passed through a rapidly rotating air bearing scanning mirror to produce a picture image with a wide-range grade tone. The high speed scanning mirror deflects the beam to reproduce a horizontal finely focused scan, less than 210 thousands of an inch in diameter across a nine and one half inch film. Moving in a vertical scan direction at a constant speed, the film is exposed to a series of adjacent scan lines which form the high fidelity picture. The resultant image is in focus across the entire area and has extremely high geometric accuracy. The calibration system automatically controls the contrast and intensity of light during the exposure of the film. The film transport enables the LBIR to generate photographs in single frame images, a rapid frame sequence of images, or a continuous wave image."

Atmospheric Propagation

One difficulty with the use of the laser for communication has been the loss of transmission through atmospheric propagation of the beam especially in bad fog or rainy weather. Occasionally, so-called "hot spots" are produced by water droplets acting as lenses for focusing the laser. Some time ago, a laser telephone line in Russia was said to have a range of only a mile.

Fiber optics wave guides, as mentioned previously, are very important for transmission. Other optical materials for transmission include lenses, prisms, and films. Goodwin of Optical Communications Research of Hughes Research Laboratories[8] has described a CO_2 laser for communications, effective over a 5-mi path even in rain or fog. The transmitter is reported to be a 1 W air-cooled CO_2 laser. Frequency modulation is done by an intracavity cadmium telluride. The receiver is optical heterodyne detection with a mercury cadmium telluride (HgCdFe) detector cooled with liquid nitrogen in a Dewar. The transmitter vs. oscillator efficiency is increased by 10% with the use of CaTe rather than GaAs modular portion. To block transmission with this apparatus, a rain field of 1 in./hr over the entire 5-mi path is needed.

In 1971 Whitmer[17] described a 200-mW argon

FIGURE 8-3. Laser distance measurement 1 m to 5 km. (Kollman Instrument Corp.)

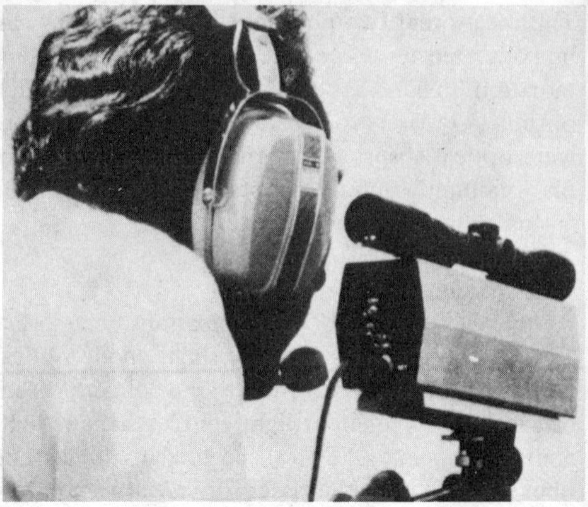

FIGURE 8-4. Laser portable communicator – transmits and receives up to 6 mi. (From *Laser Technol.*, November 1969. With permission.)

laser of Lockhead that was used to find data over a 1.2-mi path at the rate of 300 megabits/sec. This high transmission rate was made possible by a frequency modulation of the laser beam. This modulation was done by a microwave transmitter, a microwave band pass optical transmitter to operate at both microwave and optical frequencies. Whitmer[17] believes that this type of modulation makes possible the high transmission rate, but also makes the beam less susceptible to atmospheric disturbances even in rain and fog. Accuracy is improved by a 0.51-μm wavelength of argon matched to the silicon photodiode detector in the receiver. Applications considered for this include:

1. Cable television, in conjunction or substitute. Also for cable television where the user is located at a distance of a mile or so from the distribution center, as in universities.

2. Computer-to-computer hookup between buildings.

The hazards of atmospheric propagation of laser systems will be reviewed in the section on Safety.

Liquid Crystals

Another development adding to the results of laser communication is the additive effect of using laser driven liquid crystal display numbers.

The field of applications for laser-liquid crystal reactions is just beginning. Liquid crystals are organic compounds that exhibit a mesomorphic state between solid and liquid states – intermediate both in terms of temperature and molecular order. The molecules have different forms in different directions, highly anisotropic. So the "fluid takes on a measure of short-range order over a limited temperature range." Hence, the term

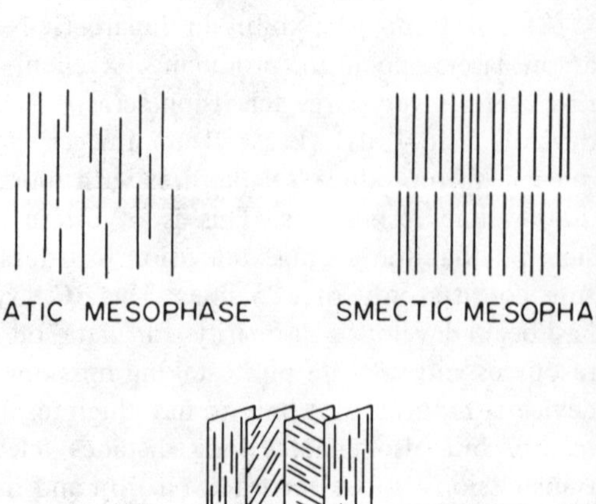

"liquid crystal." They are classified roughly as smectic – molecules parallel to each other in regular layers with molecular axes pointing the same way; nematic – molecular axes point the same way, but they no longer lie in the same layers; cholesteric – molecules all point the same way but each layer "twisted slightly relative to one above and below." However, one class may pass through another. According to Tobias,[16] smectic crystals have few applications, nematic have electrooptical properties, and cholesteric materials react to heat. Thin films are necessary and the divisions between nematic and cholesteric may be "blurred." Attempts have been made to use these crystals on the skin, on viscera such as the liver, and on papers to measure temperature changes of areas adjacent to the target of the laser impact. Except for some artistic patterns of resultant color change, liquid crystals as heat detectors were not practical. However, with more practical and stable carrying vehicles, there may be some value in measuring the intensity and spread of the thermal reactions induced by the lasers. However, these cholesteric crystals have been used as image intensifiers by a nematic film on a photoconductor. "RCA uses nematic films as light valves in a laser addressed random access memory."[16] Radiation Inc. is reported to use "dynamic scattering films for laser projection screens to get rid of laser speckle." It is still possible to have more applications with these crystals in the laser field as photochronic materials have had application through color change from UV lasers. However, Tobias' motto is well taken as it should be, for all laser research and development: *visi videro non credam* – I can't believe it unless I see it.

The liquid crystal, a transparent fluid, is reported to become "frosted" when it is exposed to heat. Since the laser driven crystals are put into a display apparatus, the cell is used in the receiving end of the laser system, coded pulses of information direct the laser beam over the medium. The liquid crystal used in the experiments at Bell Laboratories[11] was a mixture of 90% MBBA (*p*-methoxylbenzylidene-*p*-*n*-butylanaline) plus 10% cholesterol nonanoate. The liquid crystal is made as a sandwich between two fused silicone glasses coated with a transparent conducting electrode (indium/tin oxide) and an additional coating of organic material (silane) to orient the liquid crystal material. The power absorbed from the addressing beam heats the liquid crystal above its isotropic-cholestric phase and then visible light is strongly scattered by the disordered areas. This thermal riding is made possible by the neodymium-YAG laser beam 1.06 μm wavelength; He-Ne and gallium arsenide lasers could be used also.

FIGURE 8-5. Laser to measure ceiling at airport. (From *Industrial Research*. With permission.)

With CO_2 lasers it is possible to have a communicator system operating at 10.6 μm wavelength over a path 5 mi long even in the presence of "moderate rain or fog," as Hughes Aircraft Corporation describes it. In this system, the transmitter uses a sealed off air-cooled CO_2 laser with a frequency modulation made by an optical heterodyne detection with a mercury cadmium technique. This development is primarily for aerospace communication in which communications are not only necessary from a satellite-to-earth and earth-to-satellite, but also from satellite-to-satellite.

The other techniques of laser communication could produce effective laser color transmission from over the world. The future is bright for the entire field of laser communication.

The relationship of holography to communications will be reviewed later.

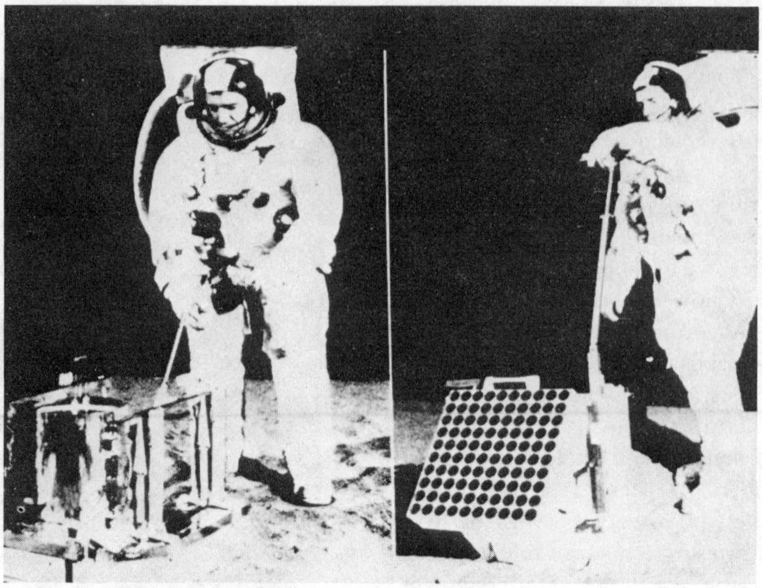

FIGURE 8-6A. Moonquake detector and laser retroreflector.

FIGURE 8-6B. Close-up lunar retroreflector. (From *Science.* With permission.)

FIGURE 8-6C. Oscillator amplifier laser for moon shots. (Spacerays Inc.)

FIGURE 8-7A. Astronaut Alfred W. Worden with scale model of SIM (Scientific Instrument Module) which includes a laser altimeter (RCA). (From Aerospace Systems Division and *Lasersphere,* July 26, 1971. With permission.)

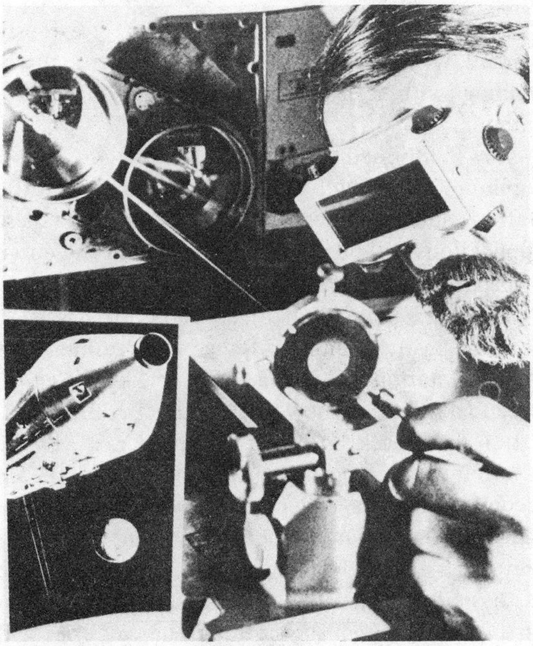

FIGURE 8-7B. Checking laser altimeter. (From *Laser Focus.* With permission.)

Space Technology

There is an increasing use of the laser in space technology as, for example, at the Smithsonian Astrophysical Observatory Agassiz Station at Harvard University. An American Optical Q-switched green laser produced 28 J in a nanosecond pulse. The telescope for these experiments is reported to be a small 8-in. diameter. This is smaller than the 60 to 120-in. telescopes used for the ruby lasers.

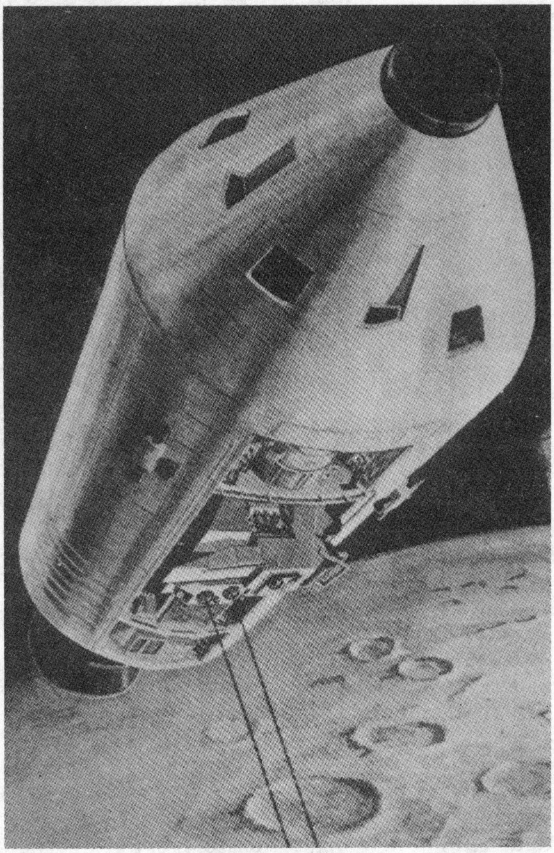

FIGURE 8-7C. Artists conception of action of laser altimeter in lunar orbit. (From *Laser Sphere,* December 1972, 3. With permission.)

Since the lunar expeditions have been terminated for the present, studies with the laser altimeter and laser retroreflector will be discontinued.

The special quartz laser retroreflector deposited on the moon surface and fortunately not obscured by moon soil was responsible for the very accurate measurements obtained. This laser ranging retroreflector of the Apollo 11 mission was a target for earth-based laser systems. The measurements chiefly concerned the earth-moon distance and the earth's rotation rate.

Recently, the Russian Luna-Vl Spacecraft put a French-made laser reflector, 18 x 8 in. weighing 8 lb and shaped like a rectangular hexagon, on the moon for study by French Space Scientists. It is reported that this is identical to the one landed on the moon by the Russian Luna-17 in November 1970.

FIGURE 8-8. Lunar ranging transmitter at Smithsonian Astrophysical Observatory Agassiz Station Q-switch 28 J/nsec with green. (American Optical Corp.)

Laser altimeters were used on the Apollo 15 and 16 missions. These altimeters revealed significant elevation differences on both sides of the moon. The laser altimeter made its third trip into the lunar orbit on Apollo 17. This was very valuable in developing measurements of high accuracy. When the mapping panoramic camera exposes a frame, the altimeter is recorded on film. The altimeter can operate at heights 50 to 70 nautical miles above the moon. The time from the sending of the light pulse and the receipt of the reflected beam is measured to calculate the altitude of the CSM above the moon.

The laser altimeter can be combined not only with a panoramic camera, but also with a gamma X-ray spectrometer and a mass spectrometer.

It is proposed that laser ranging on a high-flying spacecraft to measure changes in the gravitational constant be developed.

To detect air turbulence ahead of an aircraft, a laser Doppler clear air turbulence system is being developed at the NASA Marshall Space Flight Center.

The laser lunar trash left on the moon is estimated to be valued at $2 million. Incidentally, in space research lasers are under study for the possibility of using their tremendous power for actual launching of satellites. It is estimated that to provide a ground based laser for propulsion, levels of 1 GW would be needed. This research is directed by Dr. Philip K. Chapman[4] of Avc Everett Research Laboratory, who previously worked with lasers in NASA as an astronaut in training with a laser-coupled gravitational wave antenna.

As mentioned for space technology, laser communications will be used from earth-to-shuttle station and vice versa. It is hoped that these applications will be more successful than the initial aerospace experiments with the argon laser.

Laser tracking of artificial satellites has been done by many countries. Range data from a laser tracking station at Goddard Space Flight Center, Greenbelt, Md.[14] showed that the "orbital inclination of Beack Explorer C spacecraft can be determined to a precision of better than 0.03 arc second from data collected over six hours." The value of these data on polar motion is in their usefulness for understanding the dynamics of the earth and in particular the correlation between earthquakes and polar motion. The satellites had laser reflectors and the laser was a 1-J ruby with a pulse rate of 1/sec. The weather dependency was of significance in these experiments. Laser applications in television will allow more excellent color pictures of future satellite and shuttle experiment and also mass storage of data.

Earthquake Detection

American, Japanese, and Russian investigators are studying earthquake forecasting by means of combined laser and computer technology. Com-

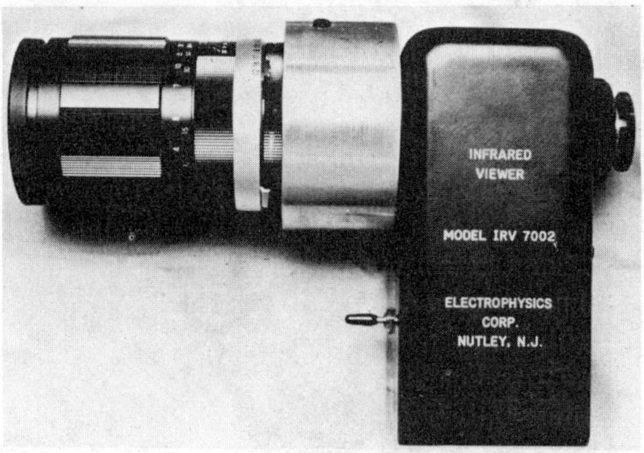

FIGURE 8-9. Infrared viewer.

FIGURE 8-10A. Laser lighthouse. (Laser im bild, Laser Elektroplik.)

FIGURE 8-10B. Laser lighthouse – Australia. (From *New Sci.*, June 3, 1971. With permission.)

puter models can develop minute fault creeping and fault failure and rock strain endurance. He-Ne beams combined with single or multiple receptors are used to detect movements of the beam. Lawrence used two beams to monitor earth movements. One experimental station for earthquake studies is near Boulder, Colo.; the other for the famed San Andreas fault is near Hollister, Calif. The use of the laser as a monitor for earthquake detection will be expanded greatly since the need for earthquake forecasting is so evident in this vein. This same technique may be used for measuring the effect of atomic explosions.

Oceanography

High output argon and tunable dye lasers are used in oceanography studies. Briefly, the laser applications in oceanography may be listed as:

1. Temperature measurements – important since relates to climatology
2. Examination of particulate matter in waters
3. Fish detection
4. Acoustical holography, including submarine detection
5. Detection of algae
6. Detection and treatment of oil pollution
7. Offensive weaponry

The laser used is usually argon, especially in pulsed systems. A pulsed argon laser 10^{-9} sec is said to penetrate 500 ft. Tunable dye lasers have also been used for investigative studies.

Computer Technology

The laser has important applications in the field

FIGURE 8-11. Laser earthquake detection.

FIGURE 8-11A. Laser over a geodetic marker.

FIGURE 8-11B. Beam on a reflector 10 mi away with travel time recorded by instruments on a helicopter with correction for atmospheric distortion. (From Matthews, S. W., U.S. Geological Survey, *Natl. Geograph.*, January 1973. With permission.)

of computer technology, in the range of machine design, layout testing, and printing. The limit of storing has not been reached. Current data indicate 10^{12} bits/cm^3. A recent laser mass memory unit, UNICON, at the University of Illinois is a trillion bit piece. This is to be used for a mass storage center for Illiac IV, the world's most powerful computer complex. An erasable recording medium of an Isomet instrument is a doped single crystal of lithium niobate ($LiNbO_3$) with a diffraction limited optical storage density of 10^{12} bits/cm^3. A local arrangement of electrons in the crystal permits the recording. The area can be erased for a new recording in this area. This crystal is superior to wet processed film for recording. Holographic recording is also possible with this material with high diffraction efficiency holograms and reported multiple image storage in depth. The storage image is believed to have a thermal lifetime of several months at room temperature. For more prolonged storage time, recycling is needed. For recording,

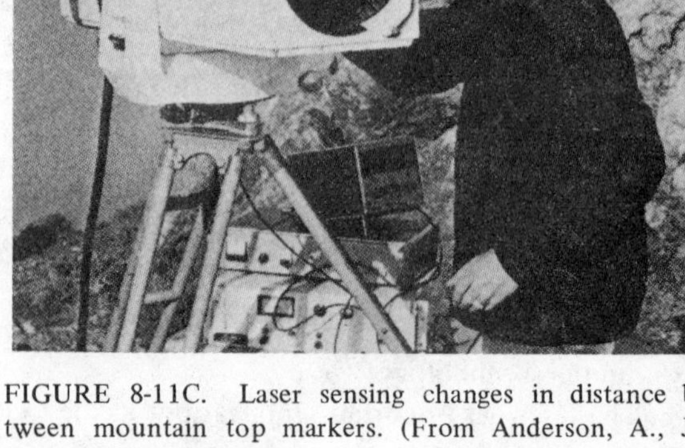

FIGURE 8-11C. Laser sensing changes in distance between mountain top markers. (From Anderson, A., Jr. *Saturday Review of The Sciences*. With permission.)

FIGURE 8-12. Modular particle sizing for sand storm. (From *Opt. Spectra.* With permission.)

FIGURE 8-12A. Instrument.

Scientists at the Air Force Systems Command's Arnold Engineering Development Center have been able to monitor a miniature sandstorm in real time, recording not only the number, but also the size of particles in the micron range (about 1/5,000th of an inch) passing a given point at supersonic speeds. The sensor system used is composed of a continuous wave laser mounted on one side of the test cell and a 32-element fiber optic array on the other. The 32 filaments are arranged in a vertical line in the sensor, right. Each fiber is connected to its own photomultiplier tube, left. As a particle passes through the laser beam it momentarily shades a number of the optic fibers, the number depending upon the size of the particle. After the particle has passed, logic circuits record it in the proper size channel. The system can count and size up to a million particles a second.

FIGURE 8-12B. Technique.

the argon ion laser is used. For readout, the phase of incident light is modulated, the split argon signal beam is turned off, and the reference beams reconstruct the pattern. Not only is adequate storage necessary, but there must be rapid accessibility.

The technology of both laser and video tape has made the computer able to disgorge extraordinary amounts of information at rapid speed. The Zenith® silent laser typewriter is said to be able to transmit 38 billion words of the *Encyclopaedia Brittanica* from New York to London in only 7 min. This is possible with modulation by an acousto-optic device. Recently, the argon laser has been used as a new film for laser recording. The argon laser breaks down the diazo molecules with the release of N_2.

In the field of communications, the transmittance of a facsimile is of importance. In addition to the above instrument, there are also others available. An example is the laser-based facsimile transmitter of Matsushita® Graphic Communication Systems, Tokyo. The speed here is 1 min/page, which is much faster than the telephone line transmitter.

There is also the nonimpact printer of RCA®, the MTR (Material Transfer Recorder). The MTR is said to print 1,000 lines/min and can also receive messages at 20,000 words/min. This printer "uses a laser to transfer digital communications alphanumerics onto ordinary paper and operates as rapidly as existing mechanical line printers." Unlike the mechanical printer, there are no keys to wear out or break. Moreover, the MTR will have a quiet operation and also will use ordinary paper.

It is evident that this reviews only some of the

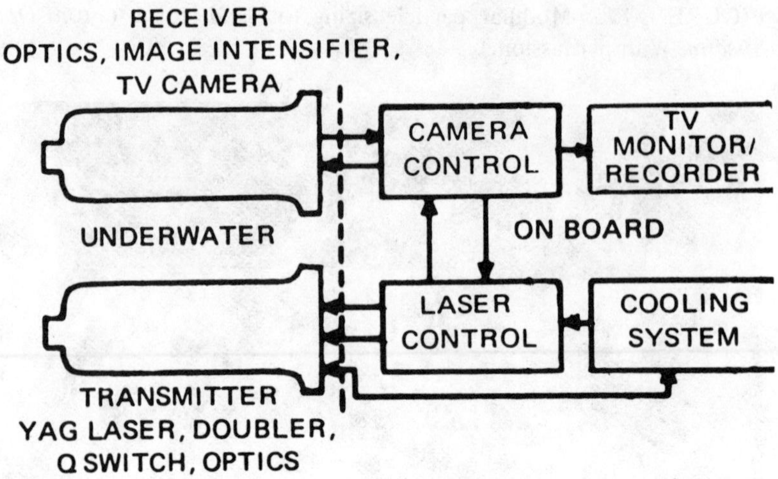

FIGURE 8-13. Outline for laser underwater survey. (From *Laser Elektro-Optik.* With permission.)

uses of the laser in the broad field of communications. For communications, high output CO_2 laser will be used in the transmitter for relatively short distances. With the efficient CO_2 laser and the reported optical heterodyne detection in the receiver, such an experimental laser communication system of Hughes is said to have a potential in excess of 300 megabit data rate or 10 television channels. Safety programs will be considered later.

With the continued development of this "new" multidiscipline of opticoelectronics, additional exciting programs soon will be possible in the field of television, other pathways of communication, and the complex systems for computer technology.

Writing and Erasing

Early in the days of the laser, it was found that the ruby laser could erase black print. Schawlow's entertaining demonstrations of the black monkey balloon inside the balloon at the San Francisco Zoo proved this property of ruby lasers. Schawlow has also developed and patented the laser typewriter eraser. The laser can reveal the electronic thin film semiconductor recording material vapor fixed to one side of a transparent electrically conducting glass sheet. Rapid readout on such a sheet can be done with a laser.

A new printing plate, the "Laser Plate," is a type of printing system which uses multiple laser beams.[12] The laser phototypesetter combines what is called matrix and CRT systems. The laser writes directly into the recording media. The first example of this experimental technique was for

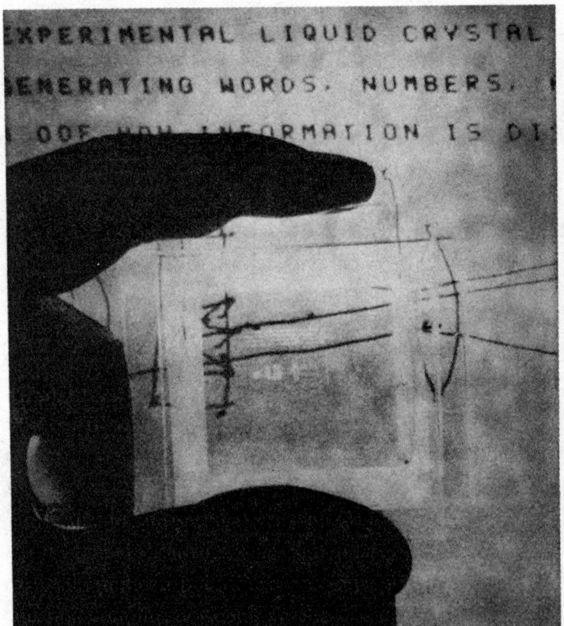

FIGURE 8-14. Laser writing information in liquid crystal slide in Bell Laboratories display system. (From *Opt. Spectra* and *Laser Weekly,* October 30, 1972, 1. With permission.)

the front page of the December 1, 1972 edition of *The Hartford Courant,* the oldest continuously published paper (1764) in the U.S. The goal in newspaper electronic production is rapid, economical automatic processing.

Label Reading

Optical label readers are now used in supermar-

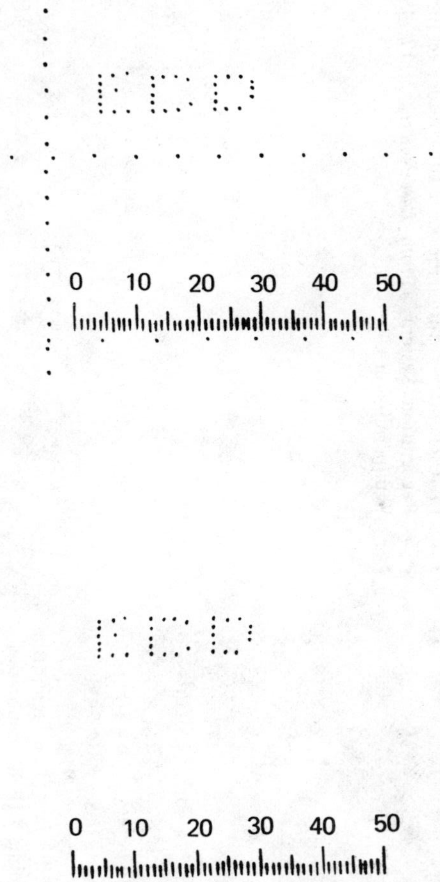

FIGURE 8-15A. Laser writing on film containing tellurium data dots written by laser on upper film erased by same laser on lower erasure of printing by pulsed ruby laser. (From *Ind. Res.*, October 1972, 28. With permission.)

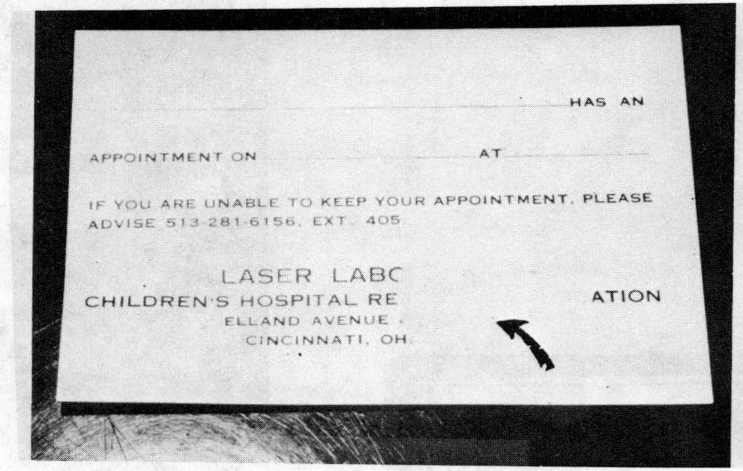

FIGURE 8-15B. Ruby laser erasure 50 J/cm² to 2.5 msec.

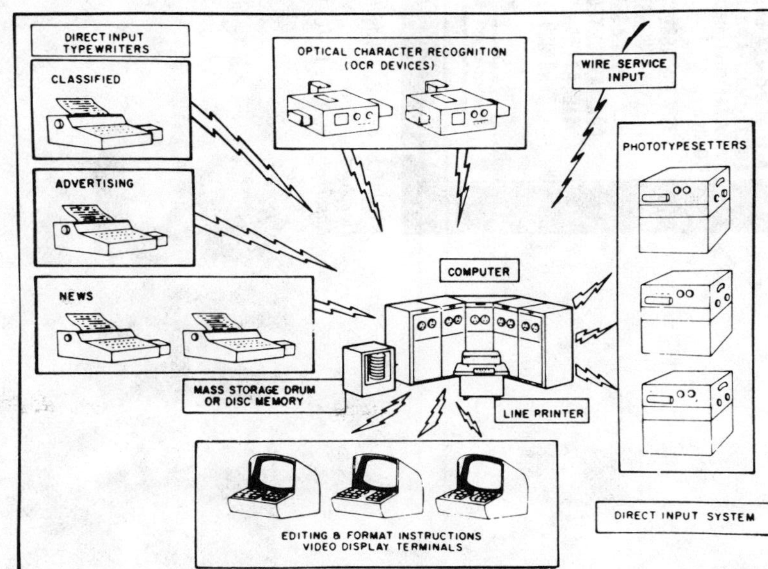

FIGURE 8-16A. Electronic newspaper production. (From American Newspaper Publishers Association Research Institute and Electronic News. With permission.)

FIGURE 8-16B. Laser phototype set – *Hartford Courant Newspaper.* (From *Lasersphere,* 2, 1, 1972. With permission.)

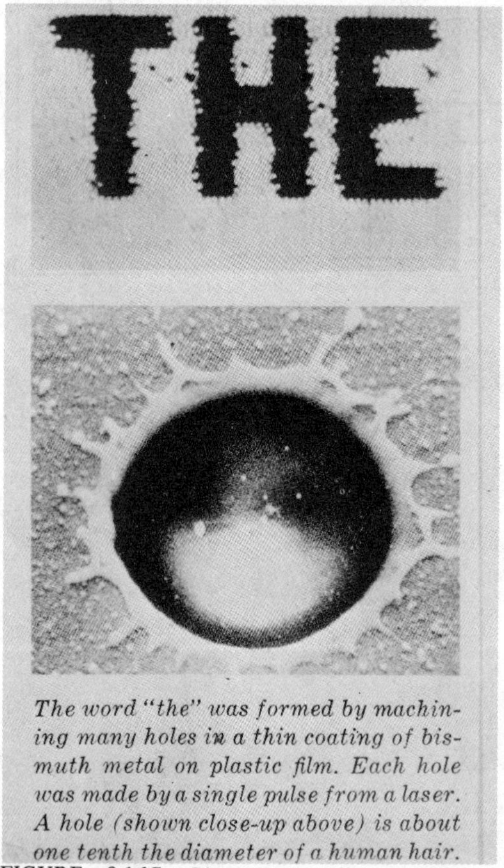

FIGURE 8-16C. Bismuth fast film microfilm copies from distant points. (Bell Laboratories Record.)

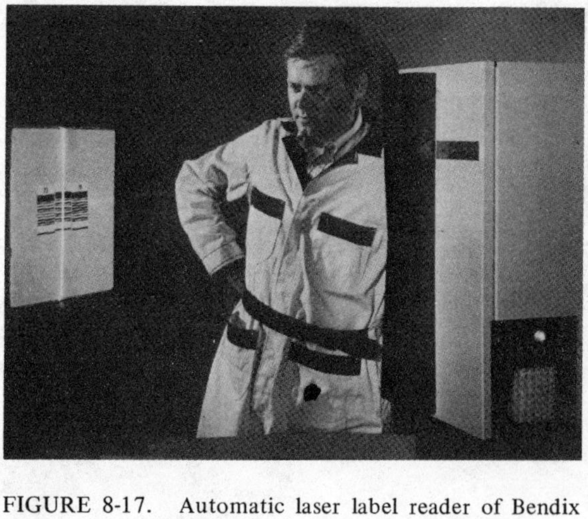

FIGURE 8-17. Automatic laser label reader of Bendix Recognition System. (From *Lasersphere,* 2, 17, 1972. With permission.)

FIGURE 8-18. Laser scanning analyser, scanning food wrapping, paper flaws, etc. (From Ferranti, *Opt. Laser Technol.*, 4, 113, 1972. With permission.)

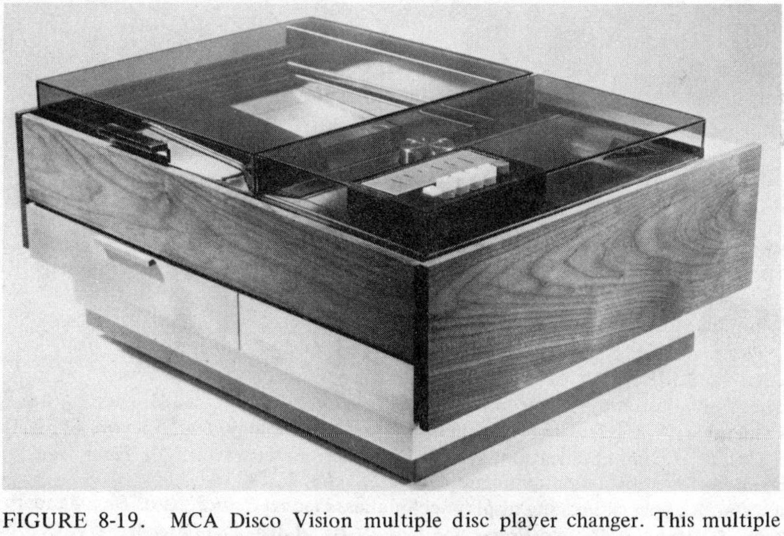

FIGURE 8-19. MCA Disco Vision multiple disc player changer. This multiple video disc player provides over 6 hr of continuous unattended playback. The playback is made by 1 mW He-Ne laser with the player attached to the antenna terminals of any standard black-and-white or color television receiver.

FIGURE 8-20. Supermarket symbol for laser check-out system.

kets for rapid cashier counter check-out. The low output He-Ne, as mentioned previously, provides no hazards to the cashier with continual exposure or for the customer including inquisitive children. Such check-out may be used in warehouses, for inventories, and for checkup as security measures. Conveyor systems are not necessary for portable He-Ne readers. In the Bendix system, the labels consist of 1/16 to 1/8 in. rings coded to represent a three-digit number. In brief, He-Ne lasers can scan:

1. Supermarket (1 mW)
2. Boxes (2 mW)
3. Bar coded labels (2 mW)
4. Railroad cars (2 mW)

Only small items of the two very important fields of communication and computer technology have been included in this brief review. As stated, this new compound discipline of electrooptics will continue to make great strides in these fields. One current example is the use of the laser in the new video discs and video cassettes to play back on T.V. sets, for information handling. These will provide communications, teaching programs, home movies, etc., all to change our pattern of life. Better lines of communication, a current cliché used to excuse many things, may be used between countries and continents to make the world smaller and hopefully better.

REFERENCES

1. **Aronsow H.**, New erasable optical recording, *Electro-Optical Syst. Des.*, July 1972.
2. Bell Laboratories Record, New crystals for communication system enters production stage, *Opt. Spectra*, 6, 32, 1972.
3. **Basignies, H.**, Communication channels, *Sci. Am.*, 227, 99, 1972.
4. **Chapman, P. K.**, Laser launched satellites, *Electro-Optical Syst. Des.*, October 1972.
5. **Constantino, C. A.**, The LBIR, *Laser Weekly*, May 15, 1972.
6. **Eberhardt, E.**, Photon-in-photon-out electron tubes, *Electro-Optical Syst. Des.*, January 1973.
7. **Glass, A. J. and Guenther, A. H.**, The laser's impact on crystal technology, *Opt. Spectra*, 6, 39, 1972.
8. Hughes Aircraft Co., A laser communications system 10.6 micrometer wavelength, *Laser Weekly*, October 30, 1972.
9. **Kahn, G.**, What's ahead for optical communication? *Opt. Spectra*, 6, 24, 1972.
10. **Maydan, D. and Kahn, F.**, Special imaging display techniques, *Electro-Optical Syst. Des.*, January 1973.
11. **Melcheor, H., Kahn, F., Maydan, D., and Fraser, D.**, Liquid crystal displays, *Laser Weekly*, October 30, 1972.
12. Newspaper laser system phototypesets, *Laser Sphere*, 2, 1, 1972.
13. **Pierce, J. R.**, Communications, *Sci. Am.*, 227, 31, 1972.
14. **Smith, D. E., Kolenkiewica, R., Dunn, P. J., Plotkin, H. H., and Johnson, T. S.**, Polar motion from laser tracking of artificial satellites, *Science*, 178, 405, 1972.
15. **Sjorgen, W. L. and Wollenhaupt, W. R.**, Lunar shape via the Apollo laser altimeter, *Science*, 179, 275, 1973.
16. **Tobias, M.**, The fluid state of liquid crystals, *New Sci.*, December 14, 1972.
17. **Whitmer, R. F.**, Laser communicator, *Electro-Optical Syst. Des.*, April 1971.
18. **Wright, P. E.**, MTR tactical line printer, *Laser Weekly*, November 13, 1972.
19. MCA Disco Vision, Video discs on TV sets, *Lasersphere*, 3, 1, 1973.
20. **Miller, S. E.**, Communication by laser, *Sci. Am.*, 24, 19, 1966.

Chapter 9

HOLOGRAPHY

The basic idea for holography was proposed in 1947 by Gabor.[7] After a long interval in 1963, the fascinating experiments by Leith and Upatnieks[22,23] and Stroke[27-30] followed. Since then, holography has been an exciting and important part of laser technology. The word holography, according to Gabor, is holo (complete) and graphy (message). Gabor, Kock, and Stroke[10] define holography as:

> "Sometimes called three-dimensional lensless photography – is the art of freezing a light wave into a photographic plate by means of another (reference) beam, and reviving it by laser or white-light illumination. The object is revived in true three dimensions, in monochrome or in natural colors. Two waves can be frozen in, one after the other, and revived together. This has opened up the new technology of nondestructive testing of vibrations and stress by holographic interferometry."

There are many books and excellent references in regard to the background of physics, optics, and technology of holography, such as the recent work by Collier et al.[2] and the monographs by Jeong.[12-16] A good current review article is the one by the three famed authorities Gabor, Kock, and Stroke.[10] In his review of holography, Jeong[16] has listed the essential major properties of the hologram:

1. The light arriving at the eyes of the observer from an illuminated hologram is precisely the same as that which would come from the original object. Therefore, this three-dimensional image can be photographed from various perspectives or scrutinized by any other ordinary optical means.

2. If a hologram is broken into many pieces, each contains a complete view of the entire object. One looks through the hologram as if it were a window, behind which the object is situated. If this window is closed, one can peep through a knothole and still see the entire scene, but with a more restricted perspective.

3. More than one independent scene can be recorded on the same photoplate. They can be viewed one at a time without cross interference by rotating or tilting the finished hologram with respect to the viewing light.

4. Although any hologram can be viewed by a monochromatic point source (not necessarily from a laser), a special kind, called the Lippmann-Bragg hologram, can be viewed by unfiltered thermal sources such as the sun or tungsten filament lamps.

5. A cylindrical hologram can be made which offers a 360° view of an object, i.e., the

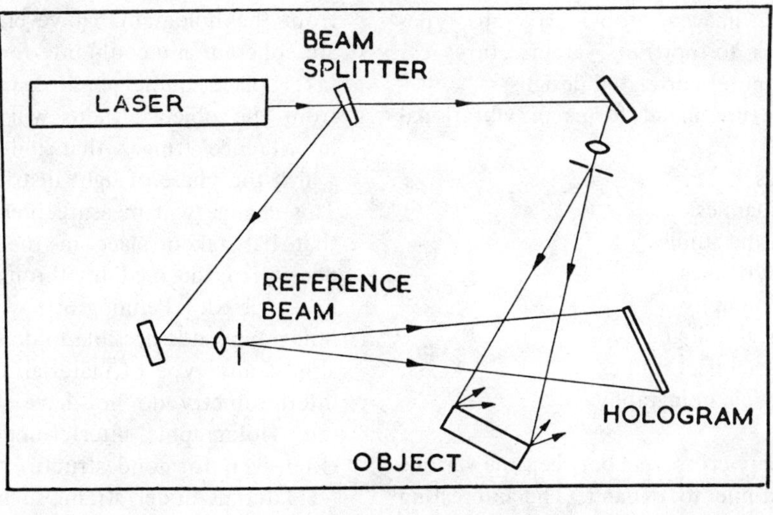

FIGURE 9-1. Laboratory setup for making a hologram. From *Holosphere.* With permission.)

observer can walk around the hologram and see the entire object from its front, sides, or back. Inverting the cylinder reveals a completely independent scene in all perspectives.

6. They can be encoded so that only the person in possession of a decoding device can see the true image.

7. Three-dimensional multicolor scenes can be recorded and reconstructed holographically.

The ruby laser, normal mode and Q-switched, He-Ne, and the argon are lasers commonly used at the present to produce the hologram. Recently, the pulsed xenon and nitrogen pulsed dye lasers have been used. The reconstruction as indicated may be done by:

1. Lasers
2. Light-emitting diodes
3. Xenon arc
4. Mercury iodide lamp

Small commercial types may be reconstructed by small flashlights and sunlight. Holograms may be made on film, glass, metal, or even ceramics. For special studies, thin film techniques with arsenic trisulfide have been used by RCA. The important safety program for the recording and reconstruction of holograms will be shown in the chapter on Laser Safety:

1. Interferometric holography
2. Computer technology, high capacity storage, and rapid accessibility
3. Specific imagery, obviously all types including designs to portraits, architecture, car models, and computer correlated designs
4. For the use in all forms of vibrational studies
5. Ballistics
6. Aerodynamics
7. Hot plasma studies
8. Hydrodynamics
9. Crystallography
10. Acoustical holography
11. Electrochemical studies
12. Microscopic holography

This list with many crossovers between the various groups will continue to expand. The interesting biomedical applications of holography will be reviewed in the chapters on Biology and Medicine.

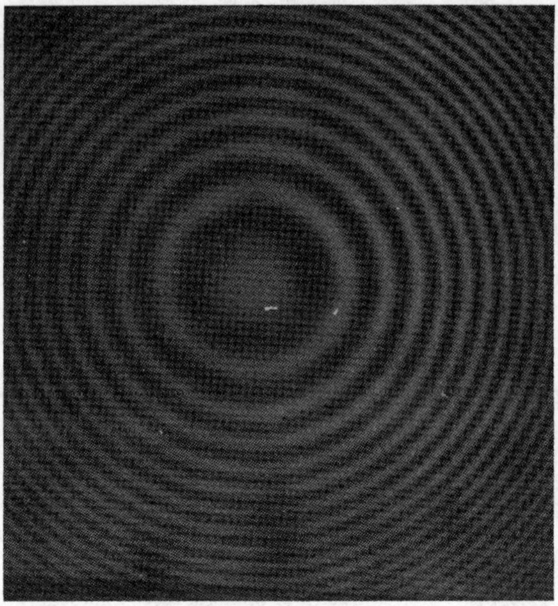

FIGURE 9-2. Hologram of a 2.5 cm diameter steel ball irradiated with 5 MHz acoustic energy and recording the backscattered field at a distance of 1.25 m. (From Perkin Elmer Corp. and *Instrument News,* 20, 2, 1970. With permission.)

Interferometry

In the reconstruction of an image from a hologram, if the hologram and the object are returned to the same positions they had during the making of the hologram and if these both are again subjected to the same direct and reference beams as before, the wave front from the object is superimposed on the reconstructed wave front from the hologram. However, these wave fronts are coherent and could interfere. If any change has taken place in the phase distribution of the light from the object "there will be light and dark interference fringes that indicate the degree to which the phase of light distribution has changed. This change will measure precisely any alteration that has taken place in the subject or in the density of the medium through which the beams have passed." Pennington[25,26] added that with holography it is possible to do interferometry with almost any type of material. Other techniques of interferometry do not have such wide ranges of use. Holographic interferometry has been used extensively for nondestructive testing.

Materials under strain, such as metals, plastic, and rubber, have been examined. Wear and tear of tires have been studied extensively.

FIGURE 9-3. Holographic nondestructive testing of a high performance turbine blade; each light fringe is a displacement approximately 12 μin. (Jodon Engineering Assoc. Inc.)

Because of current interest in hot plasma physics, the interferometric holography of laser produced gas breakdowns is important. TRW has used the ruby laser Holocamera® in these experiments with the high output Nd laser in deuterium impacts.

Holographic interferometry may also include continuous exposure as well as multiple exposures. When it becomes practical to transmit holograms at a distance, this continuous exposure interferometry will be able to monitor continuously pictures of materials under stress.

Picture Displays

There is much value in picture holograms independent of interferometry and computer generated holograms. A wide range of applications is possible. This can include panoramic types such as mosaic patterns, particle pictures of aerosol droplets, various types of displays including visual educational aids, portraiture, models, architectural designs, identification systems for personnel, materials, patterns, etc. in aerospace photographs. Media recognition could include liquid crystals, magnetic films, nonerasable plastics, etc. Here again, with the technique of holography, it would be of great value for transmission of holograms and especially for the even more difficult job, at least at present, of laser generated multicolored holographic imagery. Increased activity in the highly competitive field of video cassettes may stimulate more research and development. The fascinating application of art in holography and the important holographic interferometry studies of Wuerker on the effect of pollution on art will be mentioned in the chapter on Art.

Computer Technology

Certainly in the field of computer generated holography, the vast amount of information readily stored and readily retrieved may fulfill Gabor's[10] philosophy of holography to bring the many disciplines of science close together. In this manner, all these disciplines will work more closely together and more effectively to understand life for that ancient, oft repeated phrase, for the good of man.

After reading about the extraordinary bits of information computer generated and stored, one is justified in saying "I don't believe it until I see it." Many people walk around a cylindrical hologram and admit, "I still don't believe it." Some examples may be given. Isomet used a single crystal L_1NbO_3 as an optical recording medium to store vast quantities of digital information, analog information television picture frames, or printed page data. Optical processing can be done for credit card verification or optical character recognition systems. One beam of an argon laser, the signal beam, modulated by a test pattern transparency is focused to a 2-mm spot on the L_1NbO_3 crystal location. The recording is the interference pattern of the two beams. For the readout, the signal beam is turned off and the reference beam will

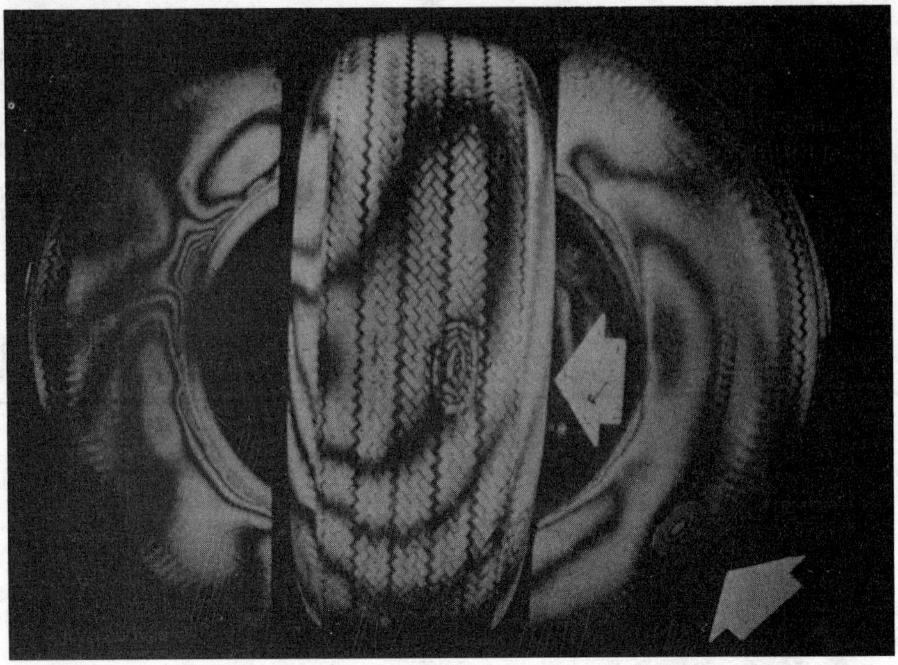

FIGURE 9-4. Holographic interferometry of a tire. (From Collier, R. A., Burkhardt, C. R., and Lin, L. H., *Optical Holography*, Academic Press, New York, 1971. With permission.)

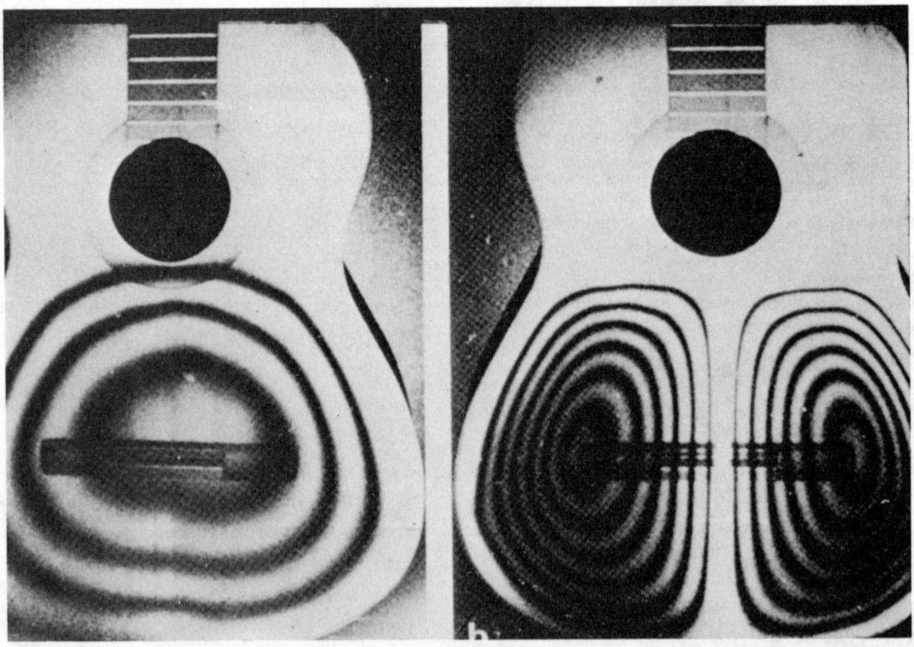

FIGURE 9-5. Time average holograms of vibrations of a guitar. (From Collier, R. A., Burkhardt, C. R., and Lin, L. H., *Optical Holography*, Academic Press, New York, 1971. With permission.)

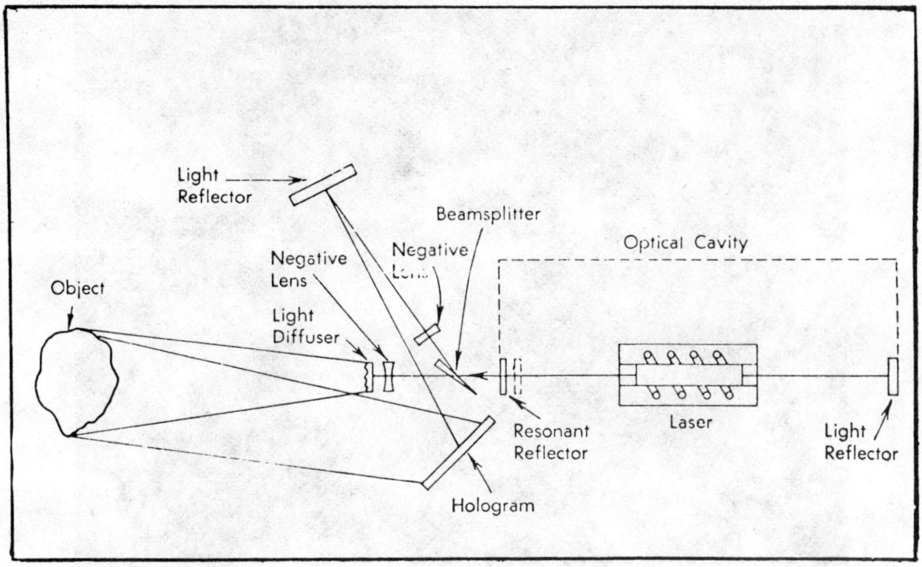

FIGURE 9-6. Holographic contouring. (From *Holosphere,* 2, 1, 1973. With permission.)

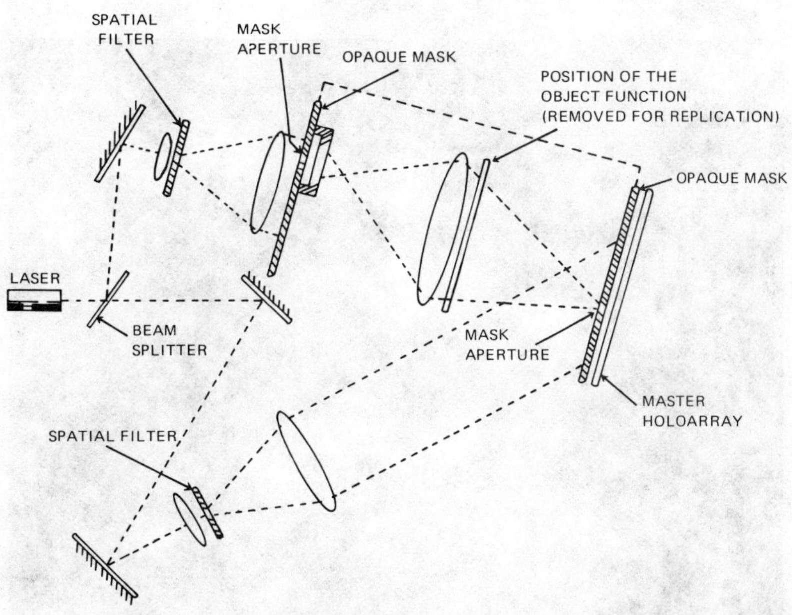

FIGURE 9-7. Technique for duplicating a hologram. (From *Laser Focus,* 5, 26, 1969. With permission.)

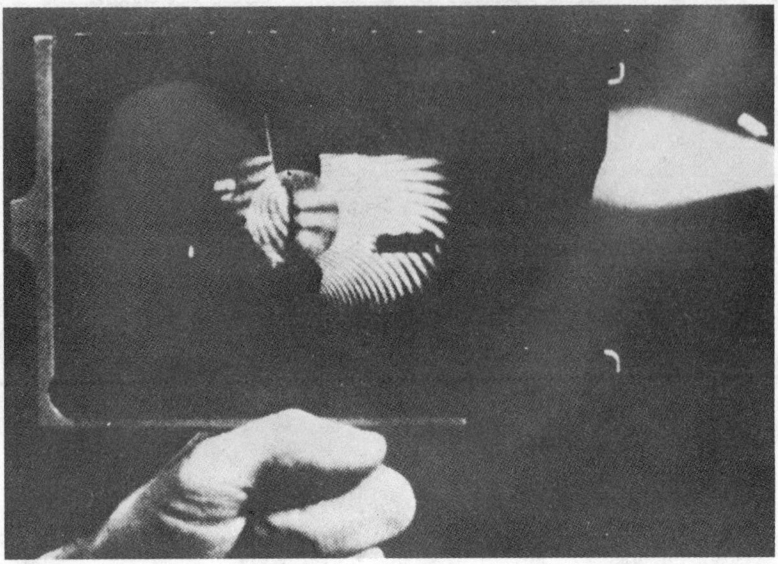

FIGURE 9-8. Holographic interferometry to test a card reader component IBM. (From *New Scientist.* With permission.)

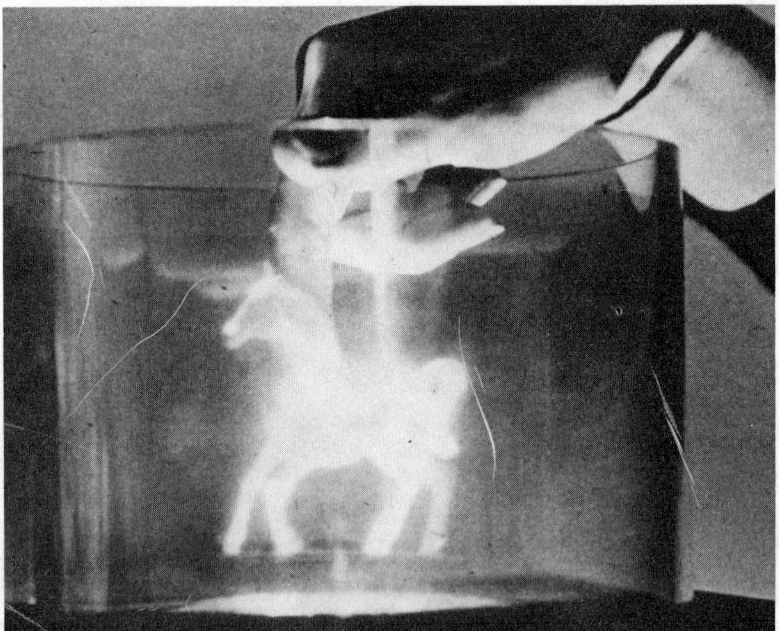

FIGURE 9-9. Cylindrical reflecting hologram 360°. (International Holographics Corp.)

reconstruct the original pattern on the screen upon passing through the hologram.

From a 4 x 6 in. microfiche card containing 60 printed pages, the Electro-Optical Center of Harris-Intertype Corp. used holography to compress all this into a few tiny strips, which can then be read by a computer.

Two wavelength holographies with computer generated holograms for interferometric testing of aspheric optical elements have been developed. The limits of computer generated holograms have not been reached and the astronomical figures for storage and retrieval will continue to expand.

Microscopic Holography

In addition to gross holographic photography, microscopic holography can also be done and important applications are offered by the conventional microscope turned into a holographic instrument. Van Lighten,[32] who has been a pioneer in research and development in this field, offers some of the following applications for observation and recording of the holographic microscope:

1. Crystallite and crystal growth
2. Bacteria movement
3. Cellular and tissue growth – very important in biology
4. Surface pattern phenomena under heat, stress, fatigue, etc.
5. Transparent materials under the influence of outside forces
6. Changes in liquid crystals under outside influences
7. Remote data gathering
8. Studies of phase objects of fog particles, aerosol, etc.

He has not only used lensless holographic microscopes but also microscopes with well corrected and uncorrected optics. Unlike conventional microscopes, holographic microscopes offer a three-dimensional image. It can be an amplitude image, a phase image, or a combination. This microscope also offers the possibility of "freezing" a picture on a two-dimensional hologram of a particular episode when many phenomena occur. Real time differential studies can also be done. As will be indicated later, there is a great opportunity for the biologist to watch three-dimensional pictures of single cell growth and the influences of internal and external factors on the growth pattern.

Acoustical Holography

Briefly, accoustical holography is a combina-

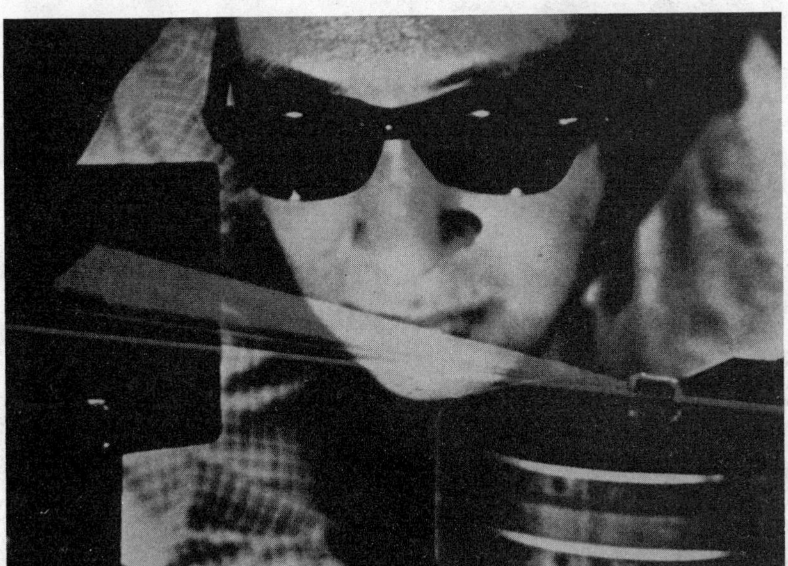

FIGURE 9-10. Computer generated holography with single small crystal in right holding 1,000 images. (Bell Laboratories.)

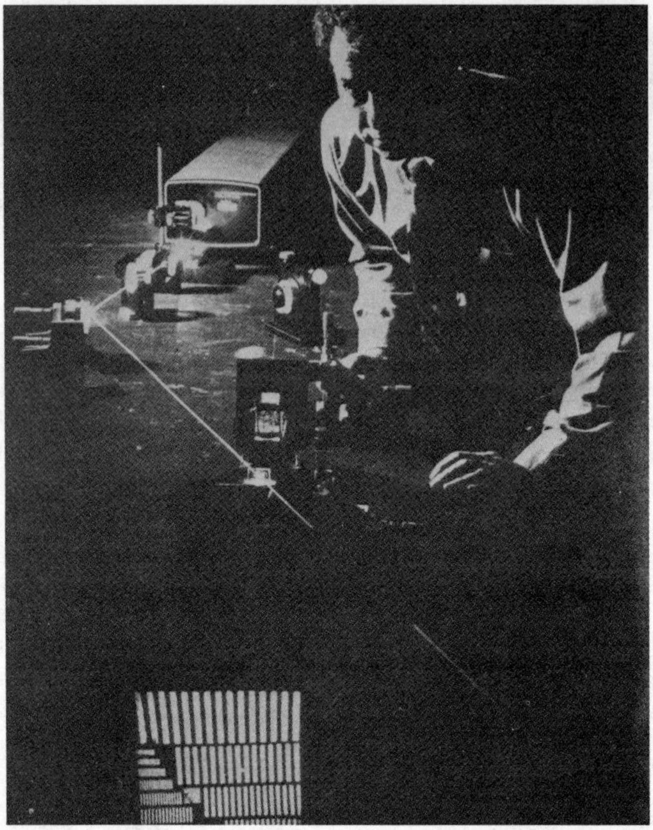

FIGURE 9-11. Argon laser beam recording test pattern in a L_1NbO_3 crystal. (From Isomet Corp. and *Lasersphere*, 2, 1, 1972. With permission.)

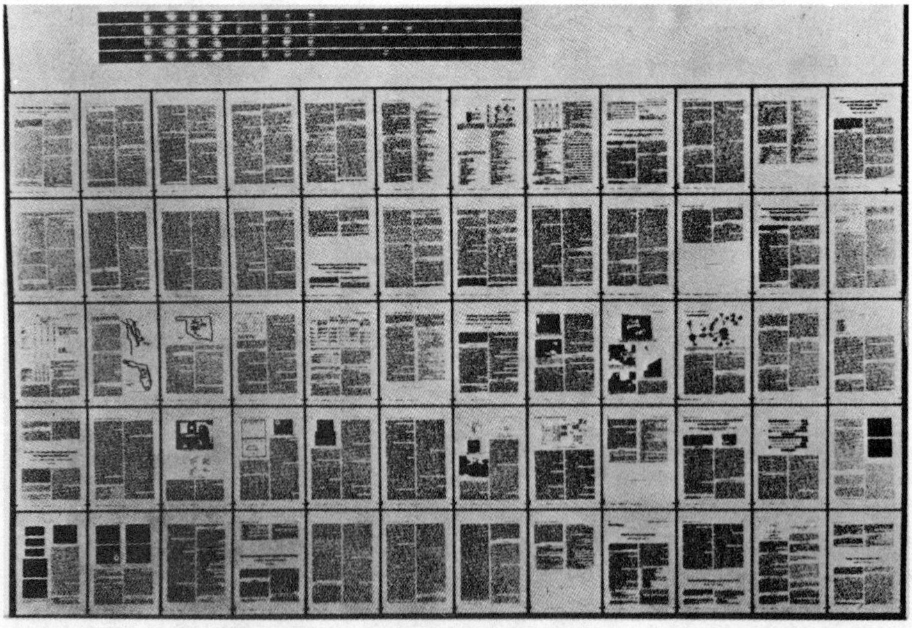

FIGURE 9-12. Microfiche 60-page-card measuring 4 x 6 in. with 4-strip holographic recording at top of the entire card. (From Electro-Optics Center – Harris Teletype Corp. and *Holosphere*, 1, 5, 1972. With permission.)

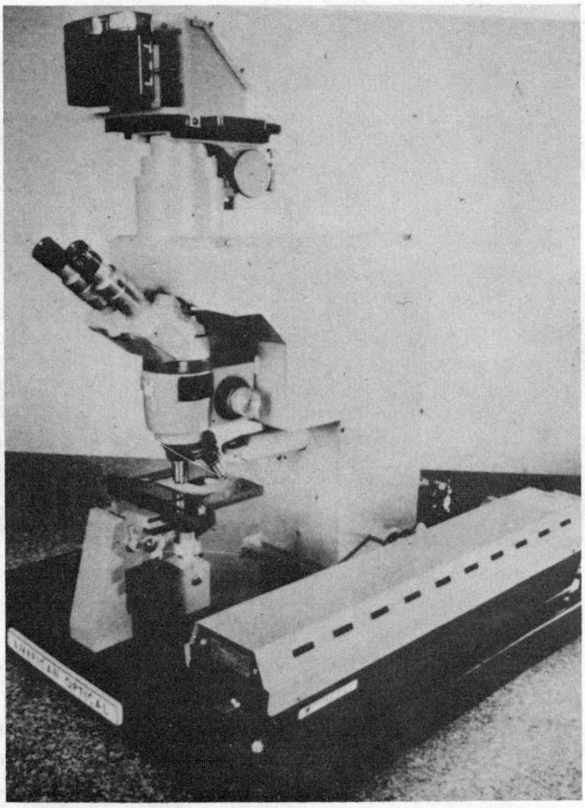

FIGURE 9-13A. Holographic microscope American Optical Co. A. Instrument.

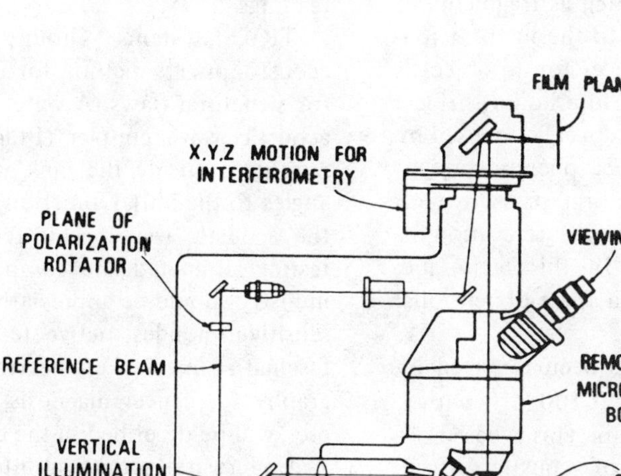

FIGURE 9-13B. Diagram.

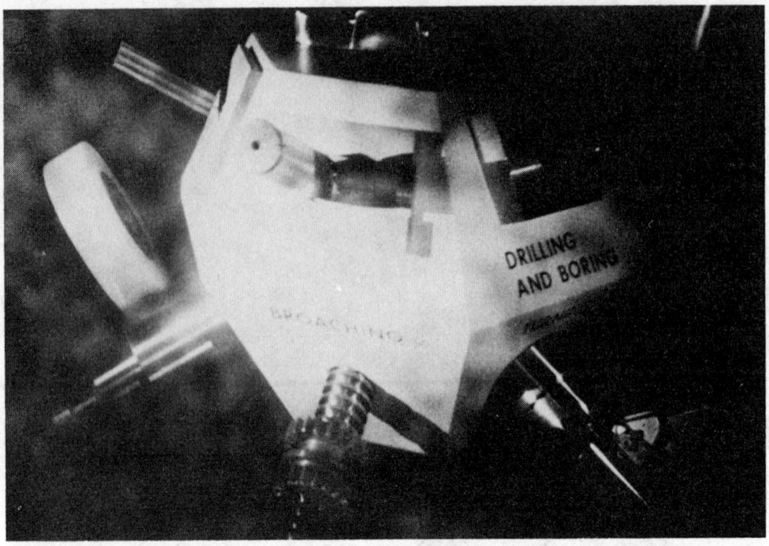

FIGURE 9-14. Composite hologram of machine tools. (Cincinnati Milacron.)

tion of acoustics and holography. In this technique the mechanical vibration form with the pulsed echo method is used. The rules governing the ultrasonic reflection are similar to those that govern reflection of light waves. An example of acoustical holography would be the internal structure analysis to produce pictures of voids, cracks, flaws, and particles in tissue such as fragments of plastics transparent to X-rays. In the instrumentation of Holosonic, the transducer for an object is focused by an acoustical lens to a liquid surface. Also, the reference beam transducer is focused by the acoustical lens. The surface phenomenon of disturbances by acoustic radiation pressure with consequent variations in surface elevation is portrayed by the hologram. In this technique, then, the hologram provides a three-dimensional image of the material.

An acoustic microscope has been reported by Korpel et al.[21] which operates at 100 MHz with a resolution in the order of 25 μm. This is based on the technique for visualization of acoustic surface waves. Acoustical optical techniques have also been used in storing and retrieval phases of computer technology.

Laser acoustical optical technology has been used in experimental spectrum analyzers with the use of the Zenith® D-70R Acoustical Optical Deflector and also used as an actual hard copy printer. The hard copy printer is reported to use a character generating system that controls the laser beam by acoustical optic devices. It is indicated that this is a nonmechanical printer for hard copy output of information such as teletyping and the computer alphanumeric output. The dye will develop red light since the paper is exposed to the laser beam and in a few seconds, it is reported, hard copy is available. Acoustical optical technology has also been developed for profile readings.

TRW Systems Group has developed an acoustooptical method for nondestructive testing for structural flaws. A water-filled tank acts as an acoustic wave coupler (Bragg cell). The laser is mounted outside the tank with its beam at right angles to the initial direction of the propagation of the acoustic waves. This technique is useful for testing laminated and composite material. Again, industry should be appreciative of these new rapid, sensitive, nondestructive testing techniques. The fascinating medical applications of acoustical holography for cancer diagnosis, studies of the fetus, etc., will be described in the chapter on medicine.

Underwater acoustical interferometry was done by Johnson and Mayer[17] with sonar transducer elements in multi-element arrays 30 m deep in a water tank 15 m in diameter. The argon laser was transmitted through a porthole with an etalon for coherence extension. The retroreflector was on the arrays covered with retroreflecting tape. The illumination beam from the same argon laser was transmitted through a second porthole used also for observation and photographic recording. The authors stated that holographic interferometry

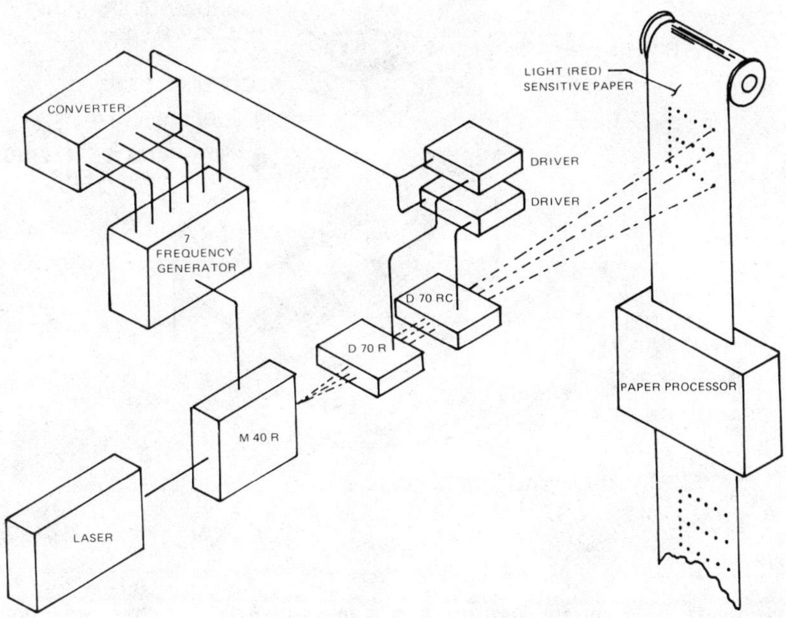

FIGURE 9-15. Development hard copy printer with laser and Acousto-optical modulation and deflector. (Zenith Radio Research Corp.)

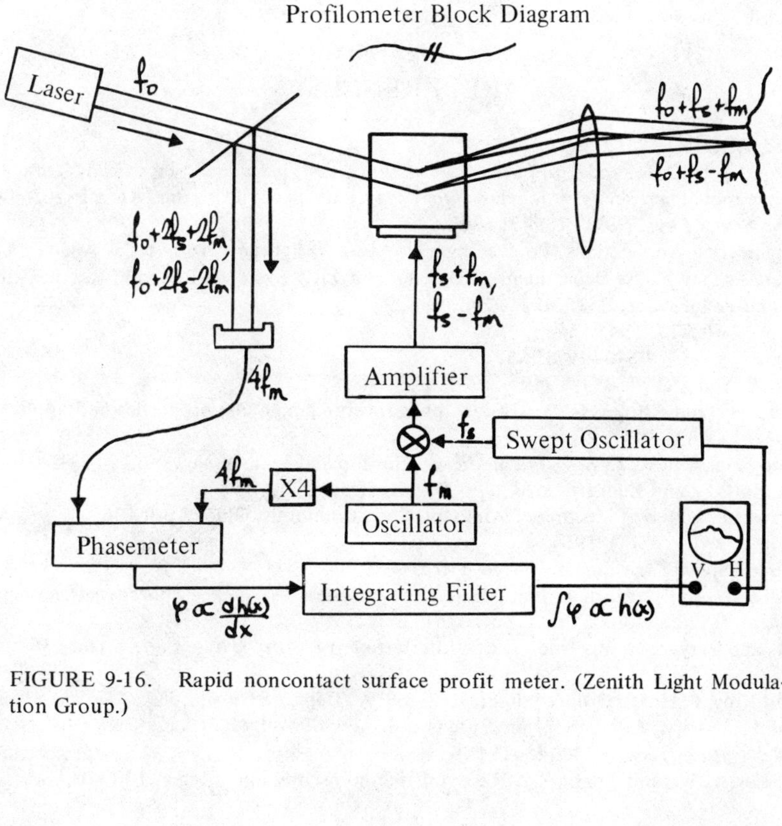

FIGURE 9-16. Rapid noncontact surface profit meter. (Zenith Light Modulation Group.)

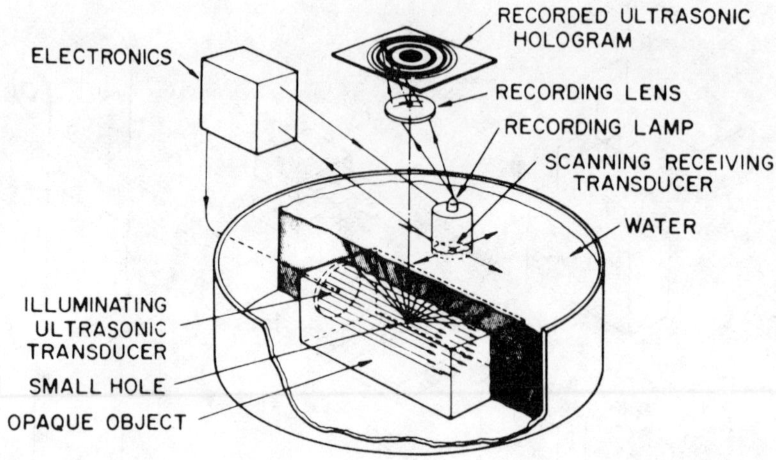

FIGURE 9-17. Acoustic hologram recorder. (Perkin Elmer Corp.)

"provides more complete data on the nature of underwater vibrations than could be obtained by any other means."

Again, holography, in all its phases, is an exciting new field of development. As better instrumentation, cheaper holograms, and better transmission techniques become available, this field will expand even more. More applications will be available in all areas of modern life and even for space technology.

REFERENCES

1. **Close, D. H.,** *Ind. Res.,* 14, 34, 1972.
2. **Collier, R. J., Burkhardt, C. R., and Lin, L. H.,** *Optical Holography,* Academic Press, New York, 1971.
3. **Cox, M. E.,** Holographic microscopy – a review, *Microscope,* 19, Second Quarter, April 1970.
4. **Denisyak, Yu N.,** *Soviet Phys. Dokl.,* 7, 543, 1962.
5. Holography Engineering Applications, Proceedings Symposium, February 16-17, Los Angeles, California Advanced Research Projects Agency of the Department of Defense and TRW Systems, Redondo Beach, Calif., 1972.
6. **Friesem, A. A. and Fedorowicz, R. J.,** *Appl. Opt.,* 6, 529, 1967.
7. **Gabor, D.,** *Nature,* 161, 777, 1948.
8. **Gabor, D.,** *Proc. Phys. Soc.,* B64, 449, 1965.
9. **Gabor, D.,** *Proc. R. Soc.,* A197, 454, 1949.
10. **Gabor, D. Kock, W., and Stroke, G. W.,** Holography, the fundamentals, properties and applications of holograms reviewed, *Science,* 173, 11, 1971.
11. **Goldman, L. and Rockwell, R. J.,** *Lasers in Medicine,* Gordon and Breach, New York, 1971.
12. **Jeong, T. H., Rudolf, P., and Luckett, A.,** *J. Opt. Soc. Am.,* 56, 1263, 1966.
13. **Jeong, T. H.,** Paper presented at Meeting of Am. Opt. Soc., Columbus, Ohio, April 1967.
14. **Jeong, T. H.,** *Am. J. Phys.,* 35(5), 1967.
15. **Jeong, T. H.,** Holography, Lake Forest College, 1972.
16. **Jeong, T. H.,** Holography, Dept. of Physics, Lake Forest College; *Reference Source of Demonstration Experiments in Physics,* Meiners, H., Ed., Ronald Press, New York, in press.
17. **Johnson, C. D. and Mayer, G. M.,** Underwater Interferometry, *Appl. Phys. Lett.,* 21(8), 369, 1972; *Laser Focus,* January 1973.
18. **Katyl, R. H. and King, P. H. Jr.,** Ordinary light, Holography, Dept. of Physics, Lake Forest College, 1972.
19. **Kasper, F. G. and Lamberts, R. L.,** Paper presented at Meeting of Am. Opt. Soc., San Francisco, 1966.
20. **Kogelnik, H.,** *Bell System Tech. J.,* 44, 2541, 1965.
21. **Korpel, A., Kessler, L. W., and Palermo, P. R.,** Acoustic microscopy operating at 100 MH_3, *Nature,* 232, 110, 1971.

22. **Leith, E. N. and Upatnieks, J.,** *Sci. Am.,* 212, 24, 1965.
23. **Leith, E. N. and Upatnieks, J.,** *Opt. Soc. Am.,* 52, 1123, 1962; 54, 1295, 1964.
24. **Lin, L. H., Pennington, K. S., Stroke, G. W., and Labeyrie, A.,** *Bell System Tech. J.,* 45, 659, 1966.
25. **Pennington, K. S.,** *Microwave,* October 1965.
26. **Pennington, K. S.,** *Advances in Holography, Lasers and Light,* Scientific American with introductions by Arthur L. Schawlow, W. H. Freeman Co., San Francisco.
27. **Stroke, G. W. and Labeyrie,** *Phys. Lett.,* 20, 368, 1966.
28. **Stroke, G. W.,** *An Introduction to Coherent Optics and Holography,* Academic Press, New York, 1966.
29. **Stroke, G. W., Brumm, D., and Funkhouser, A.,** *J. Opt. Soc. Am.,* 55, 1327, 1965.
30. **Stroke, G. W. et al.,** *J. Opt. Soc. Am.,* 57, 110, 1967.
31. **Vander Sluis, K. L. et. al.,** *Am. J. Phys.,* 33, 225, 1965.
32. **Van Lighten, R.,** personal communication, 1972.

Chapter 10

LASER-INDUCED THERMONUCLEAR FUSION

In 1966 (*Laser Focus,* May 15) Koopman and his colleagues[21] at the Maryland Institute of Technology generated a plasma with a pulsed ruby laser and controlled it with a magnetic field. The technique was developed for use in achieving controlled thermonuclear power. There has been much work on the laser approach to thermonuclear fusion since that time. Recently, very large appropriations have been made both for the development and exploitation of fusion power and, as shown previously, for the so-called laser H-bomb. One reason for all this interest is the current energy crisis. Some estimate the 1972 U.S. budget for laser-induced thermonuclear fusion as 25 million, the Russian as 50 million, with doubling for the fiscal year of 1974.

The Energy Crisis

Weinberg,[18] the Director of the Oak Ridge National Laboratory, has written that "the present energy crisis is neither theoretical nor long range; it is real and it is immediate." Our sources of energy – coal, oil, domestic and foreign imports, and natural gas – are few and shrinking. Fossil fuels are blamed for pollution and foreign imports carry a host of political and world monetary problems. So there is a frenzied search for substitutes. One area of interest is nuclear powered fission fuel with the magnetic confinement method and new laser-induced fusion. The alternatives are listed as:

1. Solar
2. Ocean thermal gradients
3. Geothermal – heat from hot rock

Most scientists admit that the main course today is to attempt to get adequate and clean power from coal and from "practical" nuclear energy.

Plasma

One of the "what's been called slippery fish" of modern physics is hot plasma which is composed of ions, molecules, and materials at tremendously high temperatures. This occurs when there is thermonuclear fusion in the plasma; unlike other gases, there is ionized gas. Some of the atoms have had some electrons snatched away. So this mixture has neutral atoms, ions and lost electrons, and free electrons. It is often called a hot fluid-like inductive mass. As a consequence, plasma can rather easily conduct electricity, which gas cannot do. The sun and stars are examples of hot plasma on earth. Fifty miles or so above our heads is the ionosphere or plasma. Deuterium heavy hydrogen, the hydrogen isotope with twice the mass of ordinary hydrogen, has been found in outer space. Plasmas made by man include the fluorescent lamp, exhaust of the rocket engine, and fireball of a nuclear exposure.

Fission vs. Fusion

In fission atoms such as uranium or thorium are split apart, releasing energy. In fusion, light atoms such as various isotopes of hydrogen are forced together, fused to create energy. In the sun, the ionized plasma has a temperature in excess of 15 million °K. With the fusion of the hydrogen ions, helium nuclei and energy that we see as sunshine are produced. Fusion is said to be produced in the sun by brute force in huge masses. Hirsch[23] directs the AEC's controlled thermonuclear research (CTR) program.

Magnetic Confinement

Can man duplicate this fusion as he exhausts all the fuel sources of the earth? One way of producing electricity with very low pollution and great efficiency in plasma physics is by magneto-hydro dynamics (MHD). This means the interaction of a moving plasma with a magnetic field. However, in brief, this has been very difficult. Logan et al.[8] have recently attempted to trap this hot plasma for fusion to take place by using so-called magnetic mirrors. The basic requirements for this fusion are

1. Minimum temperature for a fusion reaction is 46 million °K
2. The density of the plasma must be at least 10^{15} ions/cm^3
3. The plasma must be kept at this tempera-

FIGURE 10-1. Producing spheres with lasers; oxides of aluminum, thorium, chromium, hafnium-titanium, zirconium-titanium, zirconium stabilized with calcium are heated with (CW) CO_2 250 W focused to 1 mm in diameter. The oxides after melting are spun off the rod and supercooled for radioactive fuel. (Sandia Laboratories Photo.)

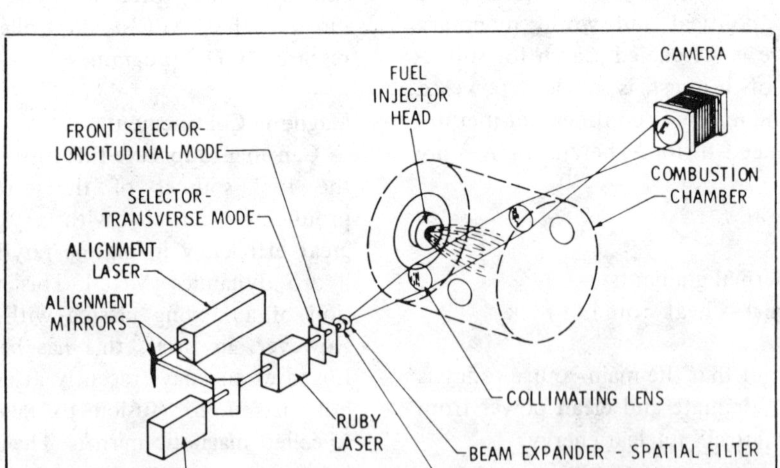

FIGURE 10-2. Holocamera for plasma analysis. (From *Holosphere.* With permission.)

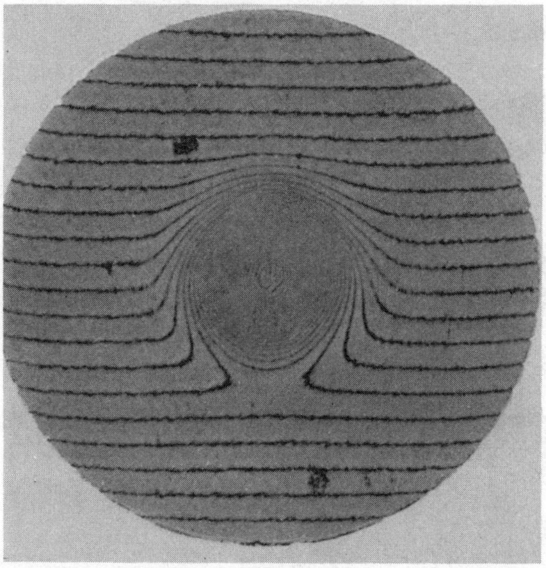

FIGURE 10-3. The University of Rochester's joint university-industry-utility Laser Fusion Feasibility Project is the only university-based program in the United States in this field. Begun in the summer of 1972, the Project builds on laser-initiated fusion studies conducted in the UR College of Engineering and Applied Science's Laboratory for Laser Energetics since 1967. Photo shows the vacuum chamber and part of the special main beam amplifier section of the laser chain used in the project.

ture and density for about 1/10 sec confinement The difficulty, then, is the task of plasma confinement holding this plasma at the necessary temperature and density long enough to start the fusion reaction.

The Russian fusion device, called the tokamak (to = toroidal, ka = chamber, mak = magnetic), has come so close to producing the necessary conditions of temperature, density, and confinement that some of the world's fusion laboratories have switched over to tokamak-type machines.

Recently the Atomic Energy Commission announced that the adiabatic toroidal compressor (ATC) at its Princeton Plasma Physics Laboratory had achieved a controlled plasma density "in the range in which a tokamak fusion reactor is expected to operate." Recently, with hot-ion injection, ATC is said to be the first tokamak device to get minimum density necessary for a fusion reactor to operate. The concept of ATC is based on studies done in Russia. Schlesinger,[6 7] of the Atomic Energy Commission, proclaimed ex-

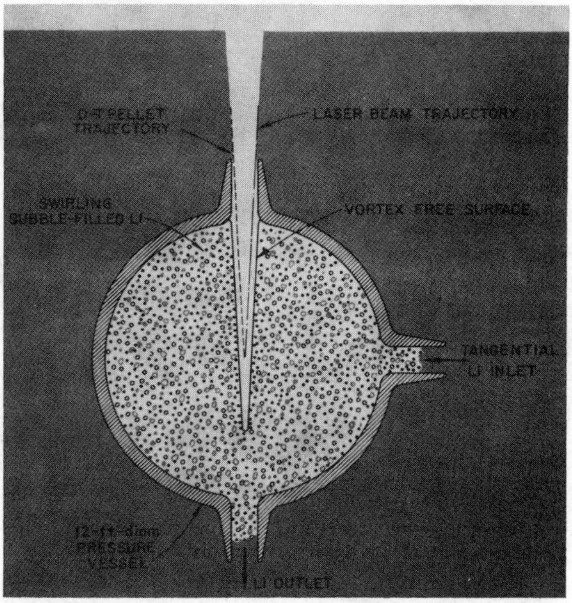

FIGURE 10-4. Laser fusion schematic. (Oak Ridge National Laboratory. Annual Report, 1970-1971.)

pansively, "These new results further brighten the prospect of realizing commercial fusion power in a tokamak confinement system. While this is a major step forward in tokamak research, I would hesitate to use the term 'breakthrough'. We still have many years of hard work ahead of us to develop this concept into a practical power system."

For the ATC cycle, an electric current is induced to flow in a doughnut-shaped plasma (high temperature gas of fusion elements). The current both confines and heats the plasma. At this stage, the ATC electron temperature rises to about 10 million °C. Then a pulsed magnetic field compresses the plasma. The temperature rise is 25 million °C and the density reaches 1×10^{14} particles/cm^3. The goal is said to be 100 million degrees. The figures[22] are given to show what the laser now must equal.

Laser Fusion

The laser has been introduced in an attempt to avoid this difficult confinement problem. A pellet of solid deuterium rather than a gas is used or deuterium plus tritium or lithium deuteride or deuterized polyethylene. According to Nuckolls and Wood,[9] the laser energy required for a practical power reactor is 10^5 to 10^6 J. Basov claims that a neodymium laser 600 J and 2 nsec is

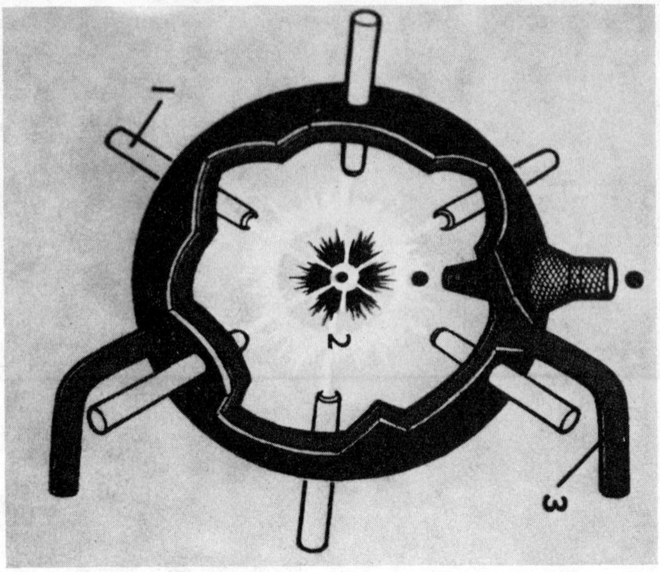

FIGURE 10-5. Laser fusion diagram. 1. array of laser beams, 2. pellets of frozen heavy hydrogen in the cross fire, 3. molten lithium to carry off to produce steam for electricity. (From Weaver, K. F. and Kristof, E., *Natl. Geograph.*, 142, 670, 1972. With permission.)

required. As their factors the Livermore group requires:

1. Deuterium pellet, 1 mg
2. The laser 10 to 15 W, 20 nsec
3. Electron temperature, 100 x 10^6 elevation in the outer layer
4. The shock wave must go to the center of the sphere of heavy hydrogen
5. The compression density must reach 10,000 times; fusion results then from this compression density plus the high temperature

Are there laser systems capable of producing such high outputs? The cynics believe that for this phase of the study, the U.S. has the biggest computers and Russia has the biggest lasers. At the Lebedev Institute in Moscow, the Russians have a device that can fire nine laser pulses at a target simultaneously. They are working on such a technique to increase it to a 27-beam system. The technique of obtaining multiple beams involves splitting each beam three times amplified, then splitting it again. After suitable amplification, the beams are focused on the hollow pellet from equidistant points in the circuit.

The Lawrence Livermore Laboratory now uses 6 Nd doped glass discs to amplify the beams. In addition to the new glass lasers, chemical lasers are said to help to develop the awesome laser output required for thermonuclear fusion. Thorne[13] suggested that theory indicates that strong magnetic fields can be created in plasmas by circularly polarized laser beams. If laser technology can be developed in such an actual and practical way, and not just calculational, then there could be almost inexhaustable supplies of pollution-free energy.

The challenge is there to attempt to produce this soon. Fraas claimed that a nuclear fusion reactor will have lower costs both for initial capital and for development. So the veil of secrecy has been lifted and the race is on in this country with Fraas at Oakridge, Nuckolls, Wood, Theissen, and Zimmerman[9] of Lawrence Livermore Laboratories, and Lubin at Rochester. This is all in addition to the work at Sandia, Los Alamos Scientific Research Laboratory, and Siegelat with KMS at Ann Arbor. *The Wall Street Journal* suggests the term of questionable nuclear fantasy regarding the work at Ann Arbor. The safety programs involving neutrons, laser beams, and heat still remain to be worked out.

Yerick[19] claimed that the control of thermonuclear fusion is probably the greatest engineering problem that has ever faced mankind. As sure examples of the applications of such total energy,

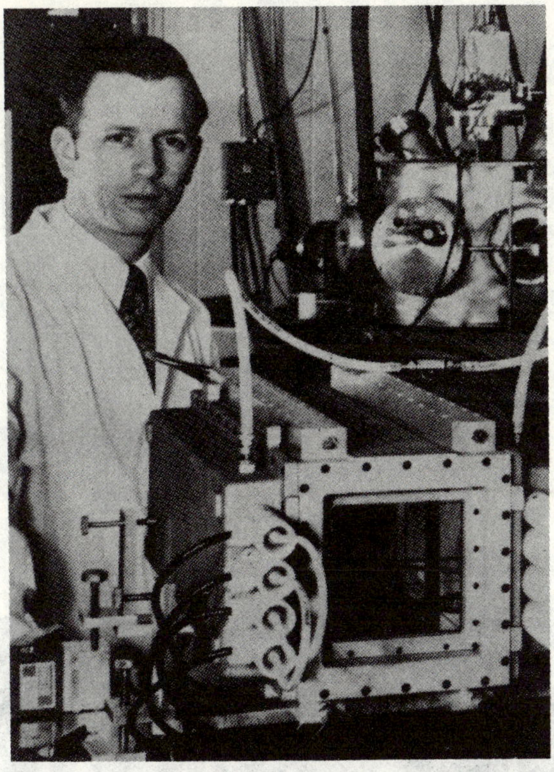

FIGURE 10-6. Moshe J. Lubin, director of the University of Rochester's Laboratory for Laser Energetics, shown beside a Zig-Zag-Face Pumped Laser module, heads the UR-based project on the feasibility of laser-produced controlled nuclear fusion. Lubin is an associate professor of mechanical and aerospace sciences and of optics in the University's College of Engineering and Applied Science. The estimated $3 million project is being conducted in collaboration with the ESSO Research and Engineering Co., General Electric Co., and Northeast Utilities.

he has suggested environmental control, mining, and highway building.

Perhaps there will be consideration of remote energy transmssion, the dream of Townes and of Hertzberg[4] of the University of Washington. This may start with the conversion of heat to coherent radiation, then the transmission of that at a distance to convert the laser energy to electrical energy and another source of power. Modulation of electron beams by the laser is still another source of work for the future.

The challenge for energy is indeed here and the need is now. Adequate funding and work and not polemics are needed in all fields of current energy work and in nuclear energy, especially between the fission and fusion experts. With regard to other outside sources of energy such as solar, it is well to heed what Weinberg[18] states: "We cannot prudently and responsibly depend on any of speculative technologies to provide the energy we need even in the long run." We would like to remove laser-induced fusion from this speculative group. Hirsch[23] is said to have indicated that the formal goal of scientific feasibility for fusion is 1980-1982. International need and cooperation are needed to achieve this great goal, therefore, a powerful stimulus for peace.

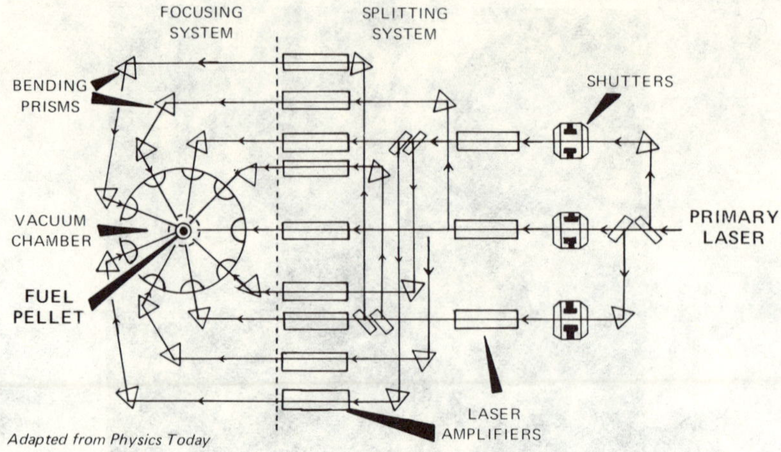

FIGURE 10-7. Soviet multipath system for achieving fusion. (From *New York Times,* October 22, 1972. With permission.)

FIGURE 10-8. Photograph of a carbon dioxide laser used in laser fusion research at the Lawrence Livermore Laboratory, operated for the Atomic Energy Commission by the University of California. This laser has a maximum energy of 100 J. Pictured is Dr. Yu-Li Pan, physicist. (Lawrence Livermore Laboratory.)

REFERENCES

1. **Bova, B.,** Physicists probe the ultimate energy source, *J. Smithsonian Inst.,* 1972,
2. **Coppi, B. and Rem, J.,** The tokamak approach in fusion research, *Sci. Am.,* 65, 1972.
3. Fusion with lasers: the wraps of secrecy come off, *New Sci.,* May 25, 1972,
4. **Herzberg, A.,** Radiant future for remote energy transmission, *New Sci.,* February 3, 1972.
5. **Kenward, M.,** Fusion power politics in the U.S., *New Sci.,* May 18, 1972,
6. Laser induced fusion, *Laser Sphere,* 2, 1, 1972.
7. Laser fusion: industry moves in, *Sci. News,* 101, 388, 1972.
8. **Logan, G., Brown, I. G., Lieberman, M. A., and Lichtenberg, A. J.,** *Phys. Rev. Lett.,* 29, 1435, 1972; *New Sci.,* 560, 1972.
9. **Nuckolls, L., Wood, L., Theissen, A., and Zimmerman, G.,** *Nature,* 239, 139, 1972; *New Sci.,* 470, 1972.
10. **Schwartz, H.,** Modulation of electron beams by laser light: new experimental and theoretical results, *N. Y. Acad. Sci.,* 172, 620, 1972.
11. **Shapley, D.,** Conservation of energy: the potential for more efficient use, *Science,* 178, 1079, 1972.
12. **Thomsen, D. E.,** Mirror, mirror on the wall, which is the fairest plasma of them all? *Sci. News,* 101, 28, 1972.
13. **Thorne, J. M.,** Controlled Fusion: Plasma Confinement with Lasers, Department of Chemistry, Brigham Young University, Salt Lake City, March 1970.
14. Tokamak plasma density now in fusion range, *Ind. Res.,* January 1973,
15. **Tuck, J. L.,** Outlook for controlled fusion power, *Nature,* 233, 593, 1971.
16. **Sullivan, W.,** Laser technique tested in search for clean power, *New York Times,* October 22, 1972.
17. **Weinberg, A. M.,** Nuclear energy, *Science,* 178, 933, 1972.
18. **Weinberg, A. M.,** Some views of the energy crisis, *Am. Sci.,* 61, 59, 1973.
19. **Yerick, G. J.,** Engineering aspects of controlled thermonuclear fusion, *N. Y. Acad. Sci.,* April 14, 1972.
20. **Yoshikawa, S.,** Fusion research in the U.S.S.R., *Am. Sci.,* 59, 463, 1971.
21. **Koopman, D.,** Quoted by *Laser Focus,* 7, 63, 1972.
22. New step to controlled fusion, *Sci. News,* 104, 36, 1973.
23. **Hirsch, R. and Metzger, N.,** America's fusion director, *New Sci.,* April 12, 1973,

Chapter 11

LASERS IN THE MILITARY

Buck Rogers may have been the first U.S. officer to plan for the use of the ray gun. A great deal of the support for current laser research comes of necessity from the military in many countries. There are obviously many needs for the laser in the military other than the public's usual opinion of lasers only as weaponry, both defensive and offensive. This review will list but a few of the main categories of research and development. As is evident, the veil of secrecy hangs heavy and low over many laser military projects. The material contained here is unclassified and is from current literature:

1. Communication
2. Range finding
3. Tracking – ground, air, sea
4. Weather reconnaissance
5. Metalworking
6. Nondestructive measurement
7. Defensive weaponry
8. Offensive weaponry
9. Laser field surgery
10. Miscellaneous

Communication

Communication systems are perhaps the most important. The lasers more commonly used are infrared types. Communications may be between ground observers, personnel or vehicular, ground to helicopter and helicopter to ground, ground to airplane, airplane to ground, and airplane to airplane, or earth to satellite and satellite to earth. Miniaturization of equipment, especially with junction diodes, has made for more flexible instrumentation.

Recent developments have made optical communications more efficient and economical. Bell Laboratories[10] has operated a room temperature junction diode laser for 3 months. Nippon Electric[10] has added a small amount of aluminum to the active gallium arsenide layer and has operated this laser for 3,000 hr. "The area actually carrying current is also greatly reduced by introducing a p type stripe made by diffusing zinc through a section of the top n type layer."[10]

Brown[10] has reported that new glass fibers, as indicated previously, have low attenuation. Bell Laboratories[10] has recently introduced a new glass fiber rod a few millimeters in diameter, "supported on a plate and enclosed within a tube 1 cm. in diameter. All three are heated and pulled out to reduce the external dimensions to a few milli-

FIGURE 11-1. Flash's laser gun combating gas cloud.

FIGURE 11-2. A laser gun.

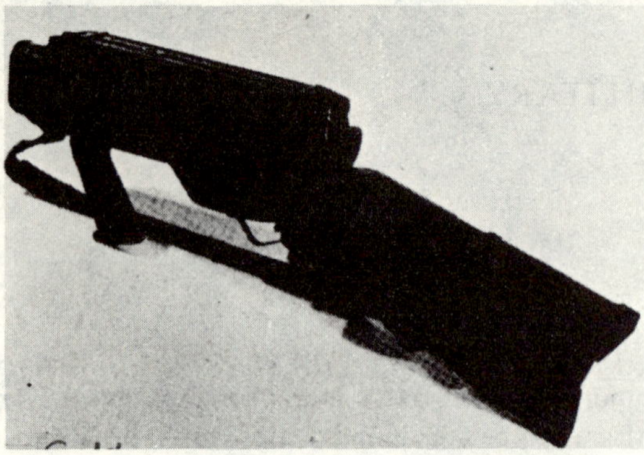

FIGURE 11-3. A distance measurement laser apparatus – International Laser Systems, Inc. (From *Laser Elektro-Optik*. With permission.)

meters. Attenuation, although higher than that of the Corning Guide (2 dB per kilometer) is still only 5 dB per kilometer." All these fibers are still in the development stage. These optical wave guides can be bent to sharp angles making installation easier not only in the military area but also in the urban area. Communications may be combined in addition to voice with computer readouts and holography. The remarkable accuracy of the laser tank range finders has been established by the use of the laser in the armed forces of many nations. The Sheridan range finder of Hughes consists of ruby laser, telescope-like optics, and control panels and electronics.

Laser holography is being used to survey terrain for carrier deck landings, especially in periods of poor weather visibility. Kruger and Grosz[13] have stated that the Germans tested remote control bombs as early as 1915. Siemens-Schuchert[13] proposed a guidance system for gliders through thin wires from the mother airships. This indeed is a far cry from the accurate laser guidance of today.

Surveillance

Infrared surveillance systems with image converters offer very excellent use for night work. Infrared lasers may also be used for lidar (laser radar) and for tracking planes. With the infrared system, the pilot and the crew are unaware that they are being tracked. The infrared system can also be used on ranging, especially popular now in tank equipment. As one considers various sensor systems, seismic with aural pick up, auditory,

FIGURE 11-4. Laser range finder. The laser is a neodymium-lutetium-YAG with a lithium neobate Q-switch peak power 1.5 MW. The range finder measures the distance to targets from 200 to 6,000 m (almost 4 mi) with a variation of no more than 10 m. A laser range finder determines distance by measuring the time it takes the laser pulse, traveling at the speed of light, to reach the target and bounce back to the finder. The readout is less than a second after its trigger button is pushed. About the size and shape of a pair of medium-sized binoculars, the rangefinder measures 6x8x3 in. Total weight, including a small rechargeable nickel-cadmium battery, is 4 lb 12 oz. The battery can be used 250 times without recharging. (United States Army Electronics Command, Ft. Monmouth, N. J., 1971.)

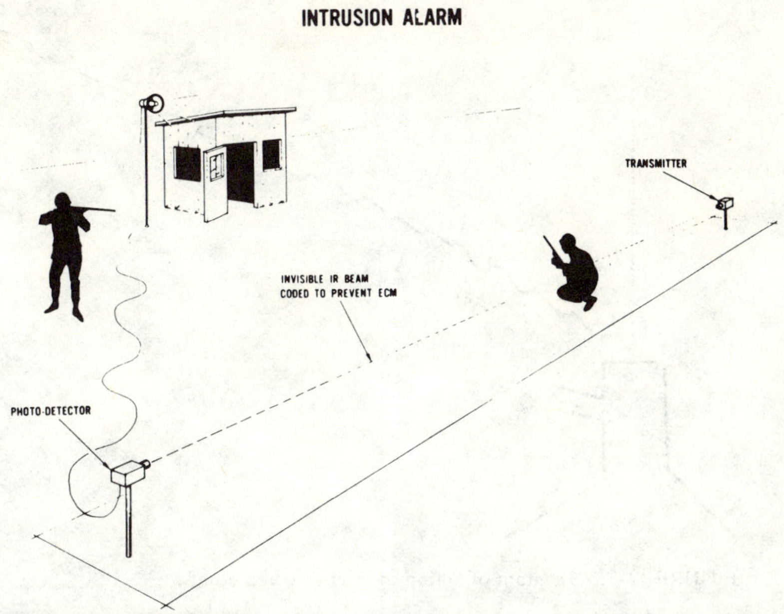

FIGURE 11-5. Ga-As lasers in military security program. (From *Opt. Spectra,* 5, 27, 1972. With permission.)

photo-detectors, ordinary infrared, etc., it is evident that laser infrared with its output and narrow beams would be more effective. Ga-As lasers make for miniaturization of such devices.

FIGURE 11-6. Laser radar transmitter – U.S. Navy Raytheon (From *U.S. News World Report.* With permission.)

Ranging and Tracking

The Army Electronics Command[12] has developed a lightweight laser finder of 4¾ lb. A laser ranger of Ferranti in England is an infrared YAG laser. Aircraft to target range is found by measuring the time interval between transmission and receiving reflected beam. The short pulse length of the laser is not given. The Ferranti laser ranger is gyrostabilized. The whole operation is shorter than other alternative radar. Ferranti also has a marked target seeker. In this concept a forward air controller on the ground directs a pulsed laser (type?) at a target. The strike aircraft seeker receives the reflected scattered beam. The seeker then locks on and tracks the energy. After stabilizing the head of the laser at the target and at the same time aligning the display for the pilots, the laser ranger is triggered automatically. Ferranti lists the following advantages over current methods of close air support:

1. Target positively identified by forward air controller
2. Pilots' search task reduced
3. Pilot sees his target earlier
4. Blind attack capability is provided

Laser tracking now can be done for aircraft at

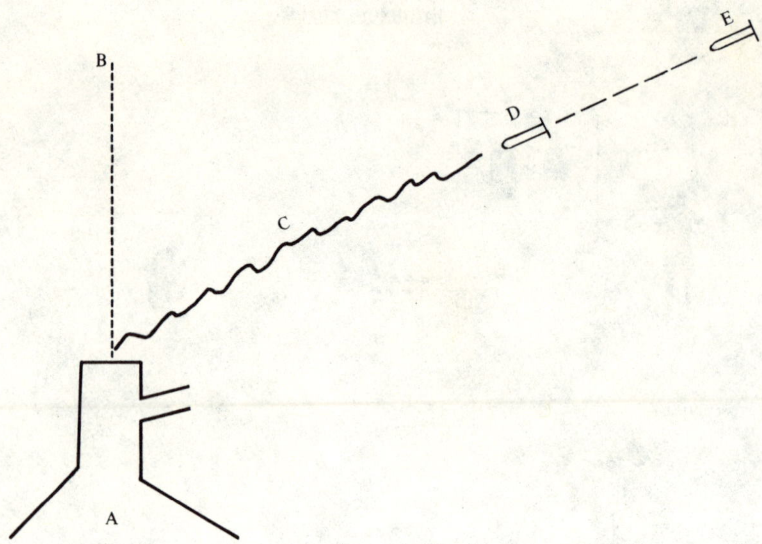

FIGURE 11-7. Diagram of action of laser-directed bomb:

A. Target
B. Designator aircraft impacts target with laser beam
C. Reflected laser beam from target
D. Bomb sensor of reflected laser energy giving data to guidance device on bomb
E. Bomb on way to target

low altitudes and long ranges. This is a combined laser-radar system. The U.S. Army is doing research and development on a Multiple Integrated Laser Engagement System (MILES). In this technique training of army personnel is done by a low power laser to simulate the firing of an aimed M16 rifle and M60 machine gun shot. High output systems are used to simulate projectiles or missiles. The transmitters are collimating lens, signal processing electronics, and semiconductor laser diodes. The laser receiver consists of an optical filter and infrared detector with electronic circuitry for signal processing and activation of a hit indicator.

A concern of the Air Force is the analysis of the laser speckle phenomenon and the effect of the speckle on tracking system from laser illuminated targets.

The recent use of the laser in Vietnam with laser-guided bombs has also shown the accuracy of the laser (LR, laser-guided). One plane or even a ground observer focuses a very thin line of laser beam on the target. The beam comes off the target in a cone of reflected energy. On release the bomb seeks out the outer edge of the cone and is guided to the pinpoint which is on target. Laser-guided bombs are estimated to cost about $3,500 a unit for the 2000 bomb and $5,000 for the 3000 bomb. The rival to the laser-guided bomb (LGBS) is the television bomb, a much more expensive type of bomb guidance program.

The army also has a laser-guided missile (THAD – Thermal Homing Accuracy Demonstration). This missile has a range of 2.8 km with laser illumination 3.0 km from target.

Weather Reconnaissance

Laser radar, or lidar, may also be used for weather reconnaissance. Laser altimeters may be used to measure ceilings. There have been previous comments on the relatively meager data available and made public on the atmospheric propagation of laser. Special infrared lidar could be used for navigation as opposed to the very effective He-Ne lighthouse in Australia for maritime traffic.

Metalworking

As indicated, laser drilling and welding have been used in the aircraft industry especially for accurate drilling in rounded surfaces. Laser interferometry has been of value in studying impacts of projectiles, stresses, and strains of metals under laser impacts. Another use of the laser has been in the puncture of armor plates to serve as a direction for the laser-guided bomb to be directed into this puncture area.

Weaponry

Laser defensive weaponry may include the detection of satellites and the possible interception by intense laser beams. Laser reflection shields, protective glasses, and clothing may also serve as some measure of protection against laser attack. One technique suggested is a small bicycle with a laser detection apparatus. Whether a program for laser civilian protection bombing and defense will ever have to be developed is unknown. If X-ray lasers are introduced into the military for weaponry, a civilian defense program may be more complex than for nuclear bomb defense. It is of interest now that the Soviet scientists are pushing for a joint Soviet-American effort to develop this X-ray laser, now called a graser (gamma ray laser). George C. Baldwin[11] of Rensselaer Polytechnic Institute is hopefully seeking American support for this cooperative work.

Laser offensive weaponry is the use of high output laser systems in more or less flexible setups to be serviceable in the field. Another type of weaponry is a single handgun used as antipersonnel or to explode ammunition dumps. Dr. John Foster, Jr., Director of Defense Research and Engineering, has recently indicated that the laser is uneconomical as a weapon. This refers to massive bombing types as well as to missile destruction. Proximity fuses triggered by the laser are said to have been developed by the Soviet Union.

The laser-induced thermonuclear fusion bomb is a type of H bomb for the have-nots, those who do not have atomic materials. The laser triggers thermonuclear fusion. This bomb is smaller, more economical, and cleaner than the present H-bombs; detonation of atomic bombs is needed for triggering these. The laser-triggered hydrogen bomb can be used as a bomb or contained in the warhead of an ICBM, Intercontinental Ballistic Missile. The studies for this bomb have been under way for more than 5 years. In the fiscal year of 1974, the Atomic Energy Commission received $34 million for this project.

FIGURE 11-8. G.T.E. Sylvania's infrared laser trackin system in a self-propelled van. (Xerox.)

Miscellaneous

The possibilities of laser field surgery with miniaturized laser scalpels will be discussed in the chapter on Medicine.

Some other uses of the laser in the military relate to the development of codes and scrambling of codes with laser fiberoptic devices. The laser may also be used as a readout device in intelligence work. One technique involves the use of the laser scanning of reconnaissance planes, usually drones; the pictures taken by the laser are reconstructed by a second laser beam and are consequently developed. Submarine detection by acoustical holography led to the developments of acoustical holography for cancer detection. Detection and transmission of conversation at the enemy headquarters are possible now with lasers.

It should not be forgotten that explosions from compact samples of liquid or solid fuels may produce explosive lasers such as the CO_2 at 1,060 nm. New lasers are constantly under investigation by the armed forces. A recent example is the work by the Air Force on new metal vapor lasers, operative at room temperature. Equilibrium vapors of various metal carbonyls transfer electrical discharge in an optical cavity.

As a corollary, applications of the lasers in the military now provide and will continue to provide applications for the good-of-man, a valuable spin-off of such work. The laser weaponry follows the ancient rule of warfare: When a new sword is introduced, a new protective shield soon follows.

REFERENCES

1. *Ferranti Avionics,* Ferranti Information Bulletin.
2. **Pincus, W.,** Tooling up for the future, *New Republic,* October 28, 1972.
3. **Compton, R. D.,** The source or where would EO be without DOD? *Electro-Optical Syst. Des.,* July 1972.
4. **Albright, N.,** Laser weapons, *Pop. Sci.,* March 1972.
5. **Ripley, A.,** Laser defense weapon of the future, *The Courier & Times,* October 17, 1971.
6. Pentagon's light ray, *U.S. News World Rp.,* October 18, 1971.
7. Death ray, *Time,* September 4, 1972.
8. **Barkan, R.,** The laser goes into battle, *New Sci.,* July 13, 1972.
9. Laser bomb work, *Cincinnati Enquirer,* January 31, 1973.
10. **Brown, R.,** Optical communications two steps nearer, *New Sci.,* 59, 19, 1973.
11. The graser, *Wall Street J.,* July 23, 1973.
12. Lightweight Range Finder Ecom, United States Army Electronics Command #2871430, Information Office, September 20, 1971.
13. **Brumfield, R.,** Viet War "smart bombs" German idea in W. W. 1, *Cincinnati Enquirer,* November 19, 1972.

Chapter 12

THE LASER IN THE LAW AND ORDER PROGRAM

Fortunately, the book *Preserve and Protect* by Alan Drury has not been as popular as his previous book *Advise and Consent.* Therefore, most of the public is unfamiliar with the assassination of the president by a laser gun in the former book. The concept of the laser as a death ray has been present since the early days of science fiction. Research in laser weaponry continues to support the idea. For those in the law and order program, the laser has a definite role and a deep concern for the future.

Lasers

Many different types of lasers may be used in the law and order program, varying from the low output He-Ne to the UV laser and even the X-ray laser. X-ray lasers would study the innermost secrets of all matter and would even furnish three-dimensional X-ray pictures of the structure of matter. They may be able to show contents of bullets and projectiles better than conventional X-ray or even neutron (N) rays.

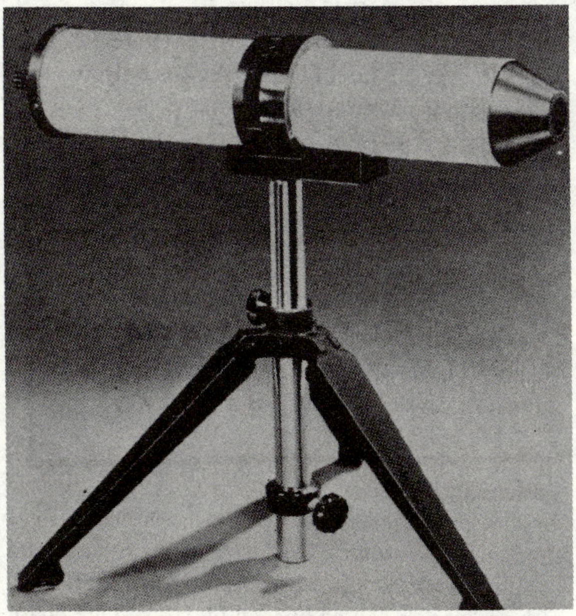

FIGURE 12-1. Miniature 2 lb 11 x 2 in. YAG laser systems, 10 pulses/sec 2 mJ/pulse peak output 14 W for use in an intrusion alarm system. (Korad.)

The Military and Industry

Some of the applications of the laser in the military relate directly or indirectly to these applications for law and order. The important applications in the military have been listed as weaponry, ranging and tracking for aircraft and missiles, computer technology, air turbulence, fog dispersion, and communications. Some of the applications in industry may also relate to the law and order program, including air pollution surveillance, communications, computer technology, metalworking, and plant surveillance and analyses. Devices have already been developed but are still in the research phase for laser radars on cars; one such device developed by the Ford Motor Co. detects any obstacle when the car is in reverse. Recent developments in holography have included the projection in the area of colored three-dimensional signs to show warnings, directions on expressways, etc. Laser holography also shows stress patterns of projectiles and metal weakness. Surface markings on bullet and projectiles can be detected by surface light scattering effects and by an apparatus such as the Zenith Profilometer®.

Forensic Laboratory Analysis

The laser has an important use in the laboratory division of the law and order program for microanalyses. Minute bits of material can be examined without the total destruction of this tiny sample. For the first time, it is possible to do a tiny spot analysis by microemission laser spectroscopy. For example, it would be possible to take different areas on a single strand of hair and do arsenic determinations at different levels on this hair shaft without complete destruction of the entire hair sample. With knowledge of the hair growth, it would be possible to determine the time of ingestion of arsenic. Bits of clothing can be examined for tiny impacts. This technique has already been used for the detection of fakery in art and for analysis of the fragile ancient glasses without breakage. Dry specimens are used. The technique is rapid and automatic with the computer. Materials could be examined and the impact spot of the laser would be so small that the analysis could not be detected. Recently, laser

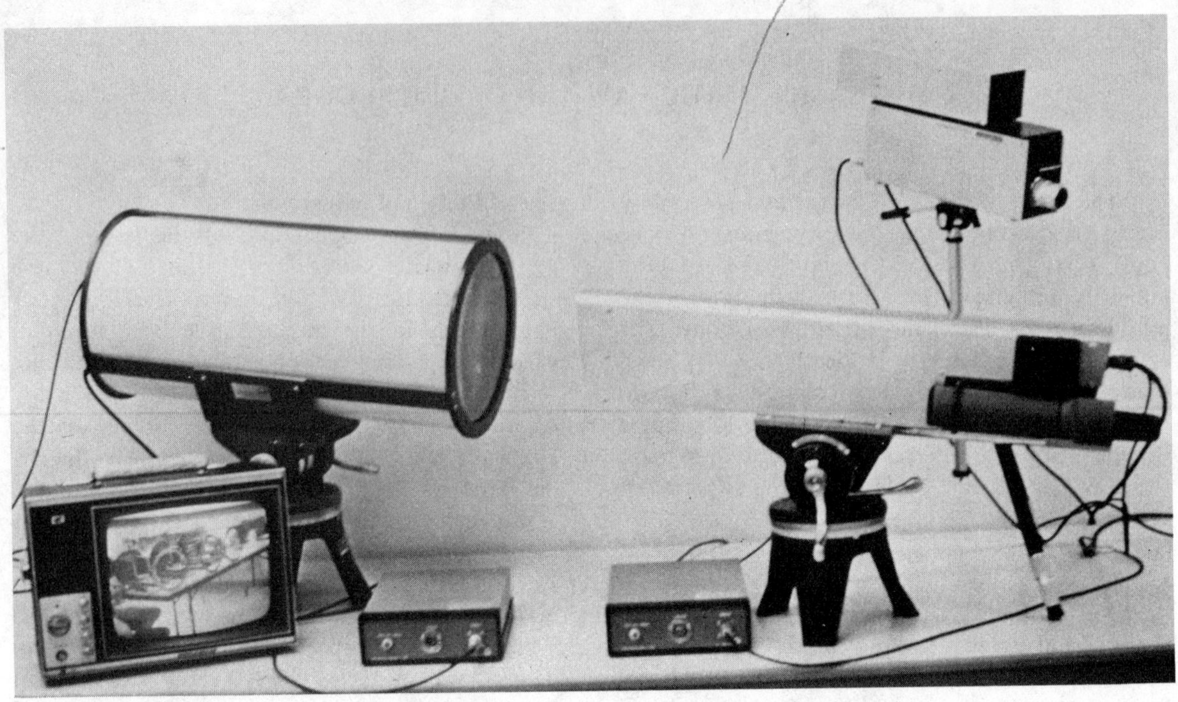

FIGURE 12-2. Laser communication system for transmitting video and modulated laser beam; video signal fed into transmitter (left), modulated laser beam directed to receiver head and receiver cabinet (right) connected to a television monitor. (Laser Communications, Inc., Cleveland.)

microemission spectroscopy has been done for analysis of many materials in human cancer, teeth, bone, and skin. Persons unfamiliar with laser technology have suggested infrared laser systems for mob and riot control. These systems are more hazardous to the eyes than tear gas. Recent experiments in England indicate that ultrasonics and infrared light, the photic driver, may possibly be the so-called nonviolent weapon.[6]

Surveillance

The laser is important in security systems, as indicated for the military. Invisible infrared lasers may be used for surveillance and for burglar alarms without obvious fixtures like the "electric eye" installations. Unlike ordinary infrared systems, laser infrared will give more rapid scanning and may be transmitted for long distances with a very narrow beam.

Recent experiments with a new laser television communications system in Cleveland by University Circle Police Department Officials are of interest and importance. The system developed by Laser Communications, Inc.[2,3] includes a modulated laser beam operating in all kinds of weather connected to special types of cameras. The laser transmitter beams the television signals back to the monitor at police headquarters almost a mile away. The police officer on duty controls the position and field of vision and can zoom if necessary, without repeaters, LCI estimates, a range of 4 mi. Where a direct line of sight is not possible, triangulation of the reflected beam may be done off a mirror or a series of mirrors. The Police Department reports in one 3-month period a 46% decrease in crime in that area.

Another type of surveillance is the use of the laser for "window tapping." This technique uses the laser to pick up reflectance from vibrations of glass windows and be transformed into voice signals. The laser then would be used to pick up conversation in a distant closed room.

Communications

New developments have been made in the field of communications. These have been listed in the chapter on Communication. Extensive new television channels will be opened by lasers. One phase of the video cassette explosion that is to occur soon will be produced with laser. It is possible this can be used at stations as well as in cars. New

FIGURE 12-3. Laser surveillance application by transmission of T.V. pictures to police headquarters about a mile away. (Case Western Reserve University Campus and University Circle area – Laser Communications, Inc., Cleveland.)

lines of communications will make data available to law enforcement officers very rapidly.

Finger- and Footprints

Fingerprint identifications with computer storage and retrieval with more rapid readouts by lasers have been done.[2] Lasers have been also proposed for three-dimensional imagery to develop a classification system of fingerprinting. A credit card identification technique has been developed recently with the credit card user's fingerprint compared to the holographically coded version of the owner's fingerprint on the credit card.[1,4] Fourier transform hologram is made from the owner's finger rubbed in castor oil or soap. This is incorporated on the credit card. The customer uses an oily finger on the plate; the beam passes through the print and through the hologram of the identity card to a light detector and logic circuit system.

An outline of footprints has been done by an experimental laser camera in England. This was done by EMI Electronics Ltd. for the Police Scientific Development Group at the United Kingdom's Home Office.[5] An interference pattern is made by the laser. On film another hologram is made at intervals of 5 min and superimposed on the same piece of filter as an accidental double exposure in holography. Then this double hologram is illuminated with a laser (He-Ne?). The interference bands in the picture represent movements of 0.00001 in. or 300 nm. in 5 min. This technique is now being tried out for examination of areas of crime to detect where more evidence may be.

Because of its absorption in color, the laser has been used as an erasure of dark color. This has been patented by one of the pioneers in laser research, Arthur Schawlow of Stanford University. The device can be attached to a typewriter and erase letters. With the laser, erasure of signatures can be done, often with great difficulty for detection if the signature is black on a white background. However, with the signatures on multicolored papers and multicolored checks, the absorption of the paper by laser as well as the signature often makes the erasure obvious.

Weaponry

As reported for the military, various types of laser guns have been proposed by the military for personnel and for area offensive weaponry. Such

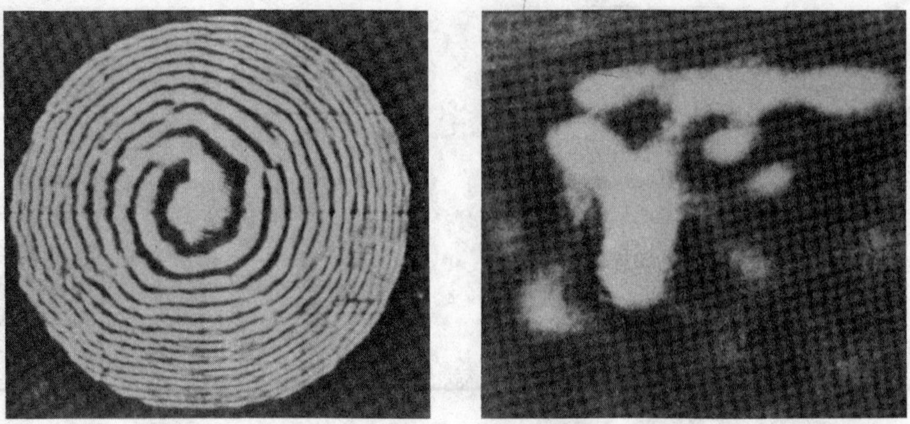

FIGURE 12-4. Plasigram of weapon detector (toy gun) with He-Ne laser illuminating plasigram (R). (University of Pennsylvania, Moore School of Electrical Engineering; *Laser*

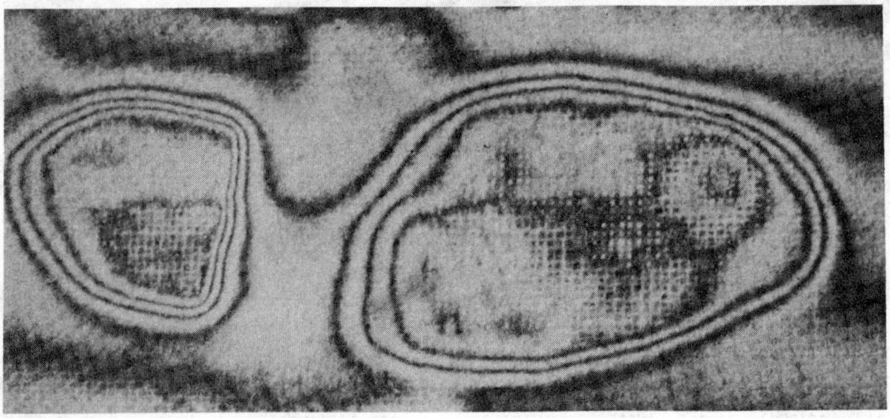

FIGURE 12-5. Footprint on a carpet visible only by laser holographic interference pattern. (EMI Electronics, Ltd.; *Holosphere.*)

guns may also be used to ignite explosives at a distance by the impacts of invisible laser rays. Such systems can be used also, unfortunately, for committing arson. At present, these devices are large and cumbersome and demand large power supplies. However, they may be supplanted in the future by what are called semiconductor lasers. These would be small and portable, easily concealed, and operate at room temperature. Unfortunately, gun magazines are advertising patterns for homemade laser guns if one could afford the ruby rods and make it up; again the eye hazards would be significant.

There has been some interest at the racetrack about the use of invisible infrared battery powered lasers whose impacts will be used illegally on horses to cause pain which would make a trotter break stride, which may even blind the horse, or for illegal communications. Security personnel on tracks should be aware of these possibilities. Actually, lasers could be used for legitimate purposes about the track, such as surveillance, security measurements, accurate measurements of time and distance, telecommunication, etc.

The high output CO_2 laser is a powerful destructive force and invisible because its frequency is far into the infrared. This vast power was exhibited by Goldfinger some years ago and now it has become reality. Presently, these installations, as indicated, are huge, cumbersome, and most of the power supply has to be in a separate installation from the laser itself. Carbon dioxide lasers are now cutting steel, clothes, and plastics; high output systems are used for cutting metals and even hard stones.

Requests have been made for consideration of the use of lasers in mob control, chiefly because of the invisible character of some lasers. However, the blinding of people in mobs is an obvious and considerable hazard.

Security Cards and Checks

A new type of security system has been introduced.[1,4] Photographic chips are embedded on either I.D. cards or tags. To the naked eye, these bits of film appear distorted. In the reader terminal, a low output He-Ne laser unscrambles this and a true image appears on the terminal screen. Similarly an unclear signature unable to be copied would be visible on a credit card and unscrambled only by the laser. An additional slip on these cards would be an optical random code from a hologram. Terminal must be used for reading this random code. Another type of I.D. card uses a laser-made hologram. The Hololock or Holocard of RCA®,[4] an argon laser, records the hologram on plexiglass blanks. The holder inserts the card in a box where the number is read by an ordinary light at the same time a verification number is punched on a keyboard. Then the machine opens a door or gives a signal. Another type of security check or actually for rapid check-out, is the laser mark on packages at the supermarket. A low output He-Ne scan automatically records computer stores and presents readout immediately to the customer. Not only is there more rapid service than manual check-out (3 to 1), but more control over cashier's error. Similar check-out techniques may be used in warehouse areas, etc.

Miscellaneous

There already have been a few instances of annoyance of neighbors with He-Ne laser beam. The eye hazards of such play are obvious. Nude 3D holograms are used also as additional obscene material.

It is obvious that knowledge and interest would do away with a lot of the fear and hysteria associated with the scare headlines. Knowledge and interest will enable one to keep up with the progress of research and developments of laser systems to see how they can fit in the order of things. The laser has a great future in the so-called law and order program not only to help man and the law forces but also as a potential source of danger to man. The authorities' knowledge and adequate lines of communication, often made more effective by laser, can again do much for the good and security of man.

REFERENCES

1. Credit card frauds, *New Sci.,* December 21, 1972, 705.
2. **Curry, A. S.,** Recent developments in forensic science, *Nature,* 235, 369, 1972.
3. **La Fleur, D. M.,** Police Communication, Laser Communications, Inc., 11000 Cedar Ave., Cleveland, Ohio.
4. **Greenway, D. L. and Russell, J. P.,** Hololock or Holocard, RCA, *New York Times,* February 27, 1972.
5. Optronics International Laser I.D., Security System, 7 Stuart Road, Chelmesford, Mass.
6. Anti-crowd weapons, *New Sci.,* 726, 29, 1973.

Chapter 13

LASER SAFETY

Knowledge and continued interest in laser safety are necessary to develop applications of lasers. Safety experts are well aware that it takes time and continued effort to educate people on the idea and practice of safety. This is especially true in laser technology. It is not uncommon to go into a laser research laboratory and find high output lasers flashing and the protective glasses remaining securely on the wall. Recently, FDA's Bureau of Radiological Health, together with State Health Agencies, surveyed 288 lasers in Colorado, Florida, Illinois, Montana, Oklahoma, Pennsylvania, and Washington and found serious defects in laser safety in high school and science classes. Bureau Director John C. Villforth[21] said that although the FDA is unaware of any injuries from improper use of lasers in schools, such products do pose a potential hazard that can be reduced by increased attention to safety procedures. All this is added material for the Bureau of Radiological Health's development of a new and rigid safety performance standard for low output lasers that are used not only in science classes but also in the construction industry and for many applications in measurement. Fortunately, the individual in the research laboratory who scorns the rules and regulations of the establishment and the careless youth of the science classes are not common representatives of the vast numbers who operate and who are exposed to lasers.

Laser safety programs are not static; they must be developed for each new laser or significant changes in parameters of the laser output. With increasing use of lasers, there is now more concern about exposure of the general public to laser beams.

The main phases of the laser safety program are first, the analysis of the laser itself, its characteristics, and transmission, next, investigations of its hazards for man, then the development of the laser area and the training, education, and protection of personnel, both those who operate and those who are exposed. It should not be forgotten that there are many lasers extending from the ultraviolet through the visible and far into the infrared. Perhaps in the very near future there may be even X-ray lasers.

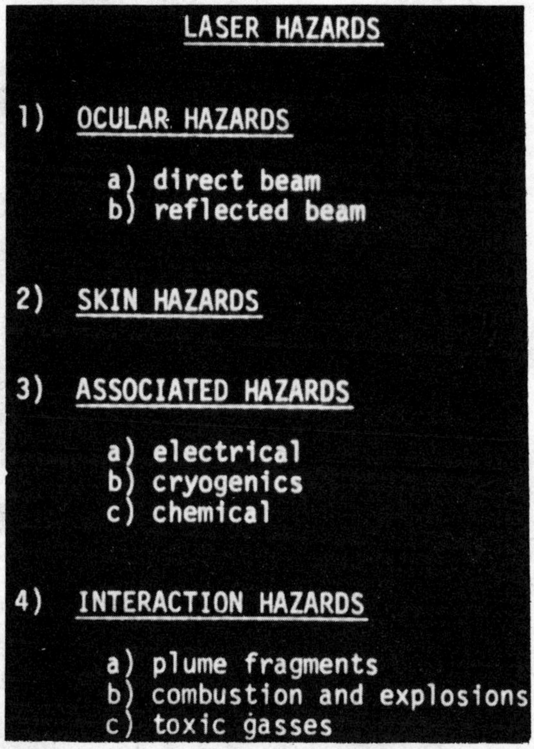

FIGURE 13-1. Laser hazards. (Laser Laboratory Safety Course.)

There are many techniques for transmitting the laser beam from its fixed position on the optical bench to its flexible use for applications for near or far use. Some of these aids include lenses, mirrors, prisms, beam splitters, fibers, etc. Some of these are part of open systems that may present hazards; some are used in closed systems, which are for the most part hazard-free.

The hazards relate to the direct and reflected beams of these various lasers that can affect eyes, exposed skin, and laser-produced air pollution related to the laser target. In addition to these hazards related to the laser, there are other nonlaser dangers such as electrical hazards, cryogenic hazards when these cryogenic materials are used, and occasionally fire hazards.

Area Control

Unfortunately, a small crowded corner of the laboratory is the only area provided when laser

ELEMENTS OF LASER SAFETY

1) **Establish the magnitude of the various hazards**
2) **Development of appropriate control measures**
3) **Establish a Medical Surveillance Program**
4) **Establish Control Program Management**

FIGURE 13-2. Elements of laser safety. (Laser Laboratory Safety Course.)

CRITERIA FOR AREA CONTROL

1) BASIC PERSONNEL PROTECTION FROM LASERS
2) ISOLATION FROM ELECTRICAL HAZARDS
3) INTERLOCKS AT ENTRANCES
4) ENCLOSURES AND BEAM TRAPS
5) INDIRECT SAFE-VIEWING SYSTEMS
6) VENTILATION SYSTEMS
7) NON-REFLECTANT SURFACES
8) AUDIBLE AND VISIBLE WARNING SYSTEMS
9) CONTROL OF CRYOGENICS AND CHEMICALS
10) LOW TRAFFIC LOCATION
11) GOOD HOUSEKEEPING

FIGURE 13-3. Criteria for area control. (Laser Laboratory Safety Course.)

FIGURE 13-4. Entrance laser laboratory.

research begins. Except for laser microscopy, and even here some eye protection is necessary, a separate area should be made available for the laser instrument even if only shields and backstops are used. When high output systems are used, a separate laser area must be constructed. Similar areas are needed for industry except when the entire laser system is enclosed, as in computers. For patient treatment installations, except for laser ophthalmology, separate laser operating facilities must be provided, both for outpatients and inpatients.

The requirements for area control have been listed repeatedly in many publications. The area control program is described in detail by Goldman and Rockwell.[13,14] These measures include, in brief, adequate space, fail-safe door interlocks to prevent entrance when either pulsed or CW lasers are in operation, and proper ventilation, especially when air contamination may result. The facilities include a diffuse reflecting wall, separate space for remote firing and monitors, a protective window for viewing treatment area, or closed-circuit T.V. for observing target area. Adequate and safe storage should be available for liquid nitrogen. As Rockwell and Epstein emphasize, "all high-voltage cables should be appropriately insulated and fastened to minimize effects of mechanical shock especially during pulsed operation."[9,27]

Personnel Control

Before there is any consideration of personnel control, it is well to list some of the concern and controversy regarding permissible exposure levels for laser radiation.

First there is the Committee-Z-136 for the Standard For The Safe Use of Lasers established by the American National Standards Institute (ANSI). The following list includes the personnel of this committee. This is public information. The make-up of this committee shows the wide range of expertise for the preparation of an important standard for a new technology.

George M. Wilkening, Chairman | **Sidney S. Charschan, Secretary**

Organization represented	Representative
Aerospace Industries Association	Robert Moore John Masek (alt.)
American Conference of Governmental Industrial Hygienists	Charles H. Powell William Kelley (alt.)
American Federation of Labor and Congress of Industrial Organizations	Edward J. Legan
American Industrial Hygiene Assoc.	Charles McHenry Isaac Matelsky (alt.)
American Institute of Chemical Engineers	Mathew M. Braidech
American Insurance Association	J. Philip Berggren Samuel A. Kaplan (alt.)
American Iron and Steel Institute	Joseph W. Miller
American Mutual Insurance Alliance	Richard I. Bleier
American Optometric Association	Paul W. Lappin
American Society for Testing and Materials	Hayes A. Lee J. A. Dwyer (alt.)
American Society of Mechanical Engineers	Erwin G. Loewen
Association of American Railroads	John L. McNabb
Association of State and Territorial Health Officers	Thomas Cashman
Automobile Manufacturers Association	Charles C. Wang, A. C. Doty, Jr. (alt.)
The Children's Hospital Research Foundation	R. James Rockwell, Jr.
Electronic Industries Association	Robert Mortensen Douglas Worth (alt.) Allen M. Wilson (alt.)
Health Physics Society	David H. Sliney
Illuminating Engineering Society	John E. Kaufman Howard Haynes (alt.)
Industrial Medical Association	Harry E. Tebrock, Dr.
Industrial Safety Equipment Association, Inc.	George Schauweker Lawrence A. Medici (alt.)
Institute of Electrical and Electronics Engineers	Samuel Fine
Instrument Society of America	Grover C. Wrenn
Institute for Advancement of Engineering, Inc.	Malcolm L. Stitch Ralph A. Lamm (alt.)
Laser Institute of America	James F. Smith Mason Cox (alt.) Sam Fine (alt.)
Medical Center of the University of Cincinnati	Leon Goldman
Medical College of Virginia	William T. Ham, Jr.
National Aeronautics and Space Administration	Rudolph M. Marrazzo Robert E. Alexander (alt.)
National Bureau of Standards	Nicholas Winogradoff
National Association of Photographic Manufacturers, Inc.	Thomas S. Ely C. E. Hilton (alt.)

Organization represented	Representative
National Constructors Association	James R. Milor J. A. Barton, Jr. (alt.)
National Electrical Manufacturing Association	H. W. Speicher D. C. Fleckenstein (alt.)
National Safety Council	John A. Janous R. C. Benashski (alt.) Juliam Olishifski (alt.) William M. Smith (alt.)
National Society for the Prevention of Blindness, Inc.	Charles J. Campbell James E. O'Neil (alt.)
Optical Society of America	Jay M. Enoch Bernard Rosett (alt.)
Polytechnic Institute of Brooklyn	Saul W. Rosenthal Glenn Heimer (alt.)
Roswell Park Memorial Institute	Edmund Klein
Society of Manufacturing Engineers	Robert Willette Glen Kett (alt.)
The Society of the Plastics Industry, Inc.	A. E. Sherr J. P. Carrol (alt.)
Underwriters Laboratories, Inc.	L. S. Homa
U. S. Army Medical Research Laboratory	Wordie H. Parr
U. S. Department of the Air Force	Donald D. Higgins Lawrence T. Odland (alt.)
U. S. Department of the Air Force, Institute of Technology	Robert W. Ebbers
U. S. Department of the Army	David H. Sliney William A. Palmisano (alt.)
U. S. Department of Health, Education, and Welfare	Richard Peterson Robert L. Elder (alt.)
U. S. Department of the Interior, Bureau of Mines	Aurel Goodwin Roy G. Stott (alt.)
U. S. Department of Labor, Bureau of Labor Standards	Floyd A. Van Atta John P. O'Neill (alt.)
U. S. Department of the Navy, Bureau of Medicine and Surgery	D. K. Heffner
NAVEL Research Laboratory	Raymond G. Nebelung Henry C. Kenedy, Jr. (alt.)
WEMA	Herb M. Dwight Eben J. Tisdal (alt.)
Virginia Commonwealth University	Walter J. Geeraets
(Individual)	John A. Carpenter
(Individual)	Leonard Gardnar
(Individual)	C. Harry Knowles
(Individual)	Grant C. Riggle
(Individual)	Howard Rosenthal
(Individual)	Hermas Swope
(Individual)	Myron L. Wolbarsht
(Individual)	H. Christian Zweng
(Alternate to Zweng)	Richard C. Honey
(Alternate to Zweng)	Art Vassiliadis

The various phases of this committee's work for the past years include, of course, the most important aspects of laser safety – the eye, then the skin, measurements, evaluation of hazards, and controls. The detailed recommendations of this committee, with permission, are reproduced in the appendix of this book.

Next is the Technical Electronic Product Radiation Safety Standards Committee of the Department of Health, Education, and Welfare.[2] This committee also will present a Proposed Standard For The Control of Radiation From Lasers. The Bureau of Radiological Health Laser Section has expanded its facilities at Rockville, Md. under the able direction of Richard W. Peterson, who is Chief of the Laser Section. Also included in the Department of Health, Education, and Welfare is the Physical Agents Branch of the National Institute of Occupational Safety and Health. Wordie H. Parr[23] is the Acting Chief of this branch.

Then there are the Safety Programs of the Armed Services with their own standards. Dunsky et al.[7] have recently revised the exposure levels for lasers for the Air Force. Sliney has been active in this program for the Army.[30,32]

The laser safety guide has been adopted recently by the British Standards Institution. An effort will be made to compare these various standards and to evaluate their effects on the use and abuse of lasers including their effect on the development of the laser industry in various areas.

For the World Health Organization[38] there is also the report of a working group convened by the Regional Office for Europe of the World Health Organization, November 15–17, 1971 at The Hague. Their recommendation was that collaborative study should be made by all concerned in the hope of establishing some international standard at least for the effect on human eyes. Whether this will be possible in the near or even distant future is not known. At least in the U.S. government and industry have agreed to have a panel of experts attempt to reconcile what is evidently "wide differences in proposed hazardous radiation standards for lasers."

Both the ANSI group and the Bureau of Radiological Health have attempted to classify lasers to evaluate their hazards. Four classes of laser hazards are proposed. The following data are taken from the Bureau of Radiological Health Laser Section Report:[2-4]

Class I Laser Product

A class I laser product is not expected to be a human hazard on the basis of current knowledge.

Class II Laser Product

A class II laser product may be unsafe for chronic viewing situations, but presents a low probability of eye injury for a single, short, accidental exposure.

Class III Laser Product

A class III laser product presents a high probability of eye injury even in the case of a single short exposure. (Note: Since all UV and IR laser emissions that are hazardous to the eye are also considered to be hazardous to unprotected skin, there is no class III laser product for these wavelengths.)

Class IV Laser Product

A class IV laser product presents an eye hazard and a hazard to unprotected skin.

The four classes of laser products are separated by laser radiation emission limits for visible laser products as seen in Appendix A of the Bureau of Radiological Health. The emission levels are presented as a function of emission duration. Each class is further modified for other wavelengths covered in the standard, as shown in Appendix B of the Bureau of Radiological Health, for the designated emission durations.

ANSI has an additional class, Class V – Enclosed Lasers and Laser Systems. Because of the construction, or what is considered as engineering controls, the levels emitted are considered to be less than the exposure considered for direct irradiation of the eye.

In addition to these specific classifications, the Bureau of Radiological Health has proposed[2] "a limit for certain collateral radiations associated with laser emission. On the basis that none of this collateral radiation is necessary outside of the laser product, emission of collateral radiations is limited to levels that are not expected to be a hazard on the basis of present knowledge. The collateral radiation limits for the 200-13,000 nm wavelength range are the same as those which define the upper limit of a class I laser product. The limit for the $1.0 \times 10^{-1} - 1.0 \times 10^{-5}$ nm wavelength range is taken from the Performance Standard for Television Receivers."

It should not be forgotten that hazards, other than irradiation, may be present. These include electrical grounding, toxic chemicals in chemical lasers, and cryogenics.

The final ANSI report (See final ANSI report in Appendix), lists the specific parameters of each class in detail.

For safety, obviously, it is also necessary to evaluate the actual risk of exposure in addition to the specific class of lasers. For example, high risk would be concerned with research and development, exposure to high output systems, laser weaponry, and in some instances, chronic exposure. Except for laser eye surgery, laser surgery is a high risk exposure. Low exposure may be associated with operation of totally closed systems without any pollution hazards, as with laser computer technology.

Now an attempt will be made to review the specific figures of permissible exposures to show the wide range of divergence among the various groups. The terms used are MPE (Maximum Permissible Exposure) and, for the Air Force, PEL (Permissible Exposure Levels).[7] Obviously, there is nothing sacred or holy about these terms. They serve only as guidelines suggested by current studies and are subject to change later, especially as the present inadequate knowledge of chronic exposure decreases. This factor will be discussed later. It is suggested that when it is "practicable," exposure levels should be maintained below the suggested permissible levels. However, for protection, legislative, and control measures, some numerical specifications should be available with due regard to their interpretation. The interest and concern of the "interpreter" should also be recorded.

The eye, of course, is the organ in man and animals that is most sensitive to the laser. Also, the inside of the eye, the retina has greater energy or power density than the surface of the cornea from the highly collimated beam of the laser. Dunsky, Fife, and Rickey[7] claimed that the size of the minimal spot in the retina is affected by:

1. Ocular chromatic aberration
2. Hot spots in the laser beam
3. Imperfection in the ocular media
4. Refractive errors in the ametropic eye

Appendix A. Laser Radiation Emission Limits for Classes of Laser Products Emitting in the Visible Spectral Region. (Total energy into a 6mm diameter circular aperture).

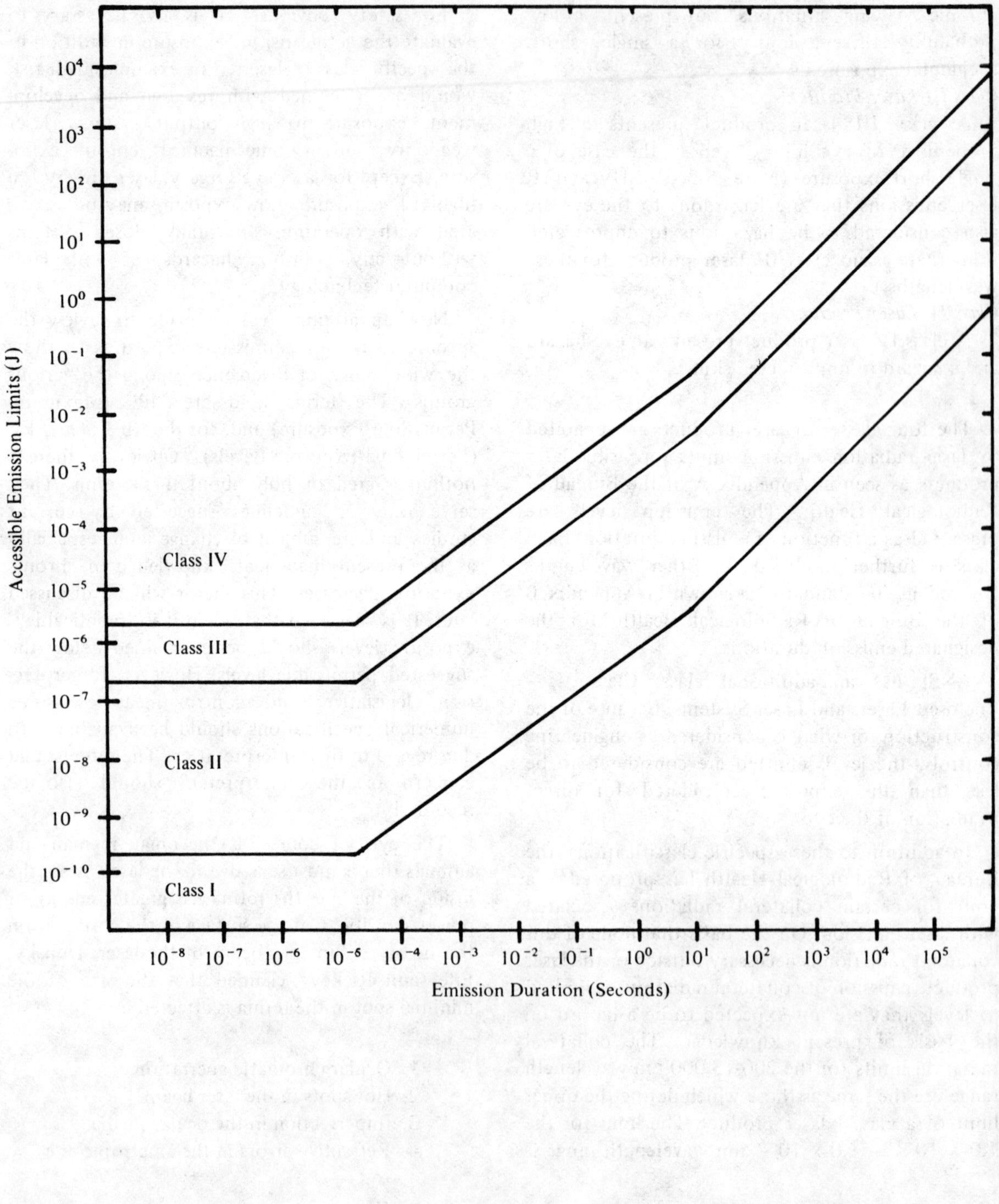

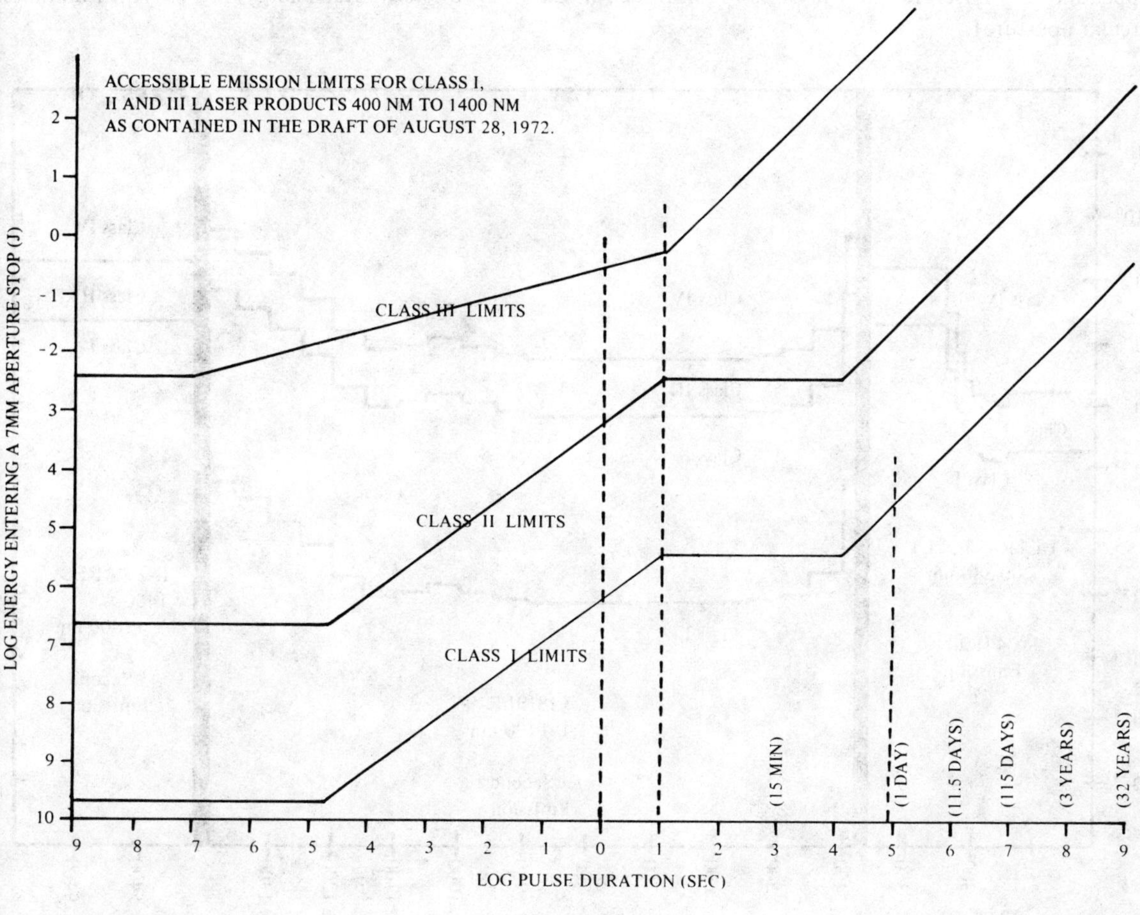
ACCESSIBLE EMISSION LIMITS FOR CLASS I,
II AND III LASER PRODUCTS 400 NM TO 1400 NM
AS CONTAINED IN THE DRAFT OF AUGUST 28, 1972.
CLASS III LIMITS
CLASS II LIMITS
CLASS I LIMITS
LOG ENERGY ENTERING A 7MM APERTURE STOP (J)
LOG PULSE DURATION (SEC)
(15 MIN)
(1 DAY)
(11.5 DAYS)
(115 DAYS)
(3 YEARS)
(32 YEARS)

Appendix B. Laser Radiation Emission Limits for Laser Products (Total energy into a 7mm diameter circular aperture)

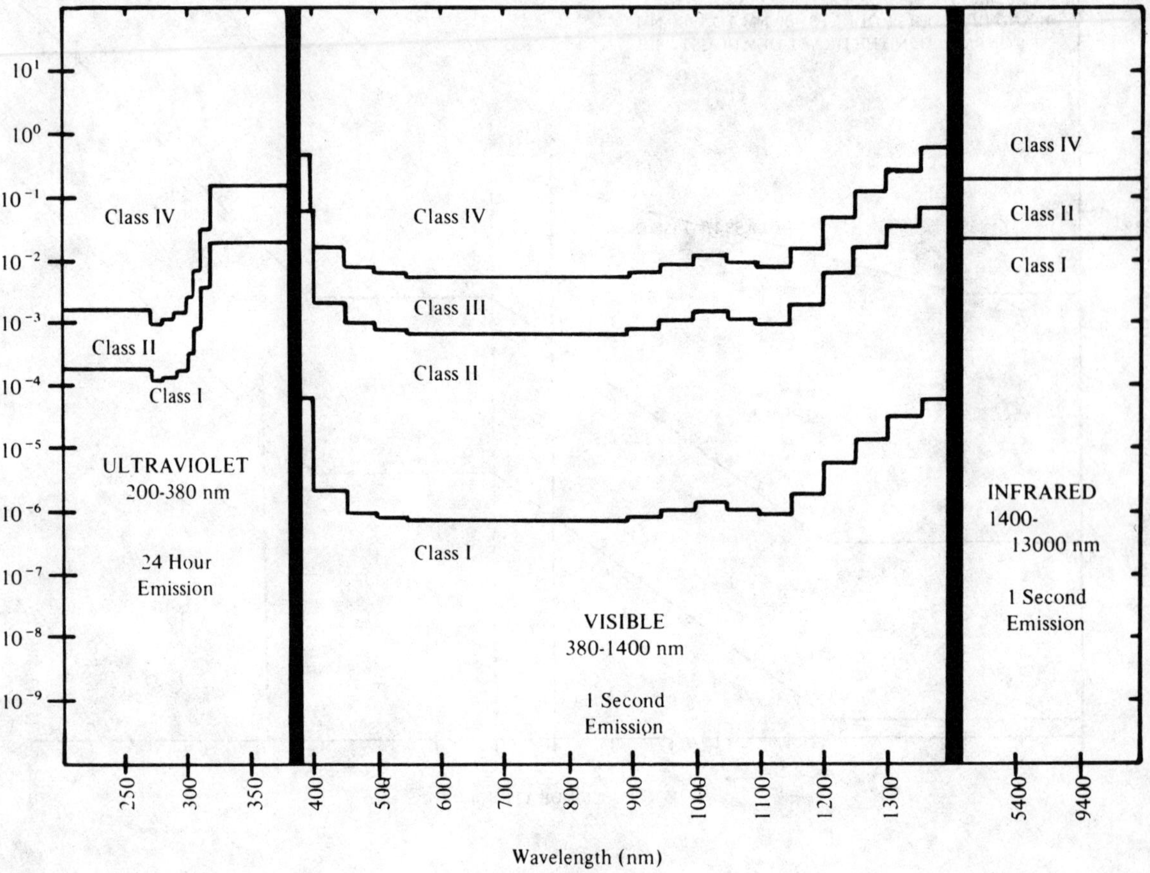

Appendix C. Comparison of Erythema Thresholds with Radiation Limits.

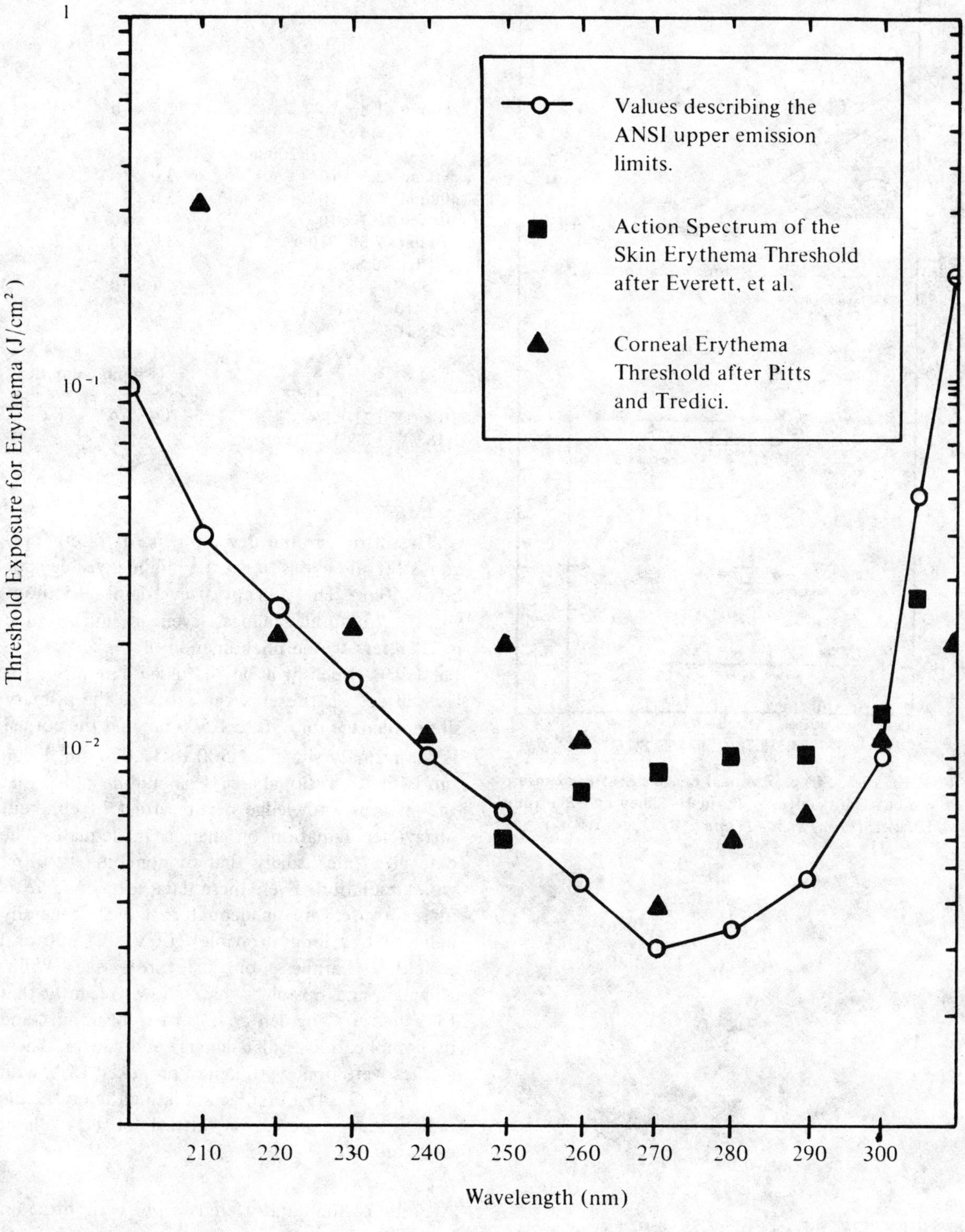

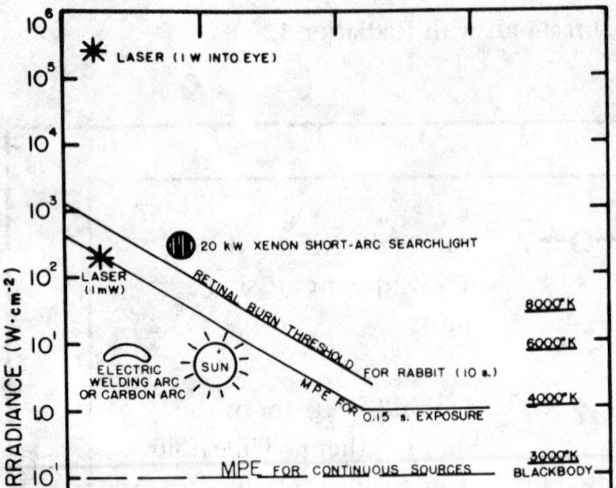

FIGURE 13-5. Typical retinal image size after exposure to various light sources – David H. Sliney and Benjamin C. Freasier. (From *Applied Optics.* With permission.)

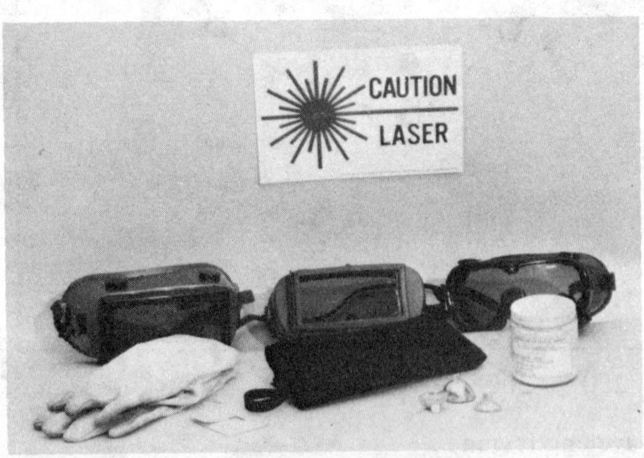

FIGURE 13-6. Laser safety devices including sign, eyecover for laser treatments about face, used in addition to protective glasses, eye shields for treatment of eyelids, protective gloves, and protective barrier cream.

Summary

A Comparison of Some Current Standards for Discussion – 1972

	ANSI	BRH
260 nm	1.8×10^{-3}	1.8×10^{-4}
	3.8×10^{-3} (200 nm)	
290 nm	3.8×10^{-3}	3.0×10^{-4}
300 nm	3.8×10^{-4}	3.0×10^{-4}
305 nm	19.2×10^{-4}	9.6×10^{-4}
310 nm	7.68×10^{-3}	3.9×10^{-4}
>312 nm	3.8×10^{-2}	2.0×10^{-2}
$<2.5 \times 10^{-5}$ S		2.1×10^{-6}
2.0×10^{-5} S-10S		
2.5×10^{-5}		
	>10S	$7.1 \times 10^{-3} + 3/4$ (+ = pulse duration)
10^{-9} 10^{-7}S	$3. \times 10^{-3}$	
10^{-7} 1S	$0.21 + 1/4$	$3.9 \times 10^{-5} + 3/4$
>1S	3.8×10^{-2} +	

The effort in the development of safety programs for the eye is to get the minimal visible spot 20 to 40µm. The concept of eye damage without visible structured change, even at microscopic level, refers to the phenomenon of evoked potential (EP), which will be discussed later. So, for laser effect on the eye, the retina is the primary site of damage for 390 to 1,500 nm and the cornea is the primary site for 1,500 to 12,000 nm; 1,500 nm is the transitional area. It should be added that the present knowledge of the effect of coherent ultraviolet radiation on man is inadequate. The data given are usually that of noncoherent ultraviolet radiation. Even these data, after long years of exposure, are inadequate. It was generally believed that long ultraviolet (LUV – > 320 nm) is either harmless or just protective. Willis, Kligman, and Epstein[37] have shown recently that LUV has an "augmentative" effect in sunburn and its complications, solar damage and cancer. These studies were done with xenon-produced LUV with Schott W G 345 Filter® eliminating the erythemic radiation of interest in ultraviolet (UV) laser exposure.

Some of the data offered to our committee on Skin Exposure and, as yet not approved or official from ANSI, can be correlated with eye exposure. For example:

Wavelength (μm)	Exposure time (sec)	Maximum permissible exposure (MPE)	Notes
0.200–0.302	$>10^{-2}$	3×10^{-3} J cm^{-2}	1 mm aperture;
0.303	$>10^{-2}$	4×10^{-3}	however in no case
0.304	$>10^{-2}$	6×10^{-3}	shall total ir-
0.305	$>10^{-2}$	1.0×10^{-2}	radiance, over all
0.306	$>10^{-2}$	1.6×10^{-2}	the wavelengths
0.307	$>10^{-2}$	2.5×10^{-2}	within the UV spec-
0.308	$>10^{-2}$	4.0×10^{-2}	tral region, be
0.309	$>10^{-2}$	6.3×10^{-2}	greater than 1 W/cm^2
0.310	$>10^{-2}$	1.0×10^{-1}	upon the cornea.
0.311	$>10^{-2}$	1.6×10^{-1}	
0.312	$>10^{-2}$	2.5×10^{-1}	
0.313	$>10^{-2}$	4.0×10^{-1}	
0.314	$>10^{-2}$	6.3×10^{-1}	
0.4–1.4	10^{-9}–10^{-7}	2×10^{-2} J/cm^{-2}	This is two times the
(Visible and	10^{-7}–10	1.1 $t^{1/4}$ J/cm^{-2}	value for eye exposures
near infrared)	710	0.2 W cm^{-2}	for the visible
1.4 –10^3	10^{-9}–10^{-7}	1×10^{-2} J cm^7	and near with 7 nm
(Far infrared)	10^{-7}–10	0.5 bt^{14} J cm^2	aperture approximately
	710	0.1 W/cm^2	same as the eye with 1
			nine aperture.

For the Air Force, Dunsky, Fife, and Rickey[7] gave the following data as of April 1972:

Wavelength Visible (400-700 nm)	Total energy/power entering the eye at the level of the cornea
Q switched (10–100 msec)	0.5 μJ
Long pulse (0.2–2 msec)	4.0 μJ
CW (2–10 msec)	2.0 mW
(100–500 msec)	1.0 mW
Neodymium 1.06 μm	
Q switched (10–100 msec)	2.5 μJ
Long pulse (0.2–2 msec)	2.0 μJ
CW (2–10 msec)	1.0 mW
(10–500 msec)	5 mW

From the Laser Microwave Division of the U.S. Army Environmental Hygiene Agency, Sliney and Freasier[31] have reviewed in detail the evaluation of optical radiation hazards.

They give the ANSI figures as of February 1972 and "indicate that the American Conference of Governmental Industrial Hygienists had proposed essentially the same exposure criteria for laser radiation."

According to *Laser Focus,*[20] the maximum permissible output of the British Standards Institution is about 2 mW. The Swiss are reported to urge a 0.1 mJ/cm for 644.3 nm.

The World Health Organization for 1972 lists the following:

General protection guides

At present, protection guides for Q-switched lasers vary between 10^{-5} to 10^{-8} J/cm^2 and for continuous irradiation between 5×10^{-3} to 10^{-6} W/cm^2.

Of greatest importance at present for protection against eye damage from lasers are the recommendations for their use, which are accepted by most countries. They include rules on installation and working areas; on the presence of adequate light in order to keep the pupil of the worker's eye small, on the avoidance of reflecting materials which might come into the beam unexpectedly; and on medical control of workers' eyes for unnoticed damage.

What does all this confusion mean as regards the application of the laser? Over 90% of the more than estimated 60,000 lasers now in use may be classified as low output. Will the adoption of rigid standards classify most all of the lasers now used as hazardous since it is assumed that the proposed guidelines of the government would be almost 1,000 times more strict than the ANSI standards? The laser industry is greatly concerned about standards that may interfere seriously with the use of lasers. Dwight,[8] of Spectra-Physics Inc., said that lasers less than 5 mW power output "expose subject to 1,000 times less radiation than that from a normal camera flashcube." Experts in photography should be able to compare the daily radiation exposure from sunlight to that of low output laser radiation. Laser experts have said that the hammer also is a hazardous instrument and have developed some fascinating reports on the hazards of the use of the hammer in the fashion of a laser safety report.

All in laser technology look forward to the Reconciliation Panel called by the Bureau of

RETINAL MAXIMUM PERMISSIBLE EXPOSURE LEVELS FOR LASER RADIATION FROM DIRECT ILLUMINATION OR SPECULAR REFLECTION

Type of laser	Q-Switched 1 ns – 1 μs pulse Energy per pulse J/cm^2	Long Pulsed 1 μs – 0.1 s pulse Energy per pulse J/cm^2	Continuous Wave (CW) Long-term exposure* Power density W/cm^2
Ruby (0.69 μm)	0.01	0.52	0.18
Neodymium (1.06 μm)	0.09	1.5	1.0
Helium-Neon (0.63 μm)	–	–	0.15
Argon (0.51 μm) (0.48 μm)	–	–	0.15

*For the CW mode only, the limit may be relaxed by two orders of magnitude for occasional (i.e., acciddental) exposure.

Maximum permissible levels at the cornea for laser radiation outside the visible light range

1. *Ultraviolet below a wavelength of 0.4 μm*
   The safe dose should not exceed 13 mJ/cm in one day, which is equivalent to 216 $\mu W/cm^2$ operating for 1 minute; the dose can be concentrated into a few seconds.
2. *Infrared above a wavelength of 1.44 μm (the CO_2 laser)*
   For a single pulse, maximum irradiation = 0.1 J/cm^2.
   For continuous exposure, average irradiance = 50 mW/cm^2.
3. *Maximum permissible levels on the skin, excluding the eye*
   All lasers including UV and IR.
   For a single pulse, maximum irradiation = 0.1 J/cm^2.
   For long-term exposure to CW lasers = 0.1 W/cm^2.

(From World Health Organization, 1972. With permission.)

Radiological Health Laser Section to attempt to satisfy all authorities (or none?). Compromises in terms of proper vivid labeling of laser instruments, learning and just teaching of laser safety, adequate personnel for surveillance for laser safety, and increased funds for basic research in laser biology and medicine will do much to offset polemics nurtured by lack of controlled experience and experiments, and by the use and abuse of data by extrapolation rather than by investigations. Apparently, the ANSI report will be generally accepted by all.

Chronic Exposure

Relatively little is known about the effect of chronic exposure to laser radiation. The chapter on Biology will present a detailed review of the data available on chronic exposure and the technique of test models to attempt to obtain data about this important phase of laser safety.

Because of the easy availability of the skin of man, chronic exposure investigations have been done on the skin at the Laser Laboratory. This can be done easily in a safe and also controlled fashion. These investigations will also be described later. Data on visible and infrared laser radiation are being made available. There is great need for studies of the acute and chronic effects of ultraviolet laser radiation.

Actually, repetitively pulsed laser impacts are not to be considered as chronic exposure. The exposures of such impulses for the skin, as regards permissible exposures, as a single pulse exposure, and the value of the average irradiation of the pulse train are to be considered in regard to exposure.

Medical Surveillance

Medical surveillance is an important aspect of a successful program on laser safety. Recent studies on new high output laser systems must be considered in current medical surveillance programs. A successful medical surveillance program means an effective program of laser application. Whereas the responsibility for the laser safety program is the laser safety officer, certain phases of the medical surveillance program may demand special expertise. The three basic aspects of medical surveillance are

1. The eye
2. The skin
3. Air pollution with possible respiratory system involvement

An ophthalmologist is required in the eye program on a parttime or consultation basis. Detailed preemployment eye exams and the check-ups and analysis of accidents with eye involvement should be the responsibility of an ophthalmologist, not an optometrist or technician. The eye surveillance program includes a special eye form chart and fundus photography. The frequency and detail of eye examinations depend on the laser exposure of the individual. A detailed eye examination is necessary upon completion of laser work.

For the skin examination, dermatologists are preferred. After instruction in laser safety and training and proper examination of the skin, other practitioners may be used in medical surveillance of the skin. Preemployment records should contain data about the reactivity of the skin and the changes of the skin especially on the exposed areas. Evidence of skin irritations, sensitivities, sun exposures, and other forms of radiation damage to the skin should be included in the record. Efforts should be made to protect the skin not only during acute exposures but also during chronic exposure since too little is known about chronic exposure of the skin to laser.

Air pollution may occur from accidents using chemical lasers, the toxic products and plume fragment from the target and of material processing. Where air pollution is a significant factor, detailed examination of the chest blood and urine must be done by trained physicians.

The Laser Safety Officer

The ancient cliche follows: If something does not work, read directions. This is one reason why laser instruments are to be labeled as hazardous since the instruction manual may not be read. In spite of the beautiful catalogues to the contrary, one must learn about lasers before one uses them, even for that which is called low output system. There must be always instruction in laser safety as well as in laser technology.

As the applications of the laser become much more extensive, there must be a laser safety officer, one responsible for laser safety, its theory, and practice. Practice means the development of area control, monitoring, responsibility for recommendations for medical surveillance and for accidents. The laser safety officer can be variously one assigned and trained, an industrial hygienist and trained, a radiation physicist and trained in nonionizing radiation, a physician, surgeon, ophthalmologist, or dermatologist. The emphasis is on training. This will be outlined later in the chapter, Training in Laser Technology. The laser safety officer must remember that programs for safety change as new developments occur.

Laser Safety Signs

Lasers and laser operating areas should be marked with bright-colored signs, as is done with ionizing radiation. A standard sign is soon to be adopted with the essential details. The word "danger" is preferred to "caution." The bright red burst lines and then the listing of the specific laser or lasers are also included.

Laser Eye Protection

Sliney and Freasier[31] stated that "laser eye protection is designed to have the greatest visual transmission along with an adequate optical density at the laser wavelengths." It is obvious then that there can be no single protective eyeglass for all lasers. The laser glasses either reflect the energy or absorb it through attenuating glass or plastic belters.

The criteria for protective eyewear are not a subject for controversy and have been given by ANSI in all the necessary phases. To develop effective protective glasses, the following factors should be considered:

1. Wavelength of laser output
2. Radiant exposure or irradiance
3. Maximum Permissible Exposure
4. Optical density of eyewear at laser output wavelength
5. Visible light transmission requirement

6. Radiant exposure or irradiance at which laser safety eyewear damage occurs

7. Need for prescription glasses

8. Comfort

9. Degradation of absorbing media

10. Strength of materials (resistance to shock)

The attenuation (O.D.) of laser protective eyewear at a specific wavelength shall be specified. Many lasers radiate at more than one wavelength; thus, eyewear designed to have an adequate O.D. for a particular wavelength could have a completely inadequate O.D. at another wavelength radiated by the same laser. This problem may become particularly serious with lasers that are tunable over broad frequency bands. Attenuation through the protective material shall be determined in all anticipated viewing angles and wavelengths.

For Visible Transmission

The adequate optical density (O.D.) at the laser wavelengths of interest shall not be sacrificed for improved visible transmission.

Identification of Eyewear

All laser protective eyewear shall be clearly labeled with optical density values at wavelengths for which protection is afforded. The glasses should be so marked that the label is readily visible.

Comfort and Fit

Protective eyewear shall provide a comfortable and snug fit so that laser radiation is satisfactorily attenuated before reaching the viewer's eyes. The snug fit must include the nose and the temple areas.

Inspection

Periodic inspections shall be made of protective eyewear to ensure the maintenance of satisfactory conditions. This shall include:

1. Inspection of the attenuator material for pitting, crazing, cracking, etc.

2. Inspection of the goggle frame for mechanical integrity and light leaks.

Responsibility of Manufacturer of Laser Safety Protective Eyewear

Manufacturers of laser safety protective eyewear shall provide the following information with each item:

1. Wavelength(s) and corresponding optical density where protection is afforded.

2. Maximum level(s) of irradiation for which protection is afforded.

Skin Protection

Protective clothing such as gloves may be used especially for the prevention of reflection from the impact of high output lasers from the target areas. Black soft leather gloves, according to photodetector measurements, afford more protection than similar gloves of white leather. In the operating room, the rubber gloves offer protection, but not for direct impact of the high output CO_2 or Nd-YAG lasers. Barrier creams containing titanium dioxide or zinc oxide also offer protection for the skin, especially for the skin adjacent to a target area. Flexible white cardboard is also an effective protectant. It is important to note that facial skin is constantly exposed about the protective glasses. This area should be considered in the evaluation of chronic exposure of the skin.

Through chronic actinic or solar exposure, the ears, nose, and lips (important areas for developing solar cancers) are also areas exposed to reflected laser energy. With the development of coherent UV radiation, these areas will have to be more adequately protected with effective UV protectants, protectants effective for solar erythema, but also for UV laser wavelength. There is no evidence today that chronic laser skin exposure is carcinogenic.

Prevention of Laser Air Pollution

Area control, ventilation, exhaust systems, and engineering controls of the laser operation are all factors concerned in the protection of the respiratory system. As indicated, environmental pollution must also be considered for those laser applications where exhaust systems protect the operator, but add to environmental pollution.

High Output Laser Systems

High output systems, at present, constitute about 10% of all lasers used. If the laser industry is not blocked, this will increase in the future. What are some of the current applications concerned with high output systems?

Materials processing, melting, heating, and

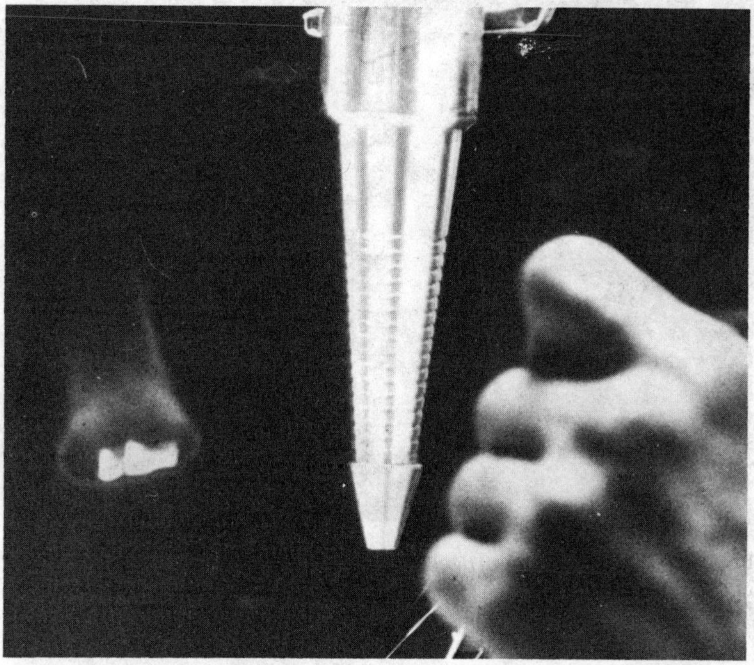

FIGURE 13-7. CO_2 laser surgery reflectance of high output laser from instrument burning cloth.

vaporization for plasma production, especially for laser-induced thermonuclear fusion, are a few of the applications for high output laser systems. In this connection, the safety program refers not only to the safety aspects of the laser itself, and this is important with high output systems, but also the safety aspects of the effects of the laser on the target materials.

Actual experiments in the author's Laser Laboratory with high output systems have included the ruby, normal mode to 20,000 J/cm^2, Q-switched to 20 J/cm^2, impulses to 4 psec. For actual work with CO_2 systems, up to 300 W have been used, for helium-neon, 75 mW, and for Nd-YAG, 200 W. Nd-YAG has been used also with Q-switched operations. In actual use today, the CO_2 lasers have included systems with outputs of kilowatts. As an example, the Bureau of Mines has recently published specifications for a 250-W output CO_2 laser TEM continuous wave 8 mm beam diameter with 1.7 mrad beam divergence. These are effective in cutting steel. The confusion about X-ray lasers has been reviewed. Planning for safety for this laser should be done now.

The hazards of laser material processing from the use of high output laser systems can be summarized briefly as affecting the eye and skin through the reflectance of laser energy from the material surface and from the vaporized plume or actually from reradiation emission induced by the gaseous plume. In addition, there may be airborne dispersion of products of these plume materials. In these instances the respiratory system may also be involved. Toxic gases in the reaction plasma or by recombination of air or from other gases in the material processing may affect the eye, the skin, and the respiratory system. In metalworking, the gases from the material may cause difficulties. Metal pollutants from current metalworking, as for example, with power densities from 250 kW/cm^2 to 50 mW/cm^2 usually include nickel, cobalt, and chromium. Curiously, very little data are available on the actual concentrations of heavy metals. In one series of experiments in the aircraft industry where often the composition of the alloys is a trade secret, with a 30-J ruby laser and with analyses done 2 ft from the impact site, the following was obtained:

Metal hastalloy	Ni (mg/m^3)	Cr (mg/m^3)
X–40	57.6	11.7
	40.2	21.9

(ACGIH standard = 1 mg/m^3 for both Ni and Cr)

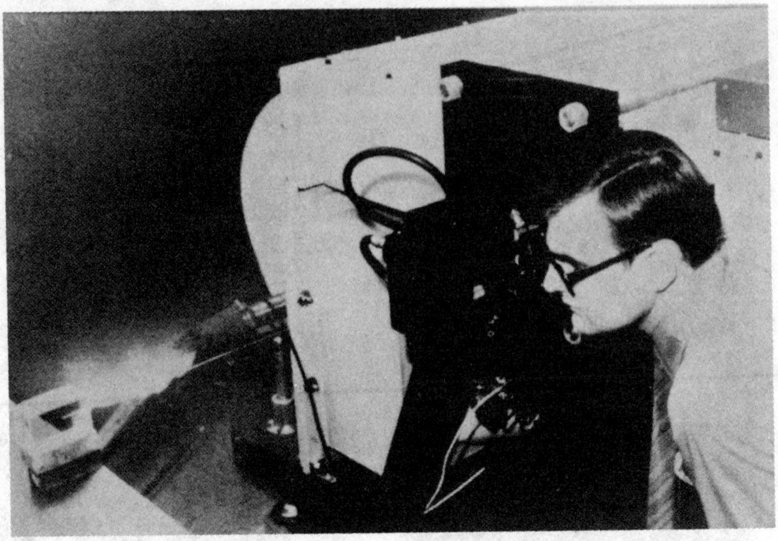

FIGURE 13-8. Gas transport laser G.T.E. Sylvania 500 W impact on plastic block. (From *Optical Spectra.* With permission.)

With the use of high output systems it is expected then that many of these are above ACGIH standards. Even in air pollution studies with the safe lasers such as high output CO_2, Nd-YAG, and soon UV for detection ranging and Raman spectroscopy, there may be hazards in atmospheric propagation of the laser beams with consequent eye damage. Also, there may be respiratory involvement from the interaction of the laser with some of the pollutants. In these experiments exposure of the public from atmospheric propagation of laser systems should be considered.

Lasers are used on plastics, for etching, drilling, for Raman spectroscopy, and for structural modification. Laser hazards in work on plastics may be developed from plastic itself in the form of actual aerosol fragments that could contribute to carcinogenicity or allergy or to toxic products from the thermal decomposition from plastics. In addition, there is always the ever present eye hazard from the spectral reflectance and from the transmission through the clear plastic materials. The toxic products from material processing of plastics may vary according to whether these are produced by slow or rapid heating of the plastic. Some thermal decomposition products that may occur from CO_2 laser processing are

1. Acrylics – irritating monomers
2. Vinyl chloride – hydrochloric acid, carbon monoxide, carbon dioxide
3. Teflon® – fluorine
4. Polystyrene (in absence of oxygen) – toxic styrene

These are but a few of the basic materials that may develop in this material processing of plastics. Studies with the use of high output ruby and Nd-YAG lasers of materials embedded in plexiglass have produced gaseous materials and explosions. As yet, lasers have not been used for processing plastic prostheses in the body.

The increasing use of the CO_2 laser in cloth cutting in single patterns or for mass cuttings is done usually with closed CO_2 systems. The hazards for laser rock cutting and drilling include eye damage through reflection and for pollution from fragments produced by the heating of the rocks. For laboratory studies on rock cutting, Carstens[3] proposed a ¼-in. lucite panel and efficient exhaust hood; however, no details of field safety programs are given. CO_2 lasers have been used in processing clays and in glazes on earthenware surfaces. If these clays and glazes contain lead and other toxic metals such as uranium yellow glazes, if they are available, there are definite air pollution hazards in this laser processing. In addition, there may be eye hazards from spectral reflectance of these glass-like materials.

As described previously, lasers are important in

Nd-YAG LASER SAFETY

1. EYE - SCHOTT KG PREFERRED TO SCHOTT BG-18.

2. GUIDE LIGHT FOR INFRARED - KRYPTON PREFERRED TO HELIUM-NEON.

3. SPECIAL NATH FLEXIBLE STERILIZABLE OPERATING PROBES.

4. AVOIDANCE OF SPECTRAL REFLECTANCE FROM SURGICAL INSTRUMENTS.

5. MAINTENANCE OF DRY OPERATIVE FIELDS TO AVOID SCATTERING OF PARTICLES OF COAGULATED BLOOD.

FIGURE 13-9. Outline high output Nd-YAG laser surgery.

present studies in thermonuclear fusion.[1] Since there is an increased optimism for controlled thermonuclear fusion in this country and less secrecy, lasers will assume a greater role in this important field of energy production. In addition, in current thermonuclear research, Q-switched holography is being used for the study of these plasmas. Although, as indicated, most of the data in thermonuclear fusion experiments are still on paper, it is necessary to plan safety programs for protection against neutron laser beams and the tremendous heat involved.

The use of the new tunable dye lasers also brings many problems for laser safety programs because of the changes in frequency and consequent changes in eye protection. Clinical investigative studies are only now underway with new high output tunable dye lasers. As a corollary, there are still too little biological data, except animal experimental work, on ultraviolet lasers. Since the ultraviolet laser will have profound effects on biological systems, more studies are needed.

The biological applications of picosecond laser systems with the instantaneous high power effects in living tissue also have to be considered, especially regarding exposure of laser personnel rather than by actual uses of such systems in biology and medicine. In our experimentation on the skin of man, repeated impacts of 4 psec have been done with Nd lasers.[39] In similar experiments the TEA CO_2 laser has also been used.[40] Specific transient thermal coagulation necrosis has been produced.

It is obvious then that laser systems are being used and will continue to be used at higher and higher outputs. Safety programs are needed not only for the operator, but also for the public, whether they are concerned with direct or indirect laser exposure.

Safety for Laser Surgery

The details of high output laser surgery with CO_2 and Nd-YAG lasers will be listed in the chapter, Laser Medicine.[10] With the CO_2 laser, there may be burns on the fingers, reflection from chrome-plated instruments, and flammability of gauze and towels. The intense light from the laser impact area may also produce eye hazards which ordinarily are not considered in the routine eye protection program with the CO_2 laser. Finally, the current optical delivery systems are difficult to sterilize. However, the new effective high output CO_2 laser of Pan Engineering, Ltd., Israel has more safety features, is more flexible, and the end pieces are easily sterilized.

Because of its compactness and flexibility, high output neodymium-YAG lasers with flexible efficient crystalline light probes are being used in biological applications of these lasers. In our laboratory, this high output system is being used with special single fiber optical delivery system of Nath on the liver of dogs, and on cancer, tattoos, and portwine marks on man.

For safety programs with high output Nd-YAG laser surgery more details are given since, unlike other laser surgery safety programs, no data are as yet available.

1. Eye safety – protective glasses Schott KG preferred to Schoot BG18
2. Guide light since output in infrared and invisible krypton preferred to helium-neon

3. Special flexible sterilizable operating probed

4. Long cotton applications for maintenance of dry fields

5. Avoidance of spectral reflectance from instruments

In brief, with planned programs of laser safety, the laser is safe for the operator and the patient.[10] For investigative laser surgery with new lasers, a definite program is followed. Instruction is given in the details of the new laser and its safety program. Controlled laser animal surgery follows. After the safety program for this has been developed, and if the laser has been shown to be effective, then planned laser surgery on man can be done. The protocol for these investigations should include details of laser safety.

Even in biology and chemistry, there are health hazards to the laser microprobe used for toxic and infective samples. These may be reduced by sterilization of the samples and by isolation chambers.

Review of the General Aspects of Some of the Hazards of Laser Material Processing

1. Eye and skin – reflection of laser energy from material surface and vaporization plume

2. Eye and skin – reradiation emission of laser induced gaseous plume

3. Eye and skin – airborne dissemination of molten fragments

4. Eye, skin, and respiratory – airborne dissemination of fragments in reaction plasma

5. Eye, skin, and respiratory – toxic gases in reaction plasma or by recombination with air or other gases in material removal processes

6. Electrical hazards from high voltage system

7. Gases and chemicals associated with laser metalworking

Holography

Since this is a laser operation, safety measures must be considered during the recording of the hologram and its reconstruction. Rockwell[27] has listed some of the hazards:

Recording

1. Accidental intrabeam viewing of a direct or specularly reflected beam at levels above the accepted MPE level

2. Viewing of a hazardous diffusely reflected beam, especially with pulsed laser holography

3. Prolonged chronic exposure experienced during a long term level occupation and holography

4. Nonbeam hazards including electronic hazards, low level light environment

Reconstruction

1. Viewing directly into the point source of the beam used for reconstruction

2. Viewing the reconstruction at unsafe intensities

3. Chronic exposure from holographic reconstruction

Holography is becoming also popular in laser art exhibits. If the laser is used to reconstruct the hologram, the public must not be able to get in the rear of the construction and into the laser beam. Such poor safety measures have been observed in a number of such exhibits on holography in art.

The question of portrait holography is also of concern in the laser safety program. This technique is becoming more popular and three-dimensional pictures have been seen with the subjects not protected with laser protective glasses during the recording. Pulsed holography of movements of the chest and of the heart will require eye protection for the patient.

Protection of Pilots

Although most of the studies on protection of pilots and aircraft have been done in the military,[11] there is need for detailed studies on the effect of atmospheric propagation of lasers on civilian aircraft and planes. At present, protection is done chiefly by avoiding flight patterns in areas of laser propagation, especially by the military. There is apparently no concern about satellite tracking, moon shots, experiments on communication, weather and air turbulence projects, air pollution studies, etc. The data available are mostly an added value from military pursuits.

Military Safety Programs

As indicated, there have been detailed and valuable studies done by the respective Armed Services in regard to laser safety programs. The Air Force especially has been active in this field. The requirements for the military differ, obviously, from those in industry and for consumer use. Laser uses are in the high risk area. Sliney[30] has emphasized that laser hazards for the military are much more difficult to define in the field than in the laboratory. More interest and concern will

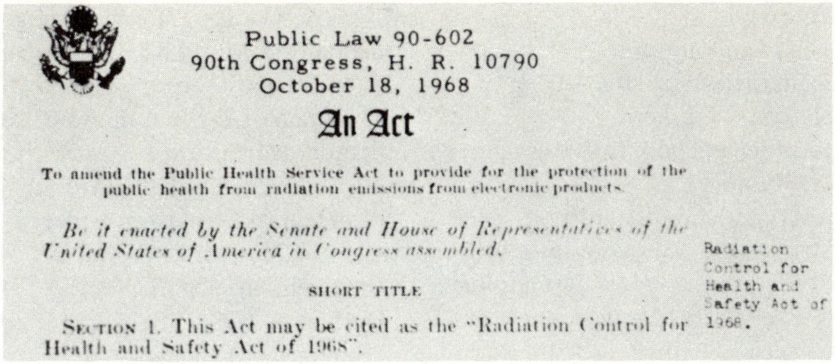

Public Law 90-602
90th Congress, H. R. 10790
October 18, 1968

An Act

To amend the Public Health Service Act to provide for the protection of the public health from radiation emissions from electronic products.

Be it enacted by the Senate and House of Representatives of the United States of America in Congress assembled,

SHORT TITLE

SECTION 1. This Act may be cited as the "Radiation Control for Health and Safety Act of 1968". Radiation Control for Health and Safety Act of 1968.

FIGURE 13-10. Public Law 90-602 for federal control of laser radiation.

involve the field use and data will be available as regards the extent of ocular hazards with the increasing use of the atmospheric propagation of laser systems. Much of the support of current laser research is done by the Armed Forces and valuable information as a result of this will be available to the public as it has been in the past. Even the current emphasis in the laser bomb will supply information in safety for laser-induced thermonuclear fusion for clean and economical nuclear power.

Safety Programs in Laser Art and Music

The laser as a new tool for the artist will be described later.[12] Following comprehensive developments of laser art at the Laser Laboratory, with consequent exhibits at a number of museums, a laser safety program has been developed.

Laser art in design, sculpture, etching, holography, photography, and music has been developed. The safety program adapted for the artist as well as for the viewing public follows a definite program developed at the Laser Laboratory. This may be listed as follows:

1. A survey of the laser design of the artist regarding safety.
2. Monitoring of the completed piece of laser or lasers is used as part of the design or in portrayal of the patterns of music.
3. After setup in the museum, there is monitoring of the piece where lasers are used again. Monitoring is done in conjunction with the local health laws and, if necessary, the local federal radiation authority.
4. If final approval is received, the exhibit opens to the public.
5. Repeated monitoring surveys occur during the exhibit where lasers are used as regards output, alignment, protective barriers, etc.
6. If a piece is purchased, and if lasers are used, it is the responsibility of the artist to instruct in the proper safety in the home or exhibit area and to inform the purchaser about registration of the laser, as determined by state and local laser registration legislation.

Art pieces, of course, created by lasers as in sculpture, etchings, and photography require no safety provisions, only a tolerant appreciation of a new art media of great interest to youth. Unless the program, as suggested above, is followed, the development of laser art may be blocked by the indifferent or rebellious artist.

Laser Safety Instruction

Actually, laser safety is a part of any instruction in lasers. The goals vary according to the specific needs, from those of the occasional user to the full-time high risk laser operator.

It has been the author's experience that a formal course in laser safety is usually necessary for adequate training in laser safety.

A laser safety course began on January 29, 1968 as a cooperative program of the Laser Laboratory, the U.S. Public Health Service, and the Department of Environmental Health of the Medical Center of the University of Cincinnati. The course was developed as a conference and workshop program to show actual laser use and safety rather than just didactic lectures. The course was designed for the following:

Occupational health specialists
Industrial health officers
Military safety officers

Health physicists
Industrial physicians and nurses
Government health officers at all levels
College and high school teachers
Professional educators and administrators
Insurance underwriters
Research scientists and administrators
Physicians and surgeons, especially in ophthalmology, investigative surgery, and dermatology
Civil engineers and construction personnel
Biomedical engineers
Laser artists and hobbyist
Small business executives

The need for such a course was shown and now it has become a semiannual affair under the direction of Rockwell,[27] sponsored by the Laser Laboratory of the Medical Center of the University of Cincinnati and the Continuing Medical Education Program of the College of Medicine of the University of Cincinnati. The faculty includes the rendent staff of the Laser Laboratory plus expertise in laser safety and laser legislation. For example, the guest lecturers for the current course on March 5–9, 1973 included:

1. Wordie H. Parr, Ph.D., Acting Chief, Physical Agents Branch, National Institute of Occupational Safety & Health, Cincinnati, Ohio
2. Richard W. Peterson, Chief, Electro-Optics Section, Division of Electronic Products, Bureau of Radiological Health, FDA, Rockville, Md.
3. John D. Pierson, Ph.D., Professor and Chairman, Laser/Electro-Optics Department, Texas State Technical Institute, Waco, Texas
4. Charles H. Powell, Sc.D., Assistant Director, Division of Criteria and Standards Development, National Institute of Occupational Safety and Health, Environmental Control Administration Public Health Service, Rockville, Md.; Chairman, Laser Safety Committee, American Industrial Hygiene Association
5. David H. Sliney, Chief, Laser Hazards Branch, Laser-Microwave Division, U.S. Army Environmental Hygiene Agency, Edgewood Arsenal, Md.; Subcommittee Chairman on "Special Considerations," Z-136 Standards Committee, ANSI
6. James Smith, Ph.D., Senior Associate Engineer, I.B.M. Corp., Poughkeepsie, N.Y., Past-President, Laser Institute of America (LIA)
7. George M. Wilkening, Head, Environmental Health and Safety Department, Bell Telephone Laboratories, Murray Hill, N.J., General Chairman of Z-136 Committee, American National Standards Institute
8. R. Marshall Wilson, Ph.D., Associate Professor of Chemistry, University of Cincinnati
9. Myron Wolbarsht, Ph.D., Director, Research in Ophthalmology, Duke University, Durham, N.C.; Subcommittee Chairman on "Biological Effects – Eye," Z-136 American National Standards Institute

The following topics, to the author's experience, serve as basic information to enable one to develop a program for laser safety:

1. Laser characteristics and devices
2. Effects on skin and internal organs
3. Effect on the eye
4. Criteria for maximum permissible exposures
5. Photobiological and nonlinear effects
6. Review of the ANSI Z-136 Standard
7. The role of BRH, DOL, and NIOSA
8. Measurement technique
9. Laboratory practice in measurements of laser output levels
10. Evaluation of safe laser environment
11. Institutional laser safety programs
12. Typical calculations required for laser hazard analysis
13. Personnel and procedural controls
14. Control measures workshop with demonstrations and class participation
15. Medical surveillance
16. Patient treatments
17. Survey of laser research laboratories
18. Safety program management
19. Legal aspects of laser safety
20. The future

This is combined with the excellent teaching films now available, as for example:

1. *Laser Light* – Scientific American
2. Prize-winning films of Upjohn Company. These are from the popular film series of the Upjohn Vanguard of Medicine film series. These are available on loan from The Upjohn Company: *Clinical Applications of Laser* and *Medical Potential of Lasers.*

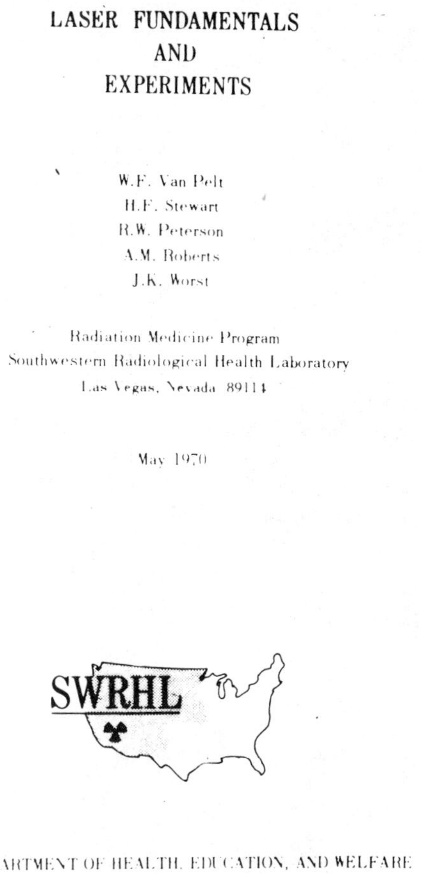
LASER FUNDAMENTALS
AND
EXPERIMENTS

W.F. Van Pelt
H.F. Stewart
R.W. Peterson
A.M. Roberts
J.K. Worst

Radiation Medicine Program
Southwestern Radiological Health Laboratory
Las Vegas, Nevada 89114

May 1970

SWRHL

U.S. DEPARTMENT OF HEALTH, EDUCATION, AND WELFARE
Public Health Service
Environmental Health Service
Bureau of Radiological Health
Rockville, Maryland 20852

FIGURE 13-11. High school laser instruction and safety booklet, Bureau of Radiological Health, Laser Section.

3. Laser Safety

All this, then, is offered as a suggestion for teaching courses on laser safety. If special needs exist, courses on laser safety can be adapted; the Laser Laboratory has done this for training army personnel in laser safety.

Conclusions

Although the basic principles of the development of a laser safety program are fixed, constant changes in laser technology will make for constant revisions. If laser technology is to continue to develop, standards of exposure must be devised by those with actual controlled laboratory and consumer and field experience. There must be provisions for recognition of the need of lasers in this new world, for protection of the operator and the public, informed and uninformed. The protection must include the possibility of the worst case exposure, but must recognize that most of the exposures will not be the worst case but the usual cases. Let us hope that the federal government will approve a safety standard for lasers which the Electronic Industries Association "claims will ban one half of the 100,000 lasers now operating in this country." There will always be disagreement about PME among experts, and those whose mothers call them experts; this should be recognized but should not interfere with the development of especially important essential applications of lasers. As one surveys the applications of the laser, except perhaps for weaponry, the laser can be used when it is needed within planned programs. Knowledge and training in laser safety always have decreased much of the fear, hysteria, and lurid headlines in the past about the hazards of the laser. Again, and in a monotonous pleading fashion, more basic research on the effect of the laser on living tissues of man will help solve this long controversy about exposures.

REFERENCES

1. **Basov, N. and Krokhin, O.,** Laser light for fusion fire, *New Sci.,* 10, 733, 1971.
2. Bureau of Radiological Health, Laser Section, Preliminary Documentation Report, Biological Aspects and Other Bases for the Performance Standard for Laser Products, Draft of October 13, 1972.
3. **Carstens, J. P. and Brown, C. O.,** Rock Cutting By Laser, Society of Petroleum, Engineers of AIME Paper #3529, October 3–6, 1971 and personal communication.

4. **Desvingnes, P., Amar, L., Bruma, M., and Village, M.,** Sur la generation d'ondes ultrasonores et formation de bulles dans le vitre d'un oeil humain par irradiation d'impulsion laser, *C. R. Acad. Sci.,* 259, 1588, 1964.
5. **Dietz, P. H.,** Probability analysis of ocular damage to laser radiation through the atmosphere, *Appl. Opt.,* 1969, 371, 1969.
6. **Ducas, T. W.,** Report from Esfahan, *Laser Focus,* December 1971.
7. **Dunsky, I. L., Fife, W. A., and Rickey, E. O.,** Determination of Revised Air Force Permissible Exposure Levels For Laser Radiation, SAM-TR-72-11, April 1972.
8. **Dwight, H. J.,** *Spectra-Physics Electronic News.*
9. **Epstein, R.,** Personal communication, 1972.
10. **Goldman, L., Rockwell, R. J., Jr., Fidler, J. P., Altemeier, W. A., and Siler, V. E.,** Investigative laser surgery: safety aspects, *Biomed. Eng.,* 4(9), 13, 1969.
11. **Goldman, L., Rockwell, R. J., Jr., Epstein, R., Kenneman, R. E., Schechter, E., and Meyer, R.,** Protection of plots against lasers, *Aerosp. Med.,* 40, 196, 1969.
12. **Goldman, L.,** Laser light, a new visual art and a new occupational hazard, *Arch. Environ. Health,* 20, 145, 1970.
13. **Goldman, L.,** Plans and goals of the laser safety conference, *Arch. Environ. Health,* 20, 148, 1970.
14. **Goldman, L. and Rockwell, R. J., Jr.,** *Lasers in Medicine,* Gordon and Breach, New York, 1971, Chap. 10.
15. **Goldman, L.,** Draft Section Review, Z-136 Standards Committee on Safe Use of Lasers, American National Standards Institute, November 19, 1971.
16. **Goldman, L., Rockwell, R. J., Jr., Naprstek, Z., Siler, V. E., Hoefer, R., Hobeika, C., Hishimoto, K., Polanyi, T., and Bredemeier, H. C.,** Some parameters of high output CO_2 laser experimental surgery, *Nature,* 228(5278), 1344, 1971.
17. **Goldman, L.,** A review for the concerned public of safety programs for current high output systems and new systems, *Am. J. Public Health,* in press.
18. **Ham, W. T., Jr., Williams, R. C., Mueller, H. A., Guerry, D., Clarke, A. M., and Geeraets, W. J.,** Effects of laser radiation on the mammalian eye, *Trans. N.Y. Acad. Sci.,* Ser. II, 28, 517, 1966.
19. **Hayes, J. R. and Wolbarst, M. L.,** Thermal model for retinal damage induced by pulsed layers, *Aerosp. Med.,* 39, 474, 1970.
20. Laser Standard, British, *Locus Focus,* December 1972 and *Laser Focus,* January 1973.
21. Laser Survey, High school and college science classes, HEW News Release, Bureau of Radiological Health, December 19, 1972.
22. **Nath, G. and Schindler, G.,** Personal communication, 1971.
23. **Parr, W. H.,** Personal communication, 1972.
24. **Parts, L., Hardy, E. E., and Rodenburgh, M.,** Laser radiation induced residue-free localized decomposition of some plastics, *I&EC Prod. Res. Dev.,* 9, 21, 1970.
25. **Peabody, R. R., Zweng, H. C., Rose, H. W., Pepper, N. A., and Vassiliadis, A.,** Corneal damage for CO_2 laser radiation, *Appl. Opt.,* 8, 377, 1971.
26. **Pitts, D. G., Prince, J. E., Butcher, W. I., Kay, K. R., Bowman, R. W., Casey, H. W., Rickey, D. G., Mori, L. H., Strong, J. E., and Tredici, T. J.,** The Effects of Ultraviolet Radiation on the Eye, Report #SAM-TR-69-10, USAF School of Aerospace Medicine, Aerospace Medical Division (AFSC), Brooks Air Force Base, Texas, February 1969.
27. **Rockwell, R. J., Jr.,** Personal communication, 1972.
28. **Salyer, I. O.,** Personal communication, 1971.
29. **Siedleicki, J. T.,** Hazardous fumes from melting of plastic toys, *J.A.M.A.,* 218, 103, 1971.
30. **Sliney, D. H.,** Evaluating health standards from military laser, *J.A.M.A.,* 214, 1047, 1970.
31. **Sliney, D. H. and Freasier, B. C.,** Evaluation of optical hazards, *Appl. Opt.,* 12, 1, 1973.
32. **Sliney, D. H.,** The development of laser safety criteria, in *Laser Applications in Medicine and Biology,* Vol. 1, Wolbarsht, M. L., Ed., Plenum Press, New York, 1971, 125.
33. **Taylor, H.,** *Electron. News,* September 13, 1971; LeBoss, B., *Electron. News,* October 1971; *Laser Report,* 7(10), 1971; *Laser Weekly,* September 27, 1971.
34. **Van Pelt, W.,** Presentation to Technical Electronic Radiation Safety Standards Committee, Office of Criteria and Standards, Bureau of Radiological Health, September 21, 1972.
35. **Vassiliadis, A.,** Ocular damage from laser radiation, in *Laser Applications in Medicine and Biology,* Vol. 1, Wolbarsht, M. L., Ed., Plenum Press, New York, 1971, 125.
36. **Vos, J. J.,** A theory of retinal burns, *Bull. Math. Biophys.,* 24, 115, 1962.
37. **Willis, I., Kligman, A., and Epstein, J.,** Effects of long ultraviolet rays on human skin, photolective or photoaugmentive? *J. Invest. Dermatol.,* 50, 416, 1972.
38. World Health Organization, *Health Effects of Ionizing and Non-ionizing Radiation,* Regional Office For Europe, Copenhagen, 1972, Chap. 9.
39. **Goldman, L., Fradin, D.W., Bloembergen, N., and Richfield, D. F.,** Studies in laser safety of new high output systems. 1. Picosecond impacts, *Opt. Laser Technol.,* 2, 11, 1972.
40. **Goldman, L., Yablonontch, E., Bloembergen, N., and Richfield, D. F.,** 2 TEA CO_2 laser impacts, *Opt. Laser Technol.,* 3, 58, 1973.

Chapter 14

THE NEW LASER BIOLOGY

Although they may not believe it, those working with lasers must be aware of the current studies in laser biology no matter what the fields of interests or applications are. In addition to the basic science of laser biology, there are related basic sciences of physics, biophysics, and chemistry. Many also do not realize that the fashionable, popular, and remunerative so-called applied research could not be possible without the background of much creative work, much seemingly of no immediate value. We will have progress in laser technology only if we do basic research. This especially refers to laser biology, the effects of the laser on individual living cells, on groups of cells in a community of cell masses, and, finally, the body as a whole. From all these detailed studies, laser safety programs are developed to expand the safe and controlled uses of the laser in all fields; applications are found especially for biology, medicine, and dentistry. The biologist can develop visions of the future as he develops understanding of lasers. Irvin Page wrote that "research workers are expected to be unworldly and slightly kooks. But the moon shots and other recent spectaculars are changing all this. . . This is discovery of new truths that never previously had been imagined. The best of it is to discover the impossible." It is a pity that those who dole out meager funds for basic research in lasers favor applied, so-called "targeted research" over basic research. Much will be said later of the social conscience of the laser scientist who can do much for the good-of-man with lasers if he has the support for his far-reaching basic research.

One of the new fields of modern dynamic biology is the new discipline of laser biology, initiated by such laser pioneers as Rounds,[57] Berns,[8,9] Bessis,[11] and their associates. Korn[70] has also recently developed a laser biological microscope. This new field is exciting for young investigators, especially those who work with the study of cells, called cytology, and all its divisions.

Several factors are responsible for the development of this new biology. The laser has opened up this new field of the study of the so-called nonionizing radiations on living tissue. Ionizing radiation is that produced by X-rays. The class of nonionizing radiation includes microwaves, ultrasonics, and the ultraviolet in addition to lasers. None of these other modalities can reach the tremendous energy and power densities of the laser.

Color Absorption

There is another exciting factor about the laser, namely, the fact that lasers in the near ultraviolet, visible, and near infrared have color dependency. This means that they can be absorbed more intensely in living tissues that are colored. These lasers, then, would include the ruby, neodymium, neodymium-YAG, argon, krypton, helium-neon, helium-cadmium, etc. They do not include the CO_2 laser, 1,060 nM, used so frequently in the biomedical applications. Thus, one phase of basic research would be to study the influence of the natural pigments, tissues containing the following natural pigments that show more intense absorption by the laser. These include melanin, melanoid, carotene, and the hemoglobins oxy- and reduced forms.

To provide color for those tissues that do not have color, a whole series of materials have been studied. These may include the vital dyes, such as Evan's blue and toluidine blue, India ink, and recently, black magnetic iron particles, ferrofluids. Recent experiments have been done in the Laser Laboratory with ferrofluids and the exciting technique of ferrohydrodynamics. Ferrofluids are dispersions of colloids, magnetic iron particles in liquid carriers. The colors vary from deep brown red to black. The particles are small, usually on the order of 100 Å, according to Rosensweig.[69] In addition, these particles are coated to prevent sticking to each other and, consequently, a homogeneous fluid is developed even when in contact with strong magnetic fields. The vehicles for use in biomedical experiments have included water and fluorocarbons. The author's studies have attempted to move these colored particles into target areas such as cancer. By means of strong magnetic fields or injections, this increased tissue density of iron particles can increase the activity in this area two- to threefold. The ferrofluids[58] also raise possibilities of local microwave absorption and, consequently, local heat and tissue hyperthermia. Ferrofluids provide for the use of radioactive iron which provides additional local ionizing radiation. It would appear that for biological studies, considerable effort is to be extended on the value of the laser as an adjuvant with their magnetic particles.

Thermal Reactions in Tissue

Selective absorption of dyes by cancer tissue also would be of value in the enhancement of absorption of laser radiation. The basic studies in biology also furnish the studies for the effect of the laser in living tissue. Many studies have been done. The linear effects of the laser radiation on tissue are primarily thermal. The thermal coagulation, death of tissue, is primarily protein denaturation. Other processes that are said to occur, especially from high output systems, are elastic recoil and pressure-wave formation, tissue ionization, harmonic frequency generation, multiphoton absorption process, free radical formation, inverse Bremstrahlung, heating, and various "photodynamic" reactions. These are especially evident with high power systems. Recent studies in tissue have been done by the author with Bloembergen, Fradin, and DeMaria,[25] with repeated picosecond pulses and with Yablonovitch et al.[26] on TEA CO_2 laser impacts.

For most of the laser systems now in use, especially for CW laser systems, the effects are chiefly thermal. The thermal reactions in tissue from the CO_2 have been studies by Hall et al.[32] and also in the authors's Laser Laboratories. Working with the CO_2 laser and making incisions at a known constant speed on albino Wistar rats, these investigators confirmed the findings of Rockwell and his co-workers[54] of the University of Cincinnati Medical Center Laser Laboratory that an ablation temperature of 100°C can be derived for the speed of laser incision. The tissue reaction is called, in brief, a thermal coagulation necrosis. However, it is curious that low intensity laser radiation may actually stimulate tissue growth instead of destroying cells. This could be of importance in the treatment of cancer, such as the black cancer or melanoma. This stimulation may not occur if prolonged (how long?) low intensity radiation is given. There have also been some uncontrolled experiments on hair growth after low intensity laser radiation. This phenomenon has been observed in the follow-up studies of the laser treatment of tattoos. When low intensity laser radiation (5 to 10 J) was used on a culture of *Escherichia coli,* stimulation of DNA and RNA synthesis was produced in direct proportion to the dose. The effect of ultraviolet on DNA has been mentioned before. This is important for UV lasers.

Biologists have also produced mammalian tissue regeneration by electrical stimulation of cellular activity. This has been done by Becker[68] with limb regeneration in rats. Becker believed that some factors should be surveyed, including hormonal responses, the availability of data carrying channels, and the relationship between the raw amount of nerve and total bulk, in addition to finding the right amount of electrical stimulation. With the precision of the collimated beam and knowledge of the target area, the laser stimulating phenomenon could be studied, especially in embryology. Edlow et al.[73] did basic studies on laser irradiation in utero.

As indicated, another phase of the effect of pulsed laser reaction in tissue is the elastic recoil and pressure waves. To study this reaction, Borg[12] developed a mathematical-physical model to obtain the time period of the behavior in pressure-density reactions as well as the time of crater formation. In living tissue experiments, surface phenomena may be studied by surface microscopy, at moderate magnification, of the living tissue as well as replica microscopy, silicone rubber molds, and replicas studied in detail under microscopy and by scanning electron microscopy (S.E.M.). The pressure reactions may be studied by transducers. Impact studies may be observed directly by high-speed photography.

It is of interest that there have not been more studies with the laser in embryology. The pioneering studies of Barnes[6] at the University of Colorado are of great importance not only for experimental teratology, but also with microirradiation techniques for cytogenetics and cytochemistry. Some recent studies on the chick embryo germinal crescent have been done by Mims and McKinnell.[45]

Microirradiation

The developments in laser biology and cytology and microsurgery of chromosomes have been made possible chiefly by studies with the technique of laser microirradiation. In brief, the laser is

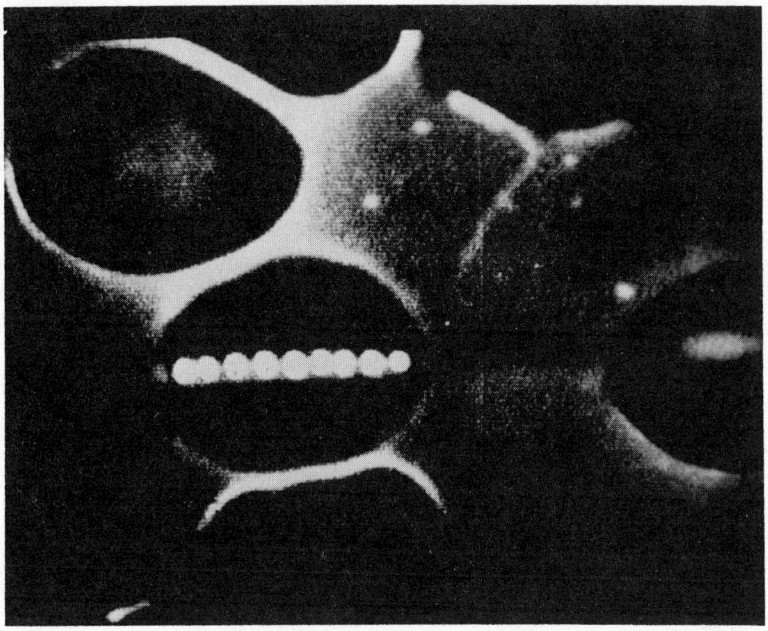

FIGURE 14–1. Nine argon laser impacts in a single 7 μm red cell. (D. Rounds.)

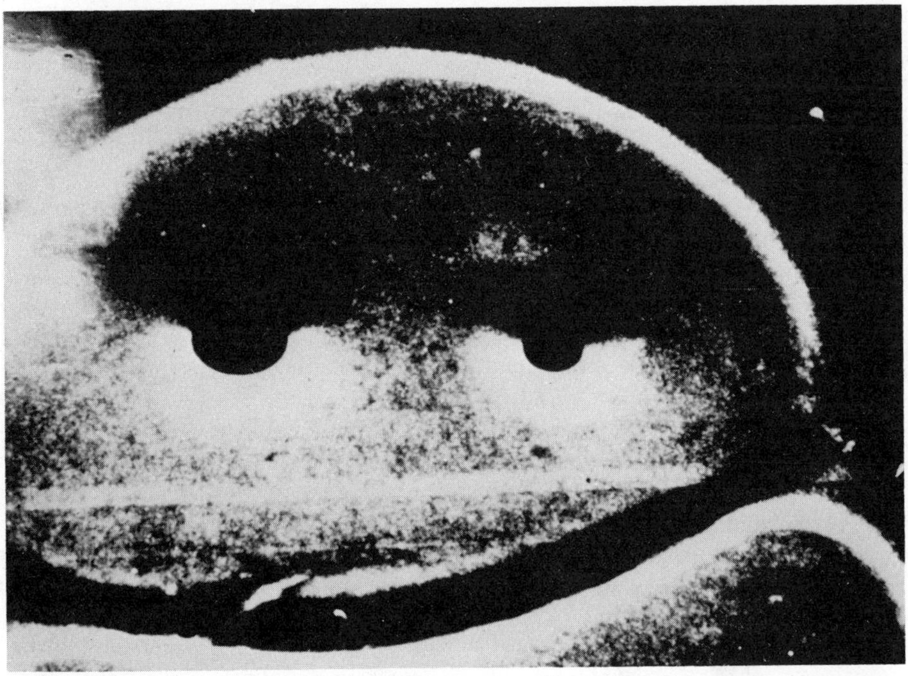

FIGURE 14–2. Normal mode ruby laser impacts of a red cell, scanning electron microscopy. (Franz Hillenkamp.)

attached to the microscope and the optics of the microscope transmit the laser beam into tissue. The spot diameters can vary from less than 1 to 50 μm.

Laser microscopy may be divided into the following general divsions:

1. Microsurgery

   A. Cells
   B. Organelles, especially mitochondria
   C. Chromosomes
   D. Melanin

2. Fluorescence microscopy, cytofluorography
   A. Scanning analysis
      a. DNA
      b. RNA
      c. Immunoglobulins
      d. Identification of malignant cells
      e. Bacterial identification
3. Scanning for viability of cells – absorbance of vital dyes
4. Infrared scanning microscopy with thermal analyses
5. Microinterferometry
6. Microholography
7. Acoustical micrography with image reconstruction by lasers[39]
8. Microemission spectroscopy

Laser microscopy has made for extensive studies on microsurgery of cells and organelles and even chromosomes. Studies on chromosomes by Berns and Rounds[9] included actual microsurgery on living large chromosomes of the salamander. Studies on fluorescence cytochemistry and S.E.M. can be done on these treated areas. After microsurgery on selective locations of the chromosomes, cells are allowed to develop and the disturbances of cellular function are studied. It is obvious, then, that laser microsurgery will have a considerable impact on studies for marking out gene mutations on the individual chromosome.

After performing argon laser radiation of the chromosomes of salamander lung epithelium (488 and 5,145 nm, 50 μsec), Berns and Floyd[8] did cytochemical analyses of chromosomal DNA and protein. Low power (1 W, 50 μJ/pulse) irradiation of acridine orange-stained chromosomes functionally and morphologically changed DNA without any effect on histone protein. On the other hand, high power irradiation (18 to 35 W, 300 μJ/pulse) of unstained chromosomes did change histone protein but did not affect DNA or nonhistone protein. Berns also used tunable dye lasers for his experiments on chromosomes. Thus, the laser has great value for structural and functional dissection of chromosomes.

We should first, perhaps, review specific cellular destruction, such as cells or paramecia stained with dye, cells containing masses of melanin, such as those of the retinal epithelium or melanoma, or just red cells themselves. These examples show absorption of the laser and consequent cell destruction. When specific elements in the cell are stained, these structures absorb laser radiation, for example, Janus Green B staining of mitochondria NADH dehydrogenase activity was used as a marker. Acridine orange-photosensitized cells showed a loss of intact lysosomes after argon laser exposure.

The toxicity of extracts of laser irradiated tissue cultures is also of interest. The toxicity refers to laser radiation effects on fresh tissue cultures. This has been shown for melanoma, but more controlled studies are needed. May, Rounds, and Cone[41] showed toxicity spread to cells adjacent to laser irradiated cells. The frequency of transmission was reported to be 15 to 45% in most tumor cell types but only 0 to 5% in nonmalignant types.

Tissue cultures have been done to develop a technique for determining the effect of subthreshold doses of chronic exposure to rabbit eyes. Tissue cultures were then taken from the cornea and the lens.

The author's studies with tissue cultures of laser radiated rabbit cornea and lens have been done to develop a biological test model system for the study of chronic exposure to laser radiation.

Routine examination of personnel exposed to laser have been done extensively by Zaret at the Bell Telephone Laboratories. However, there are little data available on such laser personnel now exposed to ultraviolet and infrared laser systems. This lack of information is important because of the increasing use of these systems and especially for the future because of continued exposure of the general public to all forms of lasers.

Relatively few studies have been done on the cornea and lens. Since the type of exposure of the general public will be chronic exposure in minimal

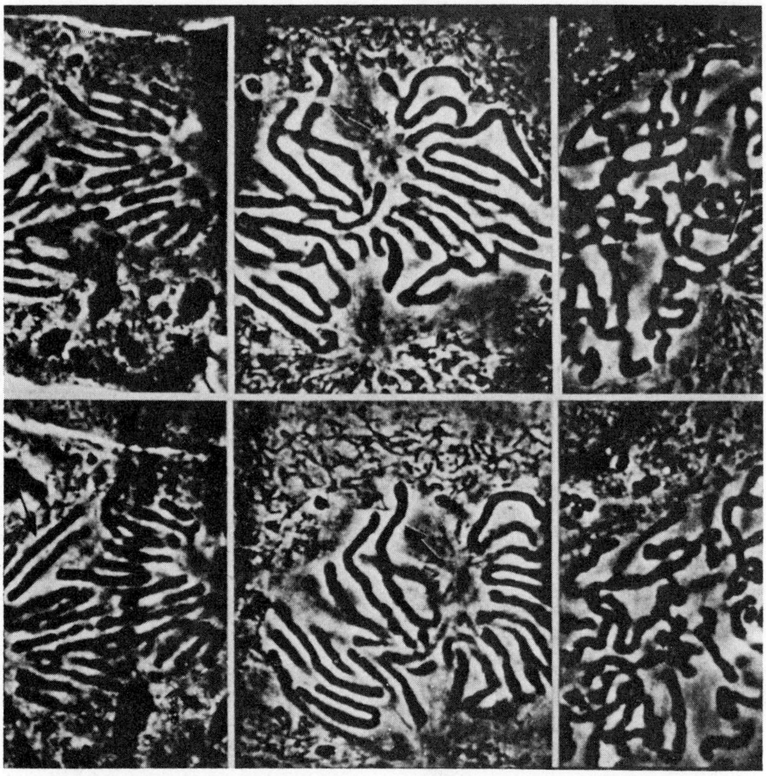

FIGURE 14–3. Laser microsurgery of chromosomes. (D. Rounds.)

doses, it is necessary to devise new techniques to evaluate exposures that may not produce changes, either visible in the fundus or detected with ordinary light microscopy, transmission electron microscopy, even scanning electron microscopy (especially of the retina), or by disturbed enzymatic reactions. We are aware of the current studies and the difficulties of interpretation of studies of functional ocular disturbances, namely, evoked potential following laser radiation in animals, not man. One technique used to record visual evoked responses in man is to use electrodes referenced to the occiput and mastoids.[71]

It is further believed that little, if any, changes occur in the cornea and the lens from laser radiation from ruby, argon, and neodymium systems since the energy or power densities are greater in the retina than on the surface of the cornea. The significant studies in relationship to these experiments have been those of Parr and Fisher.[51] Aberrant corneal epithelial cells were produced by ruby laser irradiation with dosage levels between 0.25 and 1.0 J/cm^2. Aberrant cells were found in the direct mount up to 30 days later. With an increase in energy density, more atypical cells were produced.

The other significant studies in relation to the project proposed by the Laser Laboratory at the University of Cincinnati Medical Center have been done by Cogan and Kinsey and by Pitts,[52] with studies of the effect of ultraviolet radiation on the eye. Pitts[52] has done excellent tissue cultures of the cornea but not of the lens following ultraviolet laser radiation. He indicated that the minimum energy for the production of UV photo-ophthalmia is 0.054 x 10^6 ergs/cm^2 at 270 nM. According to meager data for UV, the human eye may develop UV keratitis at two thirds the energy necessary for the rabbit eye. Similar studies are being done for infrared lasers.

Corneal damage is important in regard to CO_2 laser exposure. This has been studied extensively by Fine and his associates, by Zweng, Vassiliadis, Campbell, Leibowitz, and Peacock among others. It is important to emphasize that there are no data available, as yet, on tissue culture studies from minimal exposure to CO_2 lasers for the cornea and lens.

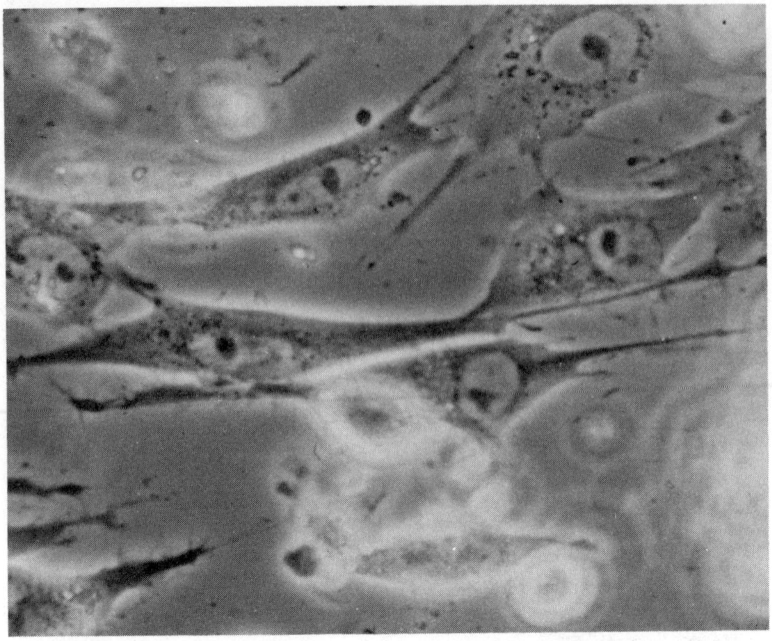

FIGURE 14–4. Subculture of corneal cells of rabbit after repeated exposures normal mode pulsed ruby laser 0.025 J/cm² showing no abnormality x 40.

Preliminary studies to determine the feasibility of the tissue culture technique were done with Dutch rabbits exposed to ruby laser, 0.5 to 0.1 J/cm². Repeated exposures were done and the eyes were enucleated immediately after initial exposure and at weekly intervals after ten impacts. Each eye had the same number of laser exposures. The nucleated eye was sectioned for routine microscopy and microscopic studies; materials were prepared for transmission electron microscopy. Tissue cultures were done of the cornea and lens from the other eye. Tissue cultures were observed for differential growth rates of cell types and intracellular changes. As a control to this technique, individual cells and colonies of cells were radiated with microradiation from the laser microscope; the changes in cell pattern were compared to tissue cultures obtained from cells irradiated *in situ* in the eye. The ruby laser impacts varied from 0.025 to 2.05 J/cm², with no changes, as yet, detected in initial or subsequent cultures. References for the eye studies are listed in the chapter about safety.

Fluorescence Microscopy

Recently, laser microscopy has had considerable application in fluorescence microscopy as cytofluography and microfluormetry.[72] In this technique, various fluorochromes stain cells or cellular components. These are scanned by lasers with a rapid and precise analysis of fluorescent areas, such as DNA, RNA, or abnormal nuclei of malignant cells.

In human serum are various molecules associated with protection against various disease states and infections. Fluorochromes can color these materials concerned with immunity. This technique is called immunofluorescence when exposed to the right wavelengths of light. Lasers have been used in these techniques, making fluorescence greater. Kaufman, Nester, and Wasserman have used this technique and recently Berquist[10,72] used a pulsed dye laser with greater illuminance. The fluorescein isothiocyanate (FITC) was excited with 7-diethylamino-4-methyl coumarin and rodamin isothiocyanate (TRITC) with sodium fluorescein.

With the variance of the wavelength of the tunable dye laser, it is possible then to tune this dye laser to many different fluorochromes. This will continue to be important for studies of chromosomes when fluorochromes other than quinacrine mustard may be used in the future. With ultraviolet laser microscopes there may be many new developments in the future for fluorescence microscopy.

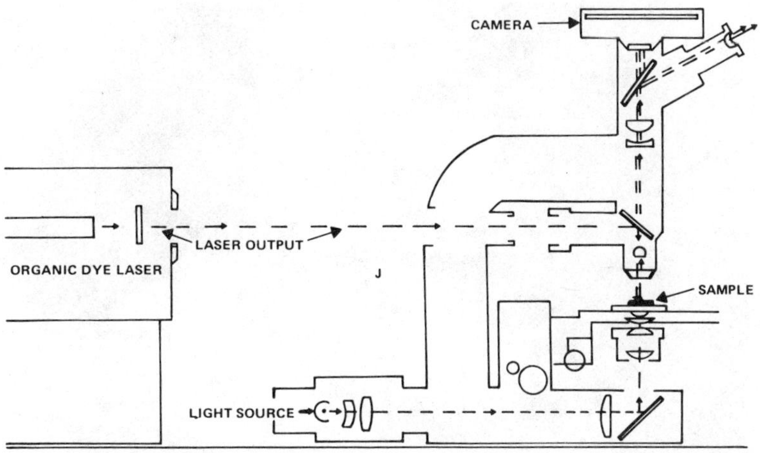

FIGURE 14–5. Pulsed dye laser microscope for immunofluorescence studies – NIR. Berquist. (From *Laser Sphere.* With permission.)

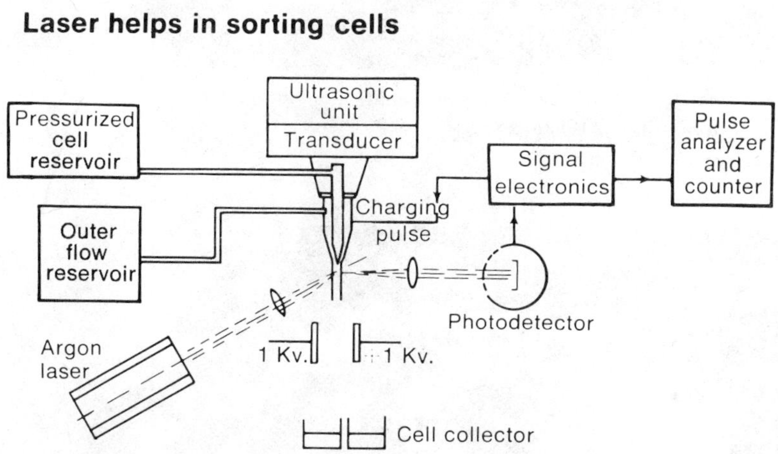

FIGURE 14–6. Fluorescent stained cells generating a signal pulse on passing through a laser beam. (Stanford University C & EN, December 13, 1972, also from Los Alamos Scientific Laboratory.)

Scanning Electron Microscopy

Scanning microscopy in the future may be used for routine blood smear analyses to identify various parasites, such as malaria and bartonellosis, Papanicolaou smears, viral diagnostic cytology, etc. Scanning electron microscopy (S.E.M.) can give detailed data of the surface changes of laser impacts. Vahl's extensive findings will be listed in the chapters about medicine and dentistry. Living cells may be differentiated from dead cells by their inability to take up vital dyes.

To determine the actual dry mass of living cells, Barer and Smith[7] have developed a rapid scanning microinterferometer based on a grating compensator. This laser illuminated system was used to measure a single dry cell mass of rat peritoneal macrophages and of nuclei isolated from the larval salivary glands of drosophila. This scanning and integrating microinterferometer can measure to within fractions of a "million-millionth" of a gram. An interference microscope of Hacker, based on the Michelson two-beam interferometry principle, is used primarily for thin-film thickness measurements and surface finish evaluation. This can be used in biology for crystallography, abrasion studies in teeth, and surface studies of bone

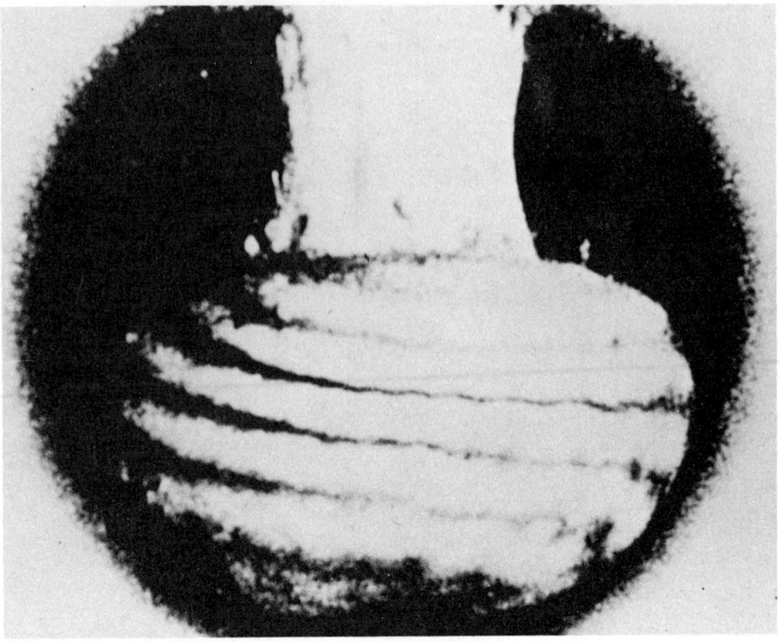

FIGURE 14–7. Interference holography – growth of a mushroom. (Franz Hillenkamp.)

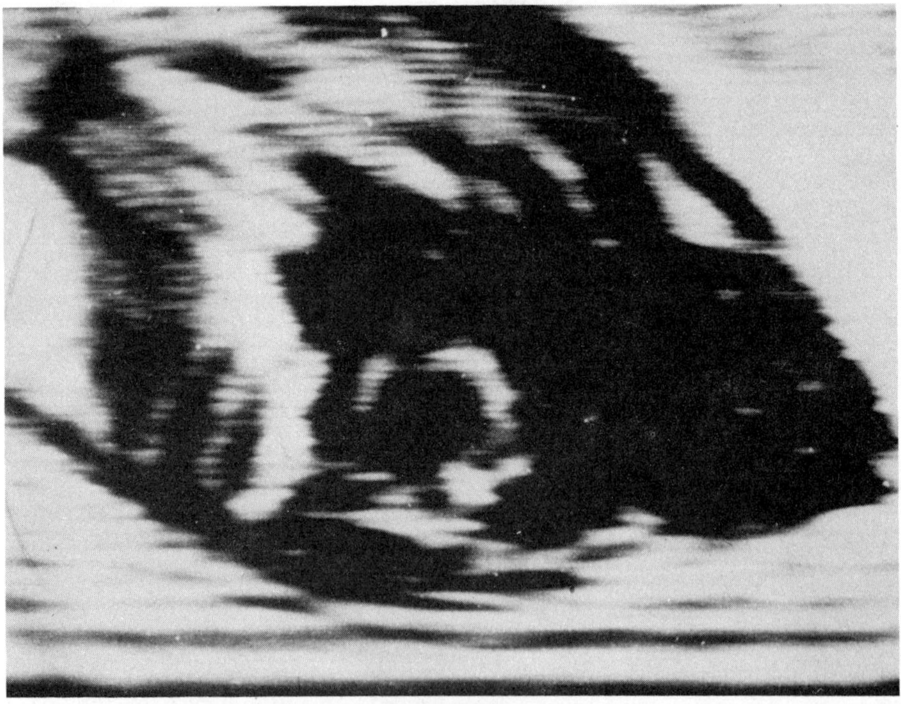

FIGURE 14–8. Acoustical holography of fish – film of Wade. (From A. Korpel.)

and cartilage. As yet, laser illumination has not been used here.

Microholography

As shown previously, holograms do give a three-dimensional, phase-faithful image. This has also been applied to microscopy by Van Lighten[67] and American Optical Co. has developed such an instrument. This introduces the exciting possibility of three-dimensional imagery of the growth, development, and stress patterns of an individual cell. Stroke,[63] one of the pioneers of laser holography, developed a holographic image deblurring method which can be used to sharpen electron micrographs. This was applied to increase the resolution of Crewe's[15] most powerful electron microscope.

Acoustical holography, the combination of ultrasonics and laser holography used early in studies of submarine detection, can be applied to biology and medicine essentially as a technique for nondestructive testing. The combination of optical and acoustical visualization has been shown in the fascinating movies of the so-called Silver Angle Fish of Wade. Correlated developments in microscopy in this field include the acoustical microscope of Korpel.[39,71]

Infrared Microscopy

With the increasing interest in infrared effects in tissue for their use in diagnostic thermography, an infrared scanning microscope has been developed. This permits thermal analysis with microscopy. Although this is to be used primarily for microcircuitry in electrooptics, it could be adapted for biology with fresh tissue slices for microthermal analyses of microcirculation dynamics, enzyme reactions as they relate to malignant cells, and, of course, thermal effects of laser impacts.

Microemission Microscopy

The other effect of the result of the attachment of the laser to the microscope is microemission spectroscopy. This has been reviewed in detail in the chapter on laser chemistry. Here, the material will be limited to biology. With this technique, microanalyses of tissue can be done without the destruction of the sample. Developments in this field have been done by Brech, Rosan,[56,57] Allemand,[1] and Glick,[20–23] and recently by Hillenkamp, Kaufmann, and Remy.[34] With this technique, for example, dry blood samples on a piece of paper have been used for cation analyses. This technique shows some of the limitations since the material must be dried. Fluids, then, cannot be analyzed.

Like its predecessors, the new laser microprobe of Brech and Allemand, now in active use, employs a focused giant pulse from a laser to evaporate a minute sample. The sample vapor is energetic enough to produce a spectrum without cross-excitation, but greater sensitivity is achieved with the use of cross-excitation. The ionized vapor short circuits an arc gap that is (or may not be) connected to a loaded capacitor bank, thus producing an electrical discharge through the sample vapor. The light emitted at this point is optimally collected and directed to the entrance slit of a fast spectrograph.

The new laser microprobe unit was designed as a system of well-matched components; the improvements were made very much along the lines suggested by Peppers et al. with some improved flexibility and the potential to be used for routine work.

One of the criteria of the new design was to ensure enough sensitivity and stability to permit the use of this unit for studies on quantitative analyses, which nowadays are more often still limited to three "bit" information like "major," "minor," and "trace," and seldom to somewhat more accurate statements in terms of relative concentration.

The instrument is divided into the excitation unit and the detection unit, which will be described separately.

The *excitation unit* consists of the laser, the microscope, and the electrical excitation and control chassis.

The *laser* has a neodymium glass rod emitting at 1.06 μm a wavelength at which photographic plates are not sensitive, and is Q-switched by means of a Pockel cell. Its output beam is either not attenuated or attenuated to 75%, 50%, 25%, and 10% of its initial value by a set of optical filters mounted on a remotely controlled turret. The beam thereafter is focused on a small aperture which is imaged on the target as an essentially round spot 5, 10, 15, 20, or 25 μm in nominal diameter, as defined by Gaussian optics. At the wavelength of 1.06 μm used here, the diffraction-limited spot diameter is in the vicinity of 5 μm. The filters and the apertures attenuate the laser output to stepwise control its total energy between 0.13 and 15 mJ, which may be further lowered by insertion of calibrated filters. The best repeatability is achieved with repetition rates of two or less shots per minute.

The *microscope* is mounted in the same housing as the laser which ensures a rigid position of the microscope relative to the laser. It is designed to be used first as a viewing microscope through which the target is positioned and then as a laser beam concentrator which aims the laser emission exactly on the chosen target which is centered by means of an illuminated reticle. The position of the reticle relative to the viewing field is adjusted to ensure an aiming precision better than 10 μm.

Magnifications up to and inclusive of those requiring oil immersion objectives are available, but for laser operation the standard magnification is 100, produced by a 10 x ocular and a 10 x objective. A scale is provided in one ocular with an apparent distance of 10 μm between two divisions when used with the mentioned optics. A 40 x high aperture reflection objective has been ordered with the hope of reducing the extension of the diffraction limit. It is expected to be delivered by Fall, 1973.

Shortly before the laser is fired, a mirror is automatically swung into the light path to direct the laser beam through the objective and, at the same time, it prevents the laser emission from escaping through the oculars and damaging the observer's eyes.

The stage of the microscope may be moved up and down both roughly for positioning purposes, and finely for focusing. It may be removed completely, if necessary, thus leaving space for bulky samples.

The energy of the *electrical cross-excitation* is taken from a capacitor the size of which is selected between 2 and 10 μF in steps of 2μF and which is loaded at 400, 600, 800, or 1,000 V. This energy is delivered to the sample vapor in the gap between two graphite electrodes located just above the sample; the site can be finely adjusted while they are in operating position. A multiple spring-loaded mechanism allows the electrodes to be positioned with a repeatability better than 10 μm in any direction and they can be moved together, or one relative to the other along the axis of the gap.

The *detection unit* comprises the relay optics, the spectrograph, and the electronic detection. The detection unit collects the light emitted by the sample and directs it through relay optics to the entrance slit of a Jarrell-Ash grating spectrograph, the output plane of which can be equipped with a photographic plate, or a polychromator housing with the associated electronic detection and display. For obvious reasons, fundamentals of the design parameters are not disclosed at this time, but the chosen solution will be described.

The *collection optics* consists of a f/1.4 Schwarzschild mirror objective and two quartz-lithium fluoride achromats. The total magnification is 8, thus matching the aperture of the collection objective to that of the spectrograph. This relatively high magnification is particularly useful when the plume size is so small that its image entirely falls in the entrance slit. Even when this is not the case, it has been shown that a high magnification is desirable for introducing as much light as possible into the spectrograph.

The lenses are corrected for chromatic aberration in the UV at 2,400 and 3,000 Å as well as for spherical aberration; this results in a sharp image that is needed for spatial studies of the emission as well as for the separation of the signal from the unwanted background.

The *spectrograph* is the Jarrell-Ash model 75-152 Fastie Grating Spectrograph® which can be operated both as a photographic and as a photoelectric instrument. Its f/8.6 aperture, together with a resolution of 0.1Å in the first order, makes it the best choice of presently available commercial grating spectrographs. With a 1,180 grooves/mm grating and a 4 x 10 in. photographic plate, it covers a wavelength range slightly over 2,000 Å at a time, the reciprocal linear dispersion being 8.3 Å mm.

The polychromator housing was modified slightly to accomodate the shielded outputs from each photomultiplier.

The *electronic detection* is a fast integrating system similar to that suggested by Peppers et al., but considerably improved.

Information concerning the shape and the duration of the electric pulses collected at the anode of a low-noise photomultiplier was obtained through preliminary measurements and guided the choice. Two ways of electronic detection are practicable: the direct oscilloscopic display with or without photographic record and the fast integration that was chosen here to equip standard instruments. An integrator is switched "on" after the Pockel cell starts "opening" and later is switched "off," leaving enough time for the entire signal to be integrated on a 1,000 pF capacitor.

Within the next 0.4 sec, the integrated value is read, displayed on a digital voltmeter, and held until the next laser shot command resets the integrator to ground some half second before the laser is fired again.

The small, half inch diameter R300 photomultiplier was chosen because of its size and its successful use in the field of Raman spectroscopy, which also requires low-noise, fast, sensitive detectors. Its power supply circuitry is arranged to deliver currents to the anode in the order of 200 mA for a short time. This peak current is 10^4 times larger than the maximum allowed current, but is not actually expected to flow long enough to damage the P.M., and it is needed to ensure proper linearity; actual currents in the tens of milliamperes are common.

Three different types of interface between the P.M. and the integrator are available:

1. A most sensitive preamplifier mounted on top of the P.M. in the same cylindrical aluminum shield as the P.M. itself. Pulses produced by a single photon in the 3,500 Å range will have a maximum peak value of 2.5 V when 900 to 1,000 V are applied to the P.M. dynode string. The rise time time is 15 nsec and the pulse pair resolution is roughly 50 nsec.
2. A similar fast amplifier, but with a 6 times smaller gain. Both amplifiers feed into 50 Ω shielded cables.
3. A direct 50 Ω output into a cable. This output is used when large signals are available that are fed directly into a transconductance amplifier preceding the integrator.

Qualitative detections in biological materials were reported by Rosan, Healy, and McNary[57] on sections of brain, pancreas, and stomach tissues. Their aim was to trace certain elements like Zn; no quantitative analysis was needed.

In past studies, relative concentrations of Mg, Al, Fe, and Zn were estimated using an external standard by Rosan, Brech, and Glick[56] in human teeth where a crater of 65 μm in diameter was produced. Other biological materials were tested for quantitative analysis by Goldman, Ruben, and Sherman,[24] who examined supra- and subgingival calculus for their content in Mg, Al, Si, Ca, P, Fe, Cu, and Zn. Sherman and his co-workers[62] carried out a similar study on tooth enamel, dentin, carious dentin, cementum, calculus, and bone. The concentrations of P ranged from 0.001 to 10%, of Mg from 0.001 to 0.1%, and so on; the crater sizes were of the order of 50 μm in diameter and 25 μm deep. Glick and Rosan[20] reviewed the application of the laser microprobe method on biological materials and reported on determinations of Ca and Mg concentrations in both dog and baby stomach tissue. The need for instrumental improvements was already pointed out in that 1966 publication. Glick[23] reported on improvements brought to a laser microprobe and tested the instrument for its analytical capability on kidney where concentrations of Ca, Fe, Zn, Cd, Cu, Mn, Al, Pb, Mo, Sn, Sr, and Cr were estimated and where detection limits were said to range from 10^{-18} g for Cr to 10^{-14} for Mg. Laser impacts were shown on samples as small as human leukocyte and rat spermatazoon head.

Many structures have been examined by laser microemission spectroscopy. Current studies have been in the field of lead pollution, with the measurement of lead in the cortex and medulla of the hair along the various regions of the hair shaft in children. These will determine the concentrations of lead correlated with growth periods of the hair. Vahl[65] did S.E.M. and studied the impact areas to determine the ultrastructure sources of lead. Controls were atomic absorption spectroscopy.

Hillenkamp, Kaufmann, and Reny[34] have used a ruby laser microprobe for mass spectroscopic analysis of inorganic and organic molecules. In red blood cells, impact areas of 0.5 μm were produced. Further developments will include a UV microbeam.

It is obvious that the laser microprobe offers considerable opportunity for trace metal microanalyses in almost any structure. This can be used for screening procedures in metabolic diseases, with disturbances of copper, sodium chloride, calcium, and phosphorus for heavy metal poisoning. The application for microanalyses for forensic medicine has been mentioned before.

Bacteria and Fungi

Next in the study of laser biology is the effect on organisms.[38] Studies have been done on color absorption, motility, cytogenetic changes, and chemotaxis. Recent reports include studies on antigenic and virulence changes after laser impacts. Nassal et al.[48] measured rapid *E. coli* motility by He-Ne laser correlation spectroscopy. Studies have

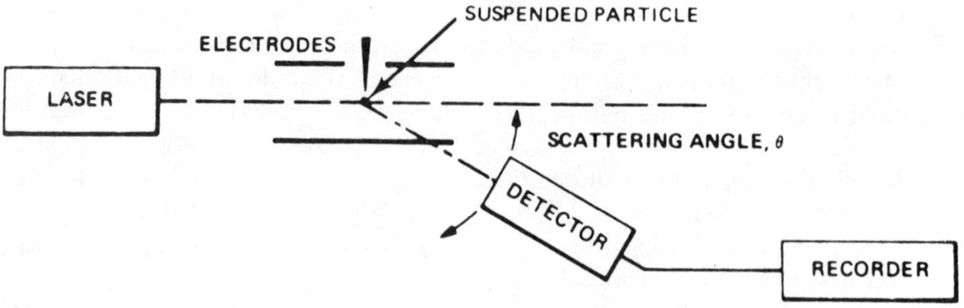

FIGURE 14–9. Size measurement of single microparticle from scattered light intensity proportional to the square of the volume of the particle. (From Science Spectrum, Santa Barbara, Calif. With permission.)

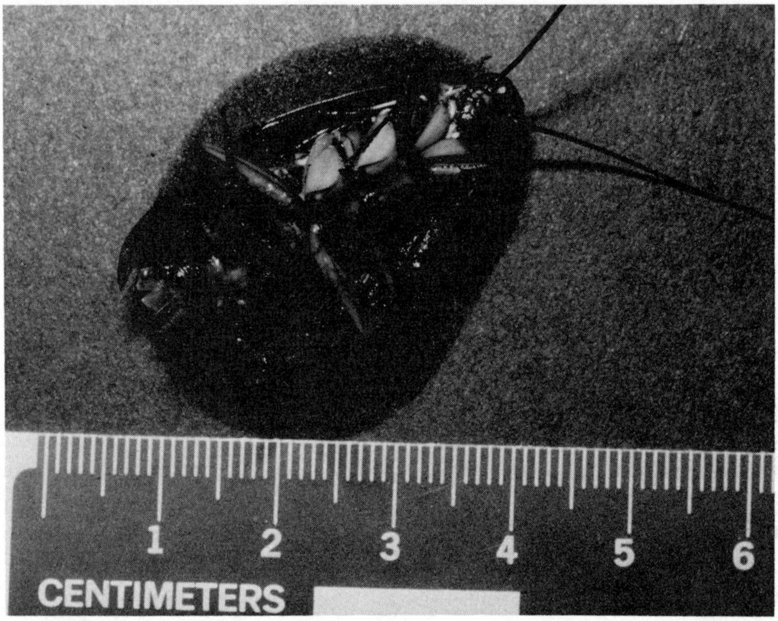

FIGURE 14–10. Ruby laser impact of roach 85 J/cm² 2.5 msec normal mode.

been done by Schwarz[61] and by Carter[14] with *Pseudomonas* and chromogenic acid fast bacilli *Serratia* and *P. acnes.*

Attempts have been made to induce plilomorphic changes in such colored fungi as *Aspergillus niger* and the pathogen *Homodendron pedrosi.* The author's studies showed all-or-none reactions with pulsed ruby laser impacts. With impacts on live organisms, safety measures must be used.

Particle Size Measurement

Laser light scattering techniques, especially with the He-Ne laser have been used for measurement of bovine plasma albumin in water and suspensions of Soule mouse leukemia virus identification of bacteria. Vaccinia virus particle measurements have been done with the Brice Phoenix® light scattering photometer, but now such determinations can be done by low output He-Ne lasers. The laser, then, can be used to determine particle size from differential light scattering data. The resultant secondary field is proportional to the polarizability of the particle, the number of electrons, or, essentially, the volume of the particle. Therefore, the scattered light intensity (i.e., the energy flow rate) is proportional to the square of the particular volume.

A development of particle studies has been the

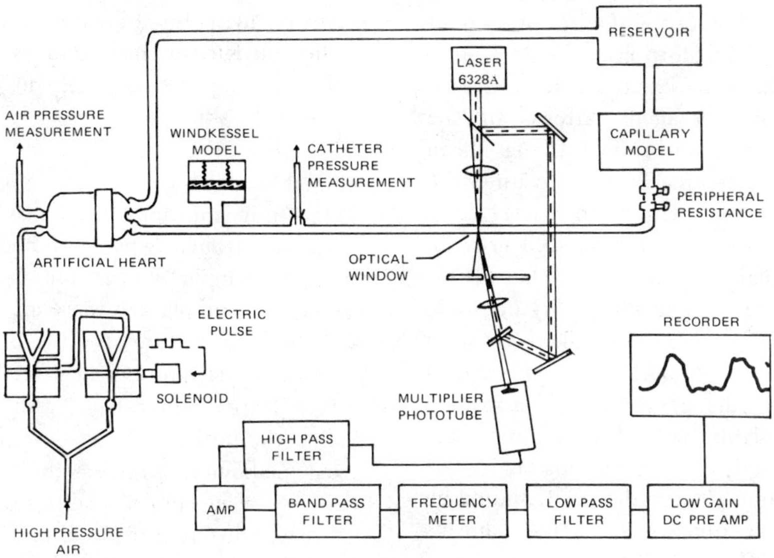

FIGURE 14–11. Laser Doppler technique. (From *IEEE Trans. Biomed. Eng.* With permission.)

study of radiation pressure, the pressure exerted by the momentum of impacts of countless photons. This has been done by Ashkin and Dziedzic[4] with a 1/4-W laser on a tiny glass sphere 20 μm in diameter. As long as the laser is focused, the particle remained trapped. The particle so levitated can be photographed by its scattered light. Laser levitation may be used to separate gases through light resonant with a particular gas. Radiation pressure may also involve atomic and molecular beams.

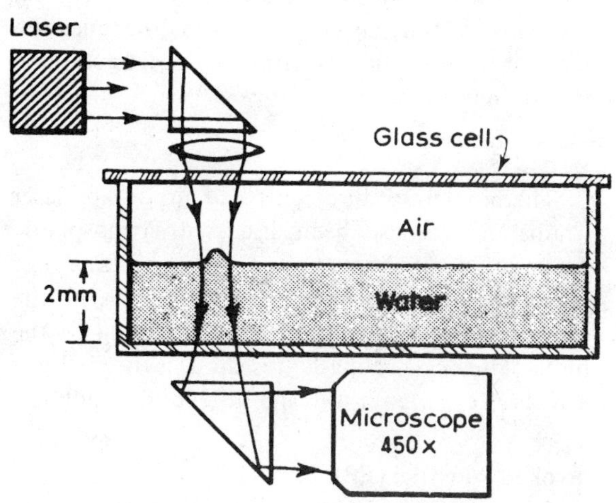

FIGURE 14–12. Light exerts outward pressure on the surface of a liquid to overcome surface tension, Ashkin and Dziedzic of Bell Laboratories. (From *New Sci.*, February 8, 1973. With permission.)

Insects

Lasers have been used in insects to observe effects of ruby and neodymium impacts on chitin and on spiders to observe changes in web patterns. Friedlander and Salet used a frequency-doubled neodymium laser, 530.0 nm and a quadrupled 265.0 nm for microbeam irradiation of neuropteran giant centrioles of unstained and acridine orange sensitized *Chrysopa carnea* cells. Spermatocytes and spermatids were irradiated. No DNA was found in giant centrioles. Mitochondria did bind more acridine orange than centrioles. They suggest that theories of centriolar autonomy should be critically evaluated.

Microsurgery in insects has been done by other investigators. Attempts have been made to do insect control with lasers in feed and grain storage areas through color differences. These are still in the investigative stage.

Botany

Laser botany is a completely undeveloped field.[50] Yet, the few studies done offer greater challenges to botanists and photochemists as well as experimental oncologists. Chloroplast membrane studies have been done with ruby laser irradiation of isolated chloroplasts. The interest is the laser absorption by chlorophyll. These studies were done by Kehani, Floyd, and Chance.

Holography shows detailed patterns of the rapid growth of the mushroom. It has been proposed that this technique be used to study the effect of pollutants in plants. Low level laser radiation by Parr[51] has accelerated seed growth.

Paleg and Aspinall[50] have shown that He-Ne impacts on the photoreceptor phytochrome system of even one quarter of a mile away were effective on the growth of barley and emphasized that the photoperiodic responses of crops are potentially controllable with lasers. Additional studies should certainly continue in this field.

On the other hand, plant pathologists should be interested in the experiments of Musayen and his colleagues[47] in Russia. A He-Ne laser was used to irradiate tomato rootlets. Chromosomal changes varied with the period of exposure. There were subsequent delays in plant growth and development from the irradiated seeds especially in floral morphology and physiology. Again, the question may be the level of irradiation as regards stimulation or inhibition.

Birds

The problem of bird control at airports by laser irradiation has also been discussed. Atmospheric propagation of laser beams must be safe for personnel in aircraft and control towers at the same time as it is disturbing or annoying to the birds. Much more basic research is needed for effects on eyes, labyrinth, and other bird structures.

Evoked Potential (EP)

The two most sensitive infrared receptors in the world are said to be those in the pit viper and boa snake. Harris and Gamon implanted electrodes to the surface of the telencephalon of snakes. The evoked potentials (EP) were measured with a signal averager (Princeton Applied Research Wave Form Educator). The stimulus was supplied by a CO_2 laser and suggested that the receptor operated on thermal principles because of the stimuli in the high infrared and the fact that the "EP latency after an infrared stimulus is comparable to that of visual EP." EP has been used also in eye of animals to study subthreshold impacts on the retina. As yet, EP often is too easily disturbed and sensitive to be used for laser studies in man. Recently, Roth et al.[71] have used a flash or changes in background intensity light for EP studies in man. Perhaps studies on Kirlian chemiluminescent photography with laser impacts may help in this area for impacts on the skin.

Animal Studies

It is now appropriate to consider animals in our progress through what has been loosely called biology. Animal studies are listed in this large category to emphasize these are still basic studies with applied features for the birth of veterinary laser medicine, another neglected field. Certainly, animal experiments have great value and also distinct limitations in the development of clinical laser medicine, surgery, and perhaps even of dentistry. The optical qualities of the target areas are often different from those of man for eyes, skin, and even viscera. The reactions in living tissue may also be different, as it is even for different species of animals. Thus, investigations of the important factor chronic exposure are difficult to extrapolate from animal to man. Finally, even with the current concern of human experimentation, we must once more repeat the tired old phrase of Goethe: the proper study of man is man.

However, in the orderly progress of research and according to rules, and in the present state-of-the-art, animal research must be done first. This gives us some idea of reactions and safety for that particular laser system as limited by all the parameters of the laser and the characteristics of the target area of that particular species. There is a vast amount of literature available on animal laser research; however, it will not be reviewed here.

Eye research studies have been discussed. Even here, the human eye corrected for color comparison is less reactive to minimal dose exposure of some laser systems than animal eyes. Studies of wound healing and secondary infection are of great importance because of the increasing use of laser surgery even for major conditions. Skin incisions with CO_2 and Nd-YAG lasers have been studied extensively in relationship to incisions as a vaporization of tissue water, as Rockwell and others have suggested. These features are important, for the ideal characteristics for laser surgery are precision, bloodless techniques, and good wound healing. The intensity and the extent of the thermal coagulation necrosis from the CW lasers relate to good wound healing. Mester et al.[42,43] have actually used low level ruby laser, 1 J/cm^2, to stimulate wound healing produced by third degree burns from high-frequency electrosurgery. The stimulation of wound healing was an increased rate of epithelial growth. This factor of laser stimulation has been mentioned above.

It is often difficult to develop the proper controls for laser surgery in man. This can be done more easily, of course, in animals. The concern is to attempt to set up comparative power levels for the development of thermocoagulation necrosis of tissue. The controls used are, of course, the scalpel, next, the high-frequency electrosurgical apparatus, next, cryosurgery with liquid nitrogen techniques, and, finally, the plasma torch. With the new developments of the plasma torch by Incopera and his associates, it may be possible to use the plasma torch safely.

Hishimoto, in the author's laboratory, has used the plasma torch in highly vascularized organs as the liver and heart. Argon gas embolism is produced with the rapid death of the animal.

From more than 10 years of experience with laser animal irradiation it has been established that the laser is not carcinogenic even with chronic exposures.

Laser surgery has been performed on practically every organ. The areas of particular interest are the skin for graft replacements, the stomach for endoscopy and arrest of bleeding of stress ulcers, and the liver for arrest of bleeding following trauma of auto accidents. Another field of special interest now is maxillofacial surgery, surgery in the mouth, pharynx, larynx, and down into the bronchi and lungs. The basic studies of Hobeika and Rockwell,[35] Högberg and Stähle,[36] Strong, and Jako have established the principles for clinical surgery in this area. The development of high output CO_2 laser surgery by Polanyi and Bredmeier has made such applications possible.

By means of laser technology, animal test models have been developed by the author's laboratory for the treatment of cerebral edema and also for study of the development of fat embolization, especially associated with bone injury.

The laser has also been used in the author's laboratory for the studies of circulation, clotting, the erasure of clots, microcirculation dynamics, etc. The laser Doppler effect has been used in studies of circulation, as it has been with ultrasonic techniques.

Studies in Cancer

The laser studies in cancer start with laser irradiation of tissue cultures of cancer cells, especially melanoma or cancer cells stained with vital dyes or substances used to induce a phototoxic effect on cells when these are exposed to lasers. Many cancers, induced and transmitted, have been used to study the value and limitations of laser cancer surgery. The concern of the spread of viable cancer tissues by impacts of pulsed lasers is real, although this has not been observed in controlled clinical studies in man. The tumors used more commonly have been the melanomas, hepatomas of the liver, and the angiosarcomas. Laser cancer studies have been done in animals not only to determine techniques of adequate removal of the cancer itself, but also for investigation of cancer immunobiology, again, especially with the melanomas. Finally, the technique of laser microemission spectroscopy can be done for analysis of cations of living cancers *in situ*. All this will help in attempting to evaluate critically the effect of laser treatment not only of itself but with the laser as an adjuvant to surgery, to cancer chemotherapy, and to radiation. Such investigations obviously are difficult in man, but at times there are definite needs for the laser as an adjuvant in the cancer treatment program.

Laser in Photobiology

As shown in the chapter on Chemistry, the laser has tremendous interest in the field of photochemistry. In the future, it will have even greater development in the entire area of photobiology.

For applied research, the laser now is of interest in the field of cutaneous photobiology. The easy availability, again, of the skin makes it possible to study the effects of radiant energy of the laser on living human skin.

It is well to review again the primary changes in the skin produced by light, usually 290 to 320 nm. Pathak[75] has listed these primary changes as:

"1.a. formation of cyclobutane dimers of pyrimidines in DNA.

b. hydration products of pyrimides.

c. cross linkage of DNA to protein.

d. cleavage of phosphate esters bonds in DNA.

2. generation of free radicals that can damage the cell membrane or other organelles.

3. a marked depression of the synthesis of DNA and of cell mitosis within 1 to 3 hours after irradiation; also there is inhibition of RNA and protein synthesis.

4. direct and indirect injury to the endothelium of blood vessels.

5. liberation of vasodilation mediators (histamine? kinins?).

6. secondary changes in proteins and RNA (e.g., inactivation of enzymes, damage to mitochondria, endoplasmic reticulum, and polysomes)."

How these primary changes will be affected by coherent radiation in these wavelengths and other wavelengths is the challenge for the future.

Conclusions

This is but a general and superficial, perhaps even for nonbiologists, review of the vast and diffuse region of laser biology. This is a new biology that is being developed too slowly through lack of lines of communication and inadequate funding for basic laser research. Many areas of research are still incomplete, some speculative, and some significant. The general dynamics of the reactions in tissue, microirradiation with all its applications for cytology and cytogenetics structural, functional, and analytic aspects, particle diagnostics, laser microbiology, entomology, and botany are all new and exciting fields. Microsurgery of chromosomes is perhaps the most significant of these at present. Laser animal research in all forms of surgery, including cancer surgery, precedes applications in man. Finally, laser reactions offer a new field of research in modern immunobiology, cytofluorography, photobiology, and genetics. With all this great need, then, how can interest and support for basic research be denied, again, for lasers for the good-of-man?

REFERENCES

1. **Allemand, C. D.,** *Appl. Opt.,* 9, 1304, 1970.
2. **Anderson, W. L.,** Holography and coherent optics in microscopy, Proc. 21st Annual Conference, Engineering, Medicine and Biology, Houston, 1968.
3. **Andreev, S. I., Symshits, Yu. I., Kaporskii, L. N., and Mustova, G. S.,** *Sov. Phys. Tech. Phys.,* 13, 657, 1968.
4. **Ashkin, A.,** The pressure of laser light, *Sci. Am.,* p. 63; *Laser Weekly,* November 1, 1971; and *Sci. News,* 309, November 6, 1971.
5. **Baba, K.,** Selective injury of mitochondria with James Green B and ruby laser light, enzyme and morphological and ultrastructural study, *Acta Pathol. Jap.,* 20, 59, 1970.
6. **Barnes, F. J.,** Personal communication.
7. **Barer, R. and Smith, F.,** New developments in laser microscopy, *New Sci.,* August 1972, 380.
8. **Berns, M. W. and Floyd, A. D.,** Chromosomal micro-dissection by laser. A cytochemical and functional analysis, *Exp. Cell Res.,* 67, 305, 1971.
9. **Berns, M. W.,** Personal communication, 1971.
10. **Berquist, N. R.,** Pulsed dye laser for microfluorometry, *Laser Sphere,* 2, 1, 1972.
11. **Bessis, M., Gires, F., Mayer, G., and Nomarski, G.,** *C. R. Acad. Sci.,* 1010, 1962.
12. **Borg, S. F.,** A mathematical physical model for the initial behavior of surface and internal tissue subjected to pressure density variations induced by laser irradiation, *Med. Res. Eng.,* 10, 20, 1971.
13. **Carlson, F. D.,** Laser in biophysical research, Int. Symp. Mod. Technol. Physiol. Sci., July 26-30, 1971.
14. **Carter, R.,** Personal communication, 1972.
15. **Crewe, A. V.,** A high resolution scanning microscope, *Quart. Rev. Biophys.,* 3, 137, 1970; *Sci. Am.*
16. **Fiel, R. J., Mark, E. H., and Munson, B. R.,** Small angle light scattering of bioparticles. III. Vaccinia virus, *Arch. Biochem. Biophys.,* 141, 547, 1970.
17. **Floyd, R. A., Keyhani, E., and Chance, B.,** Membrane structure and function. I. An electron microscopic study of the alteration induced by laser irradiation on the chloroplast lamellar membrane, *Arch. Biochem. Biophys.,* 146, 618, 1971.
18. **Floyd, R. A., Keyhani, E., and Chance, B.,** Membrane structure and function. II. Alterations in the photo-induced absorption changes after treatment of isolated chloroplasts with large pulses of ruby laser, *Arch. Biochem. Biophys.,* 146, 627, 1972.
19. **Gabor, D., Kock, W. E., and Stroke, G. W.,** Holography, *Science,* 173, 11, 1971.
20. **Glick, D. and Rosan, R. C.,** *Microchem. J.,* 10, 393, 1966.
21. **Glick, D.,** *J. Histochem. Chytochem.,* 14, 862, 1966.
22. **Glick, D.,** *Ann. N. Y. Acad. Sci.,* 157, 265, 1969.
23. **Glick, D., Marich, K. W., Carr, P. W., and Beatrice, E. S.,** *Ann. N. Y. Acad. Sci.,* 168, 507, 1970.
24. **Goldman, H. M., Ruben, M. P., and Sherman, D.,** *Oral Surg.,* 17, 102, 1964.

25. **Goldman, L., Fradin, D. W., Bloembergen, N., and Richfield, D. F.,** Studies in laser safety of new high output systems. II. Picosecond impacts, *Opt. Laser Technol.,* 2, 11, 1973.
26. **Goldman, L., Yablonovitch, E., Bloembergen, N., and Richfield, D. F.,** Studies in laser safety of new high output systems III TEA CO_2 laser impacts, *Opt. Laser Technol.,* 2, 58, 1973.
27. **Goldman, L., Vahl, J., Rockwell, R. J., Jr., Meyer, R., Franzen, M., Owens, P., and Hyatt, S.,** Replica microscopy and scanning electron microscopy of laser impacts, *J. Invest. Dermatol.,* 52, 18, 1969.
28. **Goldman, L. Rockwell, R. J., Jr., et al.,** Preliminary investigation of fat embolization from pulsed ruby impacts of bone, *Nature,* 221, 361, 1969.
29. **Goldman, L.,** Laser in biomedical research, Int. Symp. Mod. Technol. Physiol. Sci.,Munich, July 27, 1971.
30. **Goldman, L.,** *Application of Lasers in Biology and Medicine,* Springler-Verlag, New York, 1970.
31. **Goldman, L. and Rockwell, R. J., Jr.,** *Laser in Medicine,* Gordon and Breach, New York, 1971.
32. **Hall, R. R., Beach, A. D., Baker, E., and Morison, P. C. A.,** Incision of tissue by carbon dioxide, *Nature,* 232, 131, 1971.
33. **Herzenberg, L. A., Hulett, R. H., Wolf, P. L., Bonner, W. A., and Sweet, R. G.,** Lasers in cell sorting, *CLEN,* December 31, 1971.
34. **Hillenkamp, F., Kaufmann, R., and Remy, E.,** Laser microprobe, Int. Symp. Mod. Technol. Physiol. Sci., Munich, July 26-30, 1971.
35. **Hobeika, C. P. and Rockwell, R. J., Jr.,** Laser microsurgery in experimental otolaryngology, *Trans. Am. Acad. Opthal. Otol.,* 76, 325, 1970.
36. **Högberg, L. and Ståhle, U.,** The laser as a microsurgical tool. Some experiments of myelinated nerves and the labyrinth, FOA 2 Rapport C 2460.51, Forsvarets Forskningsanstalt, Stockholm, March 1971.
37. Jarrell-Ash Division of Fisher Scientific Company, Product Bulletin 45-604, 1971.
38. **Korn, M. Ya. and Chelinyya, A.,** Effect of laser microirradiation of filamentous forms of *Eschericia coli, Mikrobiologiia,* 39, 1073, 1970.
39. **Korpel, A.,** Personal communication, 1972.
40. Laser Medical Potential – Film #12 The Upjohn Vanguard of Medicine, The Upjohn Co., Kalamazoo, Mich.
41. **May, J. F., Rounds, D. E., and Cone, C. D.,** Intercellular transfer of toxic components after laser irradiation, *J. Natl. Cancer Inst.,* 46, 655, 1971.
42. **Mester, E. and Jaszsagi-Nagi, E.,** Biological effect of laser radiation, *Radiobiol. Radiother.,* 123, 377, 1971; *Laser Abstr.,* 1972–50 July 19, 1972, 1972–1302.
43. **Mester, E., Spiry, T., Szende, B., and Tota, J. G.,** Effect of laser rays on wound healing, *Am. J. Surg.,* 122, 532, 1971.
44. Microparticles Single Size Measurement, Science Spectrum, 1216 State Street, Santa Barbara, Calif.
45. **Mims, M. F. and McKinnell, R. G.,** Laser irradiation of the chick embryo germinal crescent, *J. Embryol. Exp. Morphol.,* 26, 31, 1971.
46. **Mukherjee, A. B., Blattner, P. Y., and Nitowsky, H. M.,** Quinacrine mustard fluorescence of sex chromaten in human amniotic fluid cell cultures, *Nature,* 235, 226, 1972.
47. **Musayev, M. A., Abdullayeva, T. Y., and Yegyazarov, V. V.,** The mutagenic effort of laser radiation on tomatoes, *Tsitol. Genet.* 5, 207, 1971; *Laser Abstr.,* 1972-1614, 1972-76.
48. **Nassal, R., Chen, S. H., and Lai, C. C.,** Use of Laser scattering for quantitative determinations of bacterial motility optics, *Communications,* 4, 35, 1971; *Laser Abstr.,* 1972-7, 1972-33.
49. **Ohnuki, Y., Rounds, D. E., Olson, R. E., and Berns, M. W.,** Laser microbeam irradiation of the juxtanucleolar region of prophase, nucleolar chromosomes, *Exp. Cell Res.,* 71, 132, 1972.
50. **Paleg, L. G. and Aspinall, D.,** Field control of plant growth and development through the laser activation of phytochrome, *Nature,* 228, 1970.
51. **Parr, W.,** Personal communication, 1971.
52. **Pitts, D. G.,** The Effects of Ultraviolet on the Eye, SAM-TR-69-10 Brooks AFB, Texas, 1969.
53. **Pugh, W. J. and Saunders, L.,** A laser light scattering apparatus, *J. Pharmacol. Pharm.,* 23, 855, 1971.
54. **Rockwell, R. J., Jr., Fidler, J. P., Epstein, R. A., Nafrester, Z., Franke, E. K., Hashimoto, K., and Dreffer, R. L.,** Conference on Lasers in Medicine and Biology, London, 1969.
55. **Rockwell, R. J., Jr.,** The Current Status of Laser Applications in Medicine and Biology, Laser Conference, Tel Aviv, Israel, October 1971.
56. **Rosan, R. C., Brech, F., and Glick, D.,** *Fed. Proc.,* 24, (Suppl. 14), S-126, 1965.
57. **Rosan, R. C., Healy, M. K., and McNary, W. F., Jr.,** *Science,* 142, 236, 1963.
58. **Rosensweig, R. E.,** Personal communication, 1969.
59. **Rounds, D. E.,** Personal communication, 1971.
60. **Salmeen, I., Gill, D., Rimai, L., McCormick, J. J., Maker, C. M., Arnology, W. J., and Hight, M. E.,** Application of laser beat frequency spectroscopy to the detection and characterization of an oncogenic RNA virus, *Biochem. Biophys. Res. Commun.,* 47, 1172, 1972.
61. **Schwarz, J.,** Personal communication, 1972.
62. **Sherman, D. B., Ruben, M. P., Goldman, H. M., and Brech, F.,** *Ann. N. Y. Acad. Sci.,* 122, 767, 1965.
63. **Stroke, G. W.,** Sharpening images by holography, *New Sci.,* September 23, 1971.

64. Tunable dye laser, Technical Information Bulletin, Leitz Leica, November 1972, 13.
65. **Vahl, J.,** Applicability of the scanning electron microscope to the topography of the surface of specimens after laser irradiation, Second Conference on the Laser, *Ann. N. Y. Acad. Sci.*, 168, 644, 1970.
66. **van deer Haagen, G. A. and Whitlow, D. E.,** Application of Cineholomicrography to the Study of Microcirculation, Henodynamics, U. S. Government R/D, Report AD-719, 401, February, 1971; *Laser Abstr.*, 12/22/71 1971-2756 and 1971-113.
67. **Van Lighten, R.,** Personal communication, 1971.
68. **Becker, R. O.,** Electrical current sparks mammalian tissue regeneration, *J.A.M.A.*, 223, 483, 1973.
69. **Rosensweig, R. E.,** Personal communication, 1969.
70. **Korn, M. Ya.,** A laser microirradiation attachment for a biological microscope, *Bull. Exp. Biol. Med.*, 72(10), 1226, 1971.
71. **Ford, J. M., Roth, W. T., Dirks, S. J., and Koppell, B. S.,** Evoked potential correlates of signal recognition between and within modalities, *Science,* 181, 465, 1973.
72. Microfluometry pulsed dye laser adds new dimension, *Lasersphere,* 2, 1, 1972.
73. **Edlow, J., Fine, S., Vawter, G. F., Jockin, H., and Klein, E.,** Laser irradiation: effect in rat embryo and fetus in utero, *Life Sci.*, 4, 615, 1965.
74. **Wilpizeski, C., Sataloff, J., Innis, R., and Shiner, W.,** Audiometric Effects of Pulsed Laser Irradiation of the Squirrel Monkey Inner Ear, 76th Annual Meeting of Acoustical Society of America, Cleveland, November 22, 1968.
75. **Pathak, M. A.,** Fundamentals of Photosensitization, Society for Investigative Dermatology, New York, June 22, 1973.

Chapter 15

LASER IN MEDICINE

The applications of the laser in medicine are another interesting spin-off, as it were, from studies of the hazards of the laser and the developments of the laser in industry and in the military. Possibly, the least amount of research and development and consequently the slowest progress has been in the applications of the laser in medicine. There are many reasons for this, according to the prejudices of the observer.

"In the governmental research programme we can speak with the gloomy voice of experience, for, except for the military, biomedical research on lasers is so far down on the totem-pole that it cannot be found above ground at all. Physicians have been criticised for their reluctance to adopt new advances in technology, especially electronics. The old cliché is that at least 12 years must elapse before any advances in medical electronics are adopted for the practice of medicine. However (but only in very feeble defence) it is the physician at the bedside who has to use new instruments, and who has to be careful of what he uses on his patients. There are so many cross-currents of force today, which, on one hand, demand that the physician do miracles immediately, or at least in microseconds, while on the other hand another vast and vocal group watches and condemns him for using things new and untried. The harassed physician attempts to steer a middle course; as a result, unfortunately, progress in medicine is slow.

So, just as all new drugs are tested for possible applications, all new advances in modern technology are scanned to determine their actual and potential relationships to biology and medicine. This is what is happening now in the flashing, brilliant field of laser technology."[14]

As regards biomedical applications, the laser industry is not interested.

"Many manufacturers prefer not to work in such a field, because of the poor market capabilities, tedious control of federal, state and local radiation control measures and, finally, the conservative physician as the reluctant and hesitant consumer.

The laser manufacturer meets none of these drawbacks in other market areas. In contrast, the pharmaceutical manufacturer has had the experience and the eventual profit from the tedious and expensive search for new drugs, and one recent and interesting development is that drug companies have become interested in acquiring biomedical engineering subsidiaries, in one instance, a laser company.

However, the new Federal Radiation Control Act for Health and Safety in the United States, may awaken the laser manufacturer to his responsibilities for protection of the workman and the consumer, and this may in turn make the biomedical field seem less unattractive to him.

Despite all these deficiencies in the present status of the subject, it is surprising to note that there is as yet no loud lament from the small but significant group which clings tenaciously to the almost deserted, but still the flashing, biomedical laser bandwagon."

It has been said that the two most conservative industries where the laser has had to fight its way slowly, at least in the U.S., have been the machine tool industry and medicine.

It has been stated repeatedly that there is a terrific responsibility to present a new modality for the diagnosis and treatment of patients. Consequently, progress has been slow. However, if the laser industry must realize that knowledge of the laser reactions in living tissues of man helps development of safety programs so that industry can expand for the operator and also for the consumer, the industry should realize also that effective developments of the laser, where needed, for diagnosis and treatment in medicine, would open up a considerable market in these days of expansion of health services.

To establish needs in medicine, as elsewhere in the applications of lasers, there must be, in order:

1. A study of the characteristics of a new laser system and its transmission and cost
2. A planned safety program
3. A planned protocol to include
    a. Adequate controls
    b. Where indicated, investigations at cellular level – such as with tissue cultures
    c. Animal investigative studies with due regard to the transference of data
    d. Investigative studies in man

The initial finding of such a study is, of course, difficult to obtain unless there is far-sighted vision with an appreciation of the value of basic research. An attempt will be made to show the need for the laser for today and tomorrow in the diagnosis and treatment of patients.

Although it is not fashionable to speak of research yet, there are possibilities for actual basic studies in man other than those mentioned in the chapter, Biology.

There are but few "intentional" experiments on the eyes of man, but there is a vast experience during the courses of extensive laser treatments of eye diseases in more than 10,000 patients. There is no evidence to date of precancerous or cancerous changes in the eyes of man from the thousands of laser treatments given all over the world.

So these studies, actually a phase in the controlled investigation of the application of the laser, are of value not only for the research background of the effect of lasers in man, but also they are of great value in the development of laser safety programs.

The Literature on Laser Medicine

It is not the purpose of this section to develop in great detail all the current data of laser medicines. Current literature is available through the excellent laser journals in the U.S., as listed previously, in England, as for example, *Optics and Laser Technology,* and in Germany, *Lasers Elektro-Optik.* Excellent reviews are available, as shown previously, in the *Laser Weekly* and in the abstract reviews. For laser ophthalmology, the journals of ophthalmology offer current details. Perhaps the International Ophthalmology Clinics will offer a current review of laser photocoagulation.

These are some of the current and near-current books available:

1. Annals of The New York Academy of Sciences – The Laser, May 28, 1965 (Conference Chairmen – Leon Goldman and Joseph Weber).

2. Goldman, L., *Laser Cancer Research,* Springer-Verlag, New York, 1966.

3. Goldman, L., *Biomedical Aspects of The Laser,* Springer-Verlag, New York, 1967.

4. Levine, S. N., *Advances in Biomedical Engineering & Medical Physics,* Vol. 1, Interscience, New York, 1968.

5. Annals of The New York Academy of Sciences, Second Conference on The Laser, February 10, 1970 (Conference Chairman – Leon Goldman).

6. Goldman, L., and Rockwell, R. J., Jr., *Lasers in Medicine,* Gordon and Breach, New York, 1971.

7. Wolbarst, M. L., *Laser Applications In Medicine and Biology,* Vol. 1, Plenum Press, New York, 1971.

8. Gamaleja, N. F., *Biological Effects of Laser Radiation,* (Russian) Kiev, 1972.

9. Hillenkamp, F. and Kaufmann, R., Laser in der Medizin – Gesellschaft fur strahlen and Umnvelt forschung mbH, München, 1972.

Instruction courses are other devices to teach new advances in medicine. The following is offered as a test model for the introduction of a new technology to the medical profession. This was done after almost 10 years of investigations. In April 1971, the first comprehensive course on applications of the laser was given by the Laser Laboratory in cooperation with the Department of Continuing Medical Education of the College of Medicine of the University of Cincinnati.

In addition to the start of the Laser Laboratory, the faculty and their respective assignments included:

Microirradiation	Donald Rounds
Microemission spectroscopy	Fred Brech
	Charly Allemand
Holography including microholography	Raoul Van Ligten
Laser ophthalmology	H. Christian Zweng
	Francis L'Esperance
	Lloyd Aiello
Laser otolaryngology and maxillofacial surgery	Geza Jako
	M. Stuart Strong
Laser surgery	Robert Goodale
	Thomas Polanyi
	Stanley Stellar
	John Moore
Laser dentistry	Ralph Stern

The faculty at the College of Medicine included:

William A. Altemeier
Robert Epstein
Leon Goldman
Richard J. Harris
Claude P. Hobeika
Richard W. Hoefer
William P. Mulvaney
R. J. Rockwell, Jr.
Kenneth W. Rowe
Daniel F. Richfield
Vinton E. Siler

The technical staff at the Laser Laboratory assisted in the operation and demonstrations. They included:

Marilyn Franzen
Ron Dreffer
John Johnson

There are also excellent teaching films, actually international, prize-winning medical films produced by Trubey,[43] Manager Professional Communications, The Upjohn Company. These include *Clinical Applications of Lasers* and *Medical Potential of Lasers.*

There has not been enough interest for a separate current journal of laser biology and medicine or the dignity and burden of a separate society. Yet, if one believes *Business Week,*[33] the laser has a bright future in medicine.

> "In the next two years, doctors expect to be able to filter dangerous bacteria out of the air, spot breast cancer almost as soon as it begins, and perhaps even discover how to prevent inherited diseases.
>
> Such developments may sound like science fiction. But they are just some of the results that medical researchers expect to get as they begin to realize the potential of a largely unexploited medical research tool – the laser."

Whether this is science fiction or real fiction remains to be seen. However, there is a clinical medical program for the laser today for:

1. Diagnosis
    a. For the laboratory
    b. For the field in environmental medicine
    c. For the patient
2. Treatment
3. Continuing medical education

Research

The laser is a form of so-called nonionizing radiation. Nonionizing radiation also includes microwave, ultrasound, and ultraviolet. We have but little information on the effect of these on living tissues of man, especially as regards long-term or chronic exposure. The laser offers another opportunity for precise and well-controlled work for the effect of this form of nonionizing radiation now available through most all the spectrum. The collimation, as indicated, can make for microsurgery of selected areas on the individual chromosome, all the way up to the resection of large areas of the liver for damage or for cancer. The laser, as indicated in studies in biology, has the remarkable properties for analytical spectroscopy of living tissue or even for stimulating cell growth.

In the usual order of the progress of research and development, there are now the actual studies of the effect of lasers on the living tissues of man other than for actual diagnosis or for treatment needs. These studies show how the tissues especially of the skin and eyes react to impacts of the laser below what may be called the permissible exposure levels. Such studies have been done especially on the skin of man where they are not hazardous and at a much lower level, obviously, on the eyes.

The Laser Laboratory has done extensive studies of acute and chronic exposure of the skin to many diverse laser systems including, recently, picosecond exposures and exposures to TEA CO_2 lasers. In brief, these studies have shown nonspecific reactions similar to those of mild electrical burns. Important chronic exposure studies have been done on normal skin of man, on precancerous spots, and on tattoo marks for studies on the influence of color. Studies of chronic exposure are being done on scars of previous laser reactions. The purpose of these studies is to determine the carcinogenic potential of that particular laser system, normal mode ruby, on a selected skin site, the flexor surface of the forearm, with a standardized energy density over an established period of time. Under these restrictive circumstances the laser has shown to be noncarcinogenic. This is brought out also by animal experiments, by other investigators, and by the extensive clinical experiences with hundreds of patients in the Laser Laboratory over a period of more than 10 years.

Diagnosis – Laboratory – Laser Microprobe

As mentioned previously in the sections on Chemistry and Biology, laser microemission spectroscopy can analyze for cations, either microscopic or gross samples. These are of special value in the analyses of the effect of pollution on man. Our interest in environmental pollution has been the study of lead poisoning in children, as monitored by laser microemission spectroscopy of the hair. The studies with Brech and Allemand[1] of the Fisher Scientific Company and Vahl[44,45] from the University of Münster have shown significant

FIGURE 15-1. Scanning Electron Microscopy of Hairs for Lead Studies. (Prof. Dr. J. Vahl.)

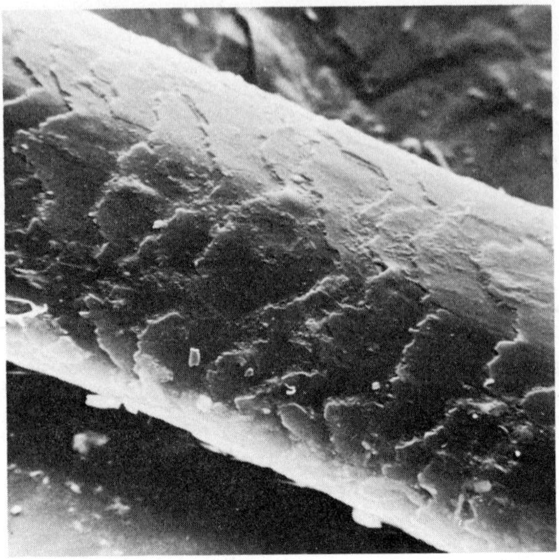

FIGURE 15-1A. Unlased portion x 106.

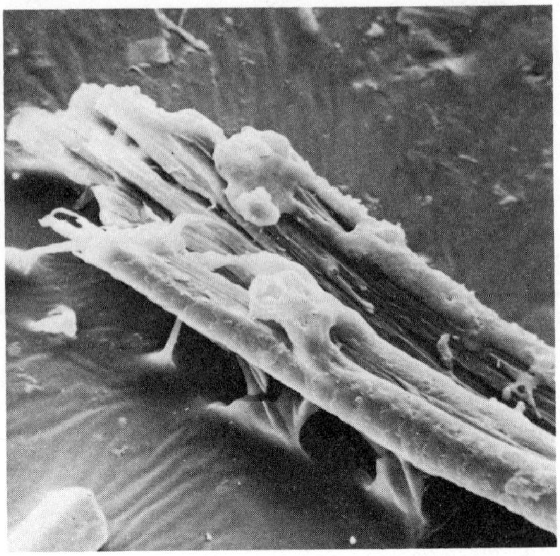

FIGURE 15-1B. Laser spot x 530.

differences in the results of microemission spectroscopy along the hair shaft of children with lead poisoning. Laser microemission spectroscopy can be limited to small impact areas along the hair shaft even in the living child. The blood levels often cannot be correlated with the levels found along the hair since the blood levels may have been reduced and the lead still stays in the hair as evidence of previous absorption. The purpose of these studies is to show that the levels along the hair shaft can be correlated with the growth patterns of the hair and may serve to indicate the periods of absorption of the lead. It is felt that the laser microemission spectroscopy of the hair is a more effective practical method of analysis than the usual mass absorption spectroscopy. In mass absorption spectroscopy the whole hair section must be used. The laser microemission spectroscopy has been quantitated and perfected so that its data are adequate and can compare favorably with the data of absorption spectroscopy. Finally, the preservation of the impact areas where the lead was actually found gives an opportunity for detailed examination of the microstructures responsible for the deposition of lead. With the help of Vahl,[45] the detailed examination of the microstructures of the impact areas has been done with scanning electron microscopy, transmission electron microscopy, and such physical methods as X-ray diffraction, X-ray fluorescence, infrared spectroscopy and microscopic element analytical research; all this will help to localize the source of origin of the lead depot and perhaps furnish information about the metabolism of lead poisoning. These additional studies would not be possible with the atomic absorption spectroscopy analysis of the hair. The sensitive laser microprobe, as described previously, can be used for the microanalyses of other environmental pollutants.

Microanalyses are also of help in learning about what are called "trace metals," i.e., copper, zinc, manganese, etc. in the body. The significance of these in regard to the life and growth of man is important.

The laser microprobe may be used on clippings from fingernails for the screening of newborn children for cystic fibrosis. These analyses would refer to high copper, sodium chloride, and phosphorus. Studies on copper metabolism could be done also for Wilson's disease.

The importance of the laser microprobe for forensic medicine has been mentioned previously for analyses of blood, skin, hair, textiles, paint, glass, etc.

However, it is in the routine hospital laboratory where the laser microprobe with its computer readout will make for considerable progress and economy in blood and tissue chemistry. As reported, the association of the laser microprobe with mass spectrometers will give data on organic materials.

Diagnosis – Patient – Holography

The laser offers interference holography for diagnosis.[47] The Holocamera of TRW® has been used to study, by double exposure holography, the respiratory movements of the chest in man.[49] This technique can also be adapted to study the impacts of the heartbeat on the chest of man. These initial studies with this technique can be done on thin subjects initially. The localization for interference holography is the precardial area. The analysis of disturbances of these impacts may give some help in the immediate recognition of abnormal heart action. These readings can be correlated with the electrocardiogram and the phonocardiogram. This correlation may be superior to the transducer now used, the stethoscope.

A fascinating phase of diagnosis, which has been mentioned before, is acoustical holography, described previously for the diagnosis of soft tissue tumors, especially cancer of the breast. The pioneering studies of Gabor, father of holography, and Byron Brenden, of Holosonics, have shown the value of this technique in the recognition of soft tissue tumors. The initial studies of Brenden have also shown the value of an ultrasonic imager to provide an internal view of internal structures. Focusing of this real time image can be done to view a particular plane within the object, as regards movements of the hand, movements of the soft tissues about the elbow and again, the early recognition of breast changes. It is emphasized that this examination is a type of laser and ultrasonic irradiation of the skin and soft tissues, and is safer than the use of soft X-ray radiation needed for the current techniques of mammography and xerography. The safety program for the operator and viewer of holography has been described in the chapter, Safety.

With Holosonics, Inc. Model 100 immersion type acoustical holographic imaging system, Holbrooke and his associates[50] at the Children's Hospital of San Francisco have done fascinating fetal studies. A large amount of detail was seen, and this technique was shown to be "non-injurious." This technique means visualization of pathology in the unborn child, such as fractures, bone deformities, circulation, etc. Gabor has received a patent for this technique, called sonoradiography. This is being developed at CBS Laboratories.

Greguss[52,53] of the Department of Ophthalmology of New York Medical College has done extensive studies in the biomedical research and clinical practice of holography. He has also developed conferences on biomedical applications of holography.

With an ultrasonic camera using 3.5 MHz, Green,[51] at the Stanford Institute, has shown detailed pictures of internal organs and soft tissues normally invisible to X-rays. The signals from the imaging camera were converted to T.V. displays and can be stored probably on videotapes.

Diagnosis – Patient – Transillumination

As reported previously, transillumination[2] is another technique that should be investigated with He-Ne, krypton, and YAG lasers and suitable film readout pictures. This soft tissue examination may be done safely through the skull of infants, bone, and brain deformities, for examination of the nasal sinuses (now done chiefly and more accurately with X-ray), and for soft tissues of the skin.

Skin examination includes the breast examination for cysts, tumors including cancer, the general skin, foreign bodies in the skin, cysts, tumors, hardening of the skin, as in scleroderma, and subcutaneous fibrosis of the lower leg for the intensity of the stasis syndrome.

Diagnosis – Patient – T.V. Diagnosis

For television diagnosis, the laser adapted to television would give specific channels for communications to the medical center. The color qualities of the laser picture television diagnosis from a distance would be appreciated. If and when the development of three dimensional movies becomes a reality, three-dimensional imagery would be available for television diagnosis.

Laser communication to hospital centers can be done for observation and monitoring operating room techniques and also for security measures. It is planned also to use such communication techniques for community medical services. Laser computer technology, with data retrieval on differential diagnosis, will also be available. This information handling by the laser will be reviewed later in regard to the educational program for the physician.

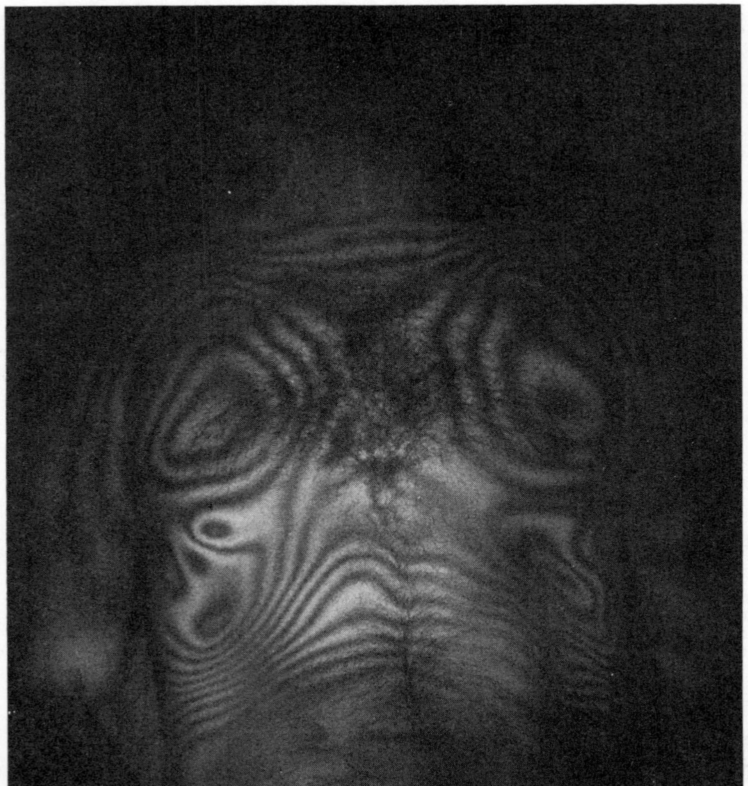

FIGURE 15-2A. Holographic image of chest during inhalation, movement of dark interference fringes one quarter wavelength movement (0.6943 μm wavelength) between two laser exposures 150 μsec apart. (TRW Holocamera, Samuel M. Zivi.)

FIGURE 15-2B. Back of patient with left lung removal, arrows show bunching of fringes about the postoperative muscle anomaly 150 μsec or interval. (TRW Holocamera, Samuel M. Zivi.)

Diagnosis – Patient – Miscellaneous

For the diagnosis of disturbances of circulation in man, laser Doppler techniques may be used. In addition, fiber optic probes, usually with incoherent light sources for observation, may be used with low output infrared systems with image converters for examination of viscera. Transillumination of viscera may be done in the operating room through visualization by fiber optics systems.

Conclusions

The use of the laser for diagnosis in medicine is still in its infancy. The few studies indicate that it is of value for rapid and economical diagnosis of the patient in the medical center, and for community medicine, especially with the current shortage of medical and paramedical personnel. These diagnostic aids are offered in the various hospital laboratories and even in the field where environmental health is concerned. Again, there is the eternal triangle of current medical research, better funding, better lines of communication, and better recognition of the need. Is this an isosceles triangle?

Treatment of the Patient

Has the treatment of the patient today reached such a superior plane of perfection that no additional improvements are necessary? The answer is obvious. What has the laser to offer as a new physical modality that is superior to current instruments? If there are some advantages, can the laser be used safely? In brief and dogmatic phrases, laser treatments are needed. There are distinct advantages with this agent and finally, as stated repeatedly, with planned programs, the laser may be used safely for the patient and also for the operator.

Advantages of the Laser as a Surgical Instrument

These are the basic properties of the laser which can be used as a surgical tool:

1. Coherency
2. Monochromatic
3. High energy and power densities
4. Selective color absorption of some laser systems, therefore, of selective value in pigmented tissues

In brief, this means that a new instrument is available which can be a tool of great precision and power and can have a selective action on certain tissues of the human body which are pigmented or, if colorless, these tissues can be colored by certain materials. There are other physical modalities used in medicine which are very powerful, but none as precise as the laser, so precise that the laser can be used for microsurgery, even in chromosomes. There are no other modalities that have selective color absorption. These other tools used for the treatment of diseases in man include the ageless scalpel, the high frequency electrosurgical instrument, the cryosurgical tool with liquid nitrogen and at the other extreme, the plasma torch[22] (only for investigative studies in animals), the ultraviolet radiation ultrasonic, and microwave instruments. These are all mentioned, for controls must be used always to determine the true value of the laser.

What can these special properties of the laser do in surgery? In brief, they can accomplish precise and bloodless surgery. Since there are many lasers, how does one select the particular laser required? Fidler[4] and Rockwell[39] suggest that the following parameters for the laser be considered:

1. Penetration ability in tissue at the laser frequency (total absorption depth)
2. Absorption selectivity of tissue substructures at the laser frequency
3. Beam size
4. Power requirements
5. Exposure time

The thermal necrosis of tissue from the laser impact has been discussed in the chapter, Biology. If the laser is to be selected, what lasers are now available for surgery? Low output laser systems are used for eye surgery, high output lasers for other forms of surgery.

The Laser Surgical Instruments

The following lasers are used for the treatment of patients. This * indicates that these lasers are used in the treatment of dermatology patients (He-Ne and krypton – usually for transillumination). The other lasers are used, thus, for surgery.

Laser	Spectral region	Wavelength-(nM)
Ruby	Red	694.3
Neodymium	Infrared	1,060.0
Neodymium-YAG	Infrared	1,060.0
Argon	Blue-green	499.5–514.5

FIGURE 15-3. Argon Laser Photocoagulator (Optics Technology).

FIGURE 15-3A. Photocoagulator with fiberoptics delivery.

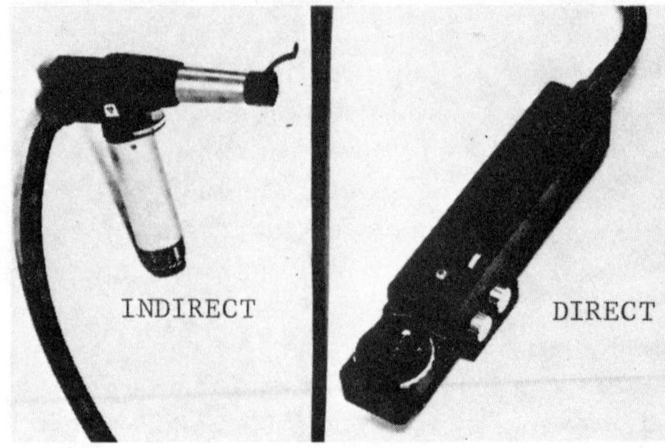

FIGURE 15-3C. The indirect and direct ophthalmoscope.

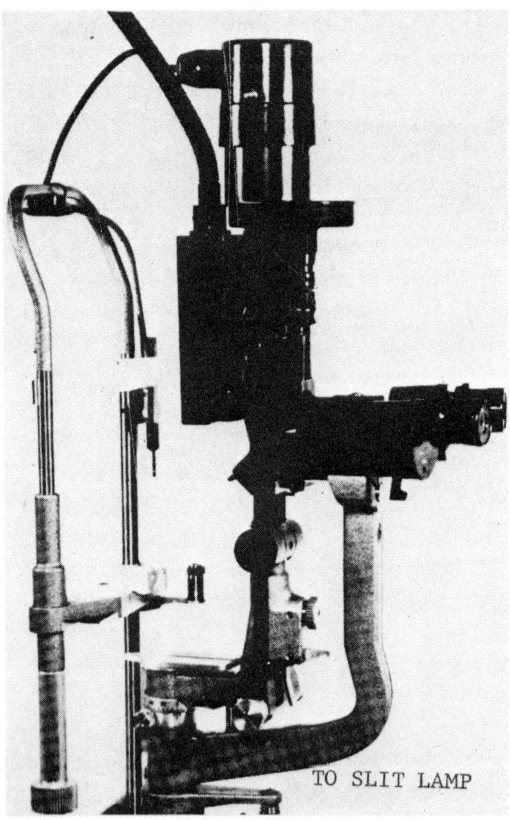

FIGURE 15-3B. To slit lamp.

*Krypton	Visible	476.2–647.1
CO_2	Far infrared	10,600.0
*Helium-neon	Red	632.8
*Ultraviolet	Ultraviolet	265.0–360.0

Laser Ophthalmology

The two lasers used in eye surgery are the red ruby laser and the bluish-green argon laser. The ruby laser is usually used in the single pulsed mode in the microsecond range with a variable retinal image diameter, according to Geeraets and Carroll, of approximately 100 to 500 μm. According to these authors, the highest percent absorption in the pigmented retinal epithelium and choroid of the human eye for light of equal intensity and incident on the cornea is between 500 and 700 nM. For direct effect on blood vessels, especially in diabetes of the eyes, the ruby is reflected and "only minimally absorbed." Vessels may be affected secondarily by heat generated in pigmented tissues on impact with the ruby laser. Absorption of the argon laser in blood vessels is approximately seven times greater than the ruby laser. Therefore, the argon laser coagulates and obliterates blood vessels easily and affects vessels of 30 to 50 μm in size. The usual retinal spot sizes for the argon laser is 50, 100, and 200 μm. Transmission of the laser in laser eye instruments may be by multiple mirrors, lenses, or recently by fiber optics. Laser photocoagulation may be used with direct or indirect ophthalmoscopes or with the slit lamp. The ruby laser also may be used with direct or indirect ophthalmoscopes.

A recent gas laser ophthalmoscope of International Research & Development for Victoria Infirmary of Newcastle-on-Tyne in Englnad is reported to be compatible with various types of gaseous lasers, including tunable dye lasers. This makes it possible to select certain wavelengths for certain specific eye disorders. Unlike the ruby, the argon laser, because of its photophobia and duration, may require retrobulbar anesthesia.

The rival to laser photocoagulation is, of course, the white light or high pressure xenon photocoagulator used for more than 10 years prior to the development of the laser. Some objections to the xenon arc photocoagulator include injury to the macula, increased heat spread, not the marked precision as found with the laser, lack of color specificity, and more prolonged exposure with excessive induced photocoagulation.

Some pioneers in clinical laser ophthalmology directly or indirectly are, in alphabetical order:

Lloyd M. Aiello
Charles C. Campbell
F. A. L'Esperance
Walter J. Geeraets
W. T. Ham, Jr.
A. Patz
Kenneth W. Rowe
A. Vassiliadis
M. L. Wolbarst
Milton M. Zaret
H. C. Zweng

Many more could be included in this list, as the clinical users of laser photocoagulation now include much more than 10,000 patients.

For the specific diseases to be treated by laser photocoagulation, one is referred to the special journals and reference works on laser ophthalmology. The ruby laser is used mainly for tears and postretinal detachments; the argon laser is used mainly for vascular disorders of the eye. Geeraets and Carroll[8] claim that for blood vessel proliferation into the vitreous humor in diabetic retinopathy, "this is the area where the argon laser has found its principle and dominant role in clinical ophthalmology." The main study is now with diabetic retinopathy and other disorders requiring reduction of retinal blood supply. These may include severe glaucomatous optic atrophy, severe myopia, retinitis pigmentosa, and large chorioretinal scars. Sixteen medical centers are now doing a critical cooperative study of photocoagulation in diabetic retinopathy. This shows that even after 15 years, photocoagulation (xenon, ruby, and argon lasers) is still unproven.

One recent interesting use for the laser photocoagulator was the treatment of a patient with a worm in her eye, a trematode using the frog as a host. This larval form, resident in the eye for more than 6 years, presumably after preparing frogs' legs for cooking, was destroyed with an argon photocoagulator.

A second patient, treated by the same group in Canada, had a fox tapeworm in her eye, *Taenia crassiceps* (dog licking face?). This was removed with a microcryoprobe. In the laser treated eye, vision improved to 20/50 then fell to 20/400 with degeneration of the macula. The patient with the fox tapeworm has no central vision.

Ophthalmologists must be trained in laser ophthalmology and are not expected to become experts simply by purchasing a laser photocoagulator. Well-controlled studies also are still needed, such as the one directed by the National Eye Institute of the National Institutes of Health. The American Academy of Ophthalmology offers training courses in laser ophthalmology. In the field of laser safety, it is the laser ophthalmologist, when available, who should direct such a program on laser eye safety.

Laser-induced cataracts have been reported in rabbits by Leibowitz and Luzzio.[35] It is assumed that the increased temperature on the anterior surface of the lens can cause a cataract. Rowe is not aware of any cataracts in man produced by the ruby laser. The small miotic pupil in the animal eye helped to develop the cataract. Rowe believes that the calcium content of a cataract is proportional to the rapidity of the development of the cataract. The technique of tissue culture of lens epithelium as done by Geeraets[7] with human, rabbit, and monkey lens epithelial explants and by the Laser Laboratory following laser exposure of rabbit eyes to ruby lasers will provide a controlled experimental technique for studies of radiation-induced cataracts.

Clinical laser photocoagulation is an accepted form of eye treatment. As suggested, more extensive use in controlled studies by laser eye expertise will evaluate this new field of laser ophthalmology.

A computer operated CO_2 laser has been used to produce tiny holes in contact lenses to make such lenses more comfortable. Unlike mechanical drilling, the CO_2 laser makes tiny holes without the need of repolishing in the contact lens. Ophthalmologists differ as to the value of such additional holes.

Laser Aids For The Blind

There are various devices to aid the blind by the production of radar-like techniques. At present, these types include:

1. Light emitting diodes (LED)
2. Ultrasonic devices (Kay Ultrasonic Sensor)
3. Laser devices
   a. Canes
   b. Spectacles
4. Tactile sensations from pictorial data (Tactile Vision Substitution System)
5. Traffic light beeper – beeping noise indicates green light for only 12 short seconds

This discussion will be limited to laser canes and spectacles. For years, the Veterans Administration has been working on electronic mobility aids for the blind. The laser cane of Bionic Industries Inc. is equipped with three laser junction diodes, one aimed downward, one aimed upward, and one straight ahead. This distance is measured by the angle of reflected light. Sensing is by vibration of the handle. The obstacle's level is indicated by the pitch. The cane can be charged for approximately 4 hr of use. The Laser Laboratory has done safety studies on this cane with its 3 gallium arsenide lasers, with 100 nsec pulses and repetition rate, and 40 pulses/sec. The radiation emitted are levels about 7.0 μm.

The laser cane is heavy and large. It weighs about 1½ lb, whereas the small aluminum cane used ordinarily weighs about 2 oz. On the city streets, the blind person prefers devices that do not make him feel too conspicuous. The Miniature Electronic Object Detection System for the Blind of Mims[37] is a pair of spectacles with the temple pieces holding infrared light emitting diodes. An ear tube is connected to a receiver housed in a tube on the other side of the glasses. This weighs only 3 oz. The laser spectacles, similar to the LED spectacles of Mims, are still in the early stages of development. Sensing for this laser type will be done by vibration or buzzing about the ears.

The Laser Laboratory has been planning a tactile sensing device for the blind using laser radar, lidar, with the production of three-dimensional imagery for tactile sensing by the sensitive fingertips of the blind.

With the rapid development of electrooptics and biomedical engineering, the challenge is for more miniaturized economical, effective laser aids for the blind.

High Output Laser Systems for Surgery

Unlike the low output lasers used for eye surgery, high output systems are needed for general and special surgery at present. As shown, four laser systems are used:

1. Ruby
2. Argon
3. Neodymium-YAG
4. CO_2

Except for certain requirements, the surgeon prefers to use CW laser systems, argon, Nd-YAG, and CO_2. The CW lasers act as a precise optical knife.

Penetration ability is listed. The mm penetration depth is "defined as that depth necessary to absorb 99 per cent of the incident radiation." Fidler and Rockwell have summarized the tissue penetration depth of the four lasers used currently:

Laser	Wavelength (nM)	Absorption coefficient (cm^{-1})	Penetration depth (mm)
Argon	488–514	55	0.84
Neodymium-YAG	1,060	11	4.2
CO_2	10,600	200	0.23
Ruby	694.3	20	2.3

A CW laser cuts through various types of tissues of different colors, consistency, and vascularity. The argon laser is effective on pigmented tissues, especially those containing blood. The Nd-YAG is a poor cutting tool, but coagulates blood vessels well. The CO_2 laser incises tissue well, but controls only capillary bleeding. The ruby laser is excellent for deeply pigmented tissues.

The Laser Laboratory has had clinical experience with all the four laser systems. For general

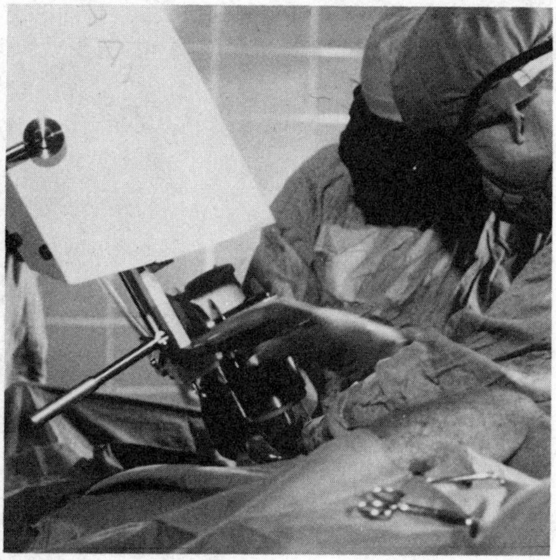

FIGURE 15-4A. Ruby laser in operating room for treatment of skin cancer.

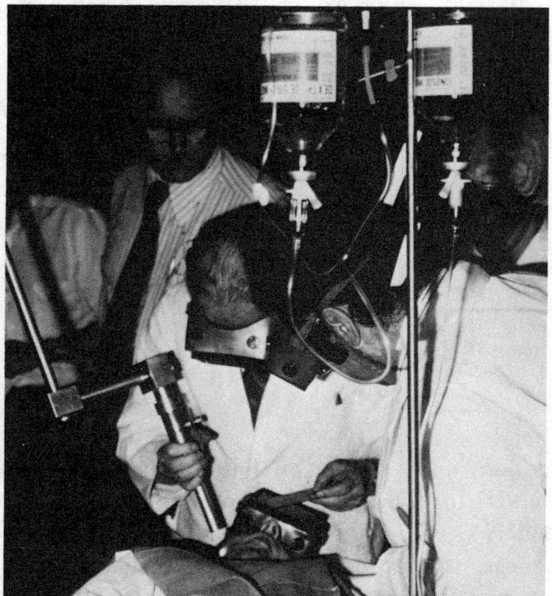

FIGURE 15-4B. Argon laser treatment of portwine mark under general anesthesia.

FIGURE 15-5. Carbon Dioxide Laser. (American Optical Co.)

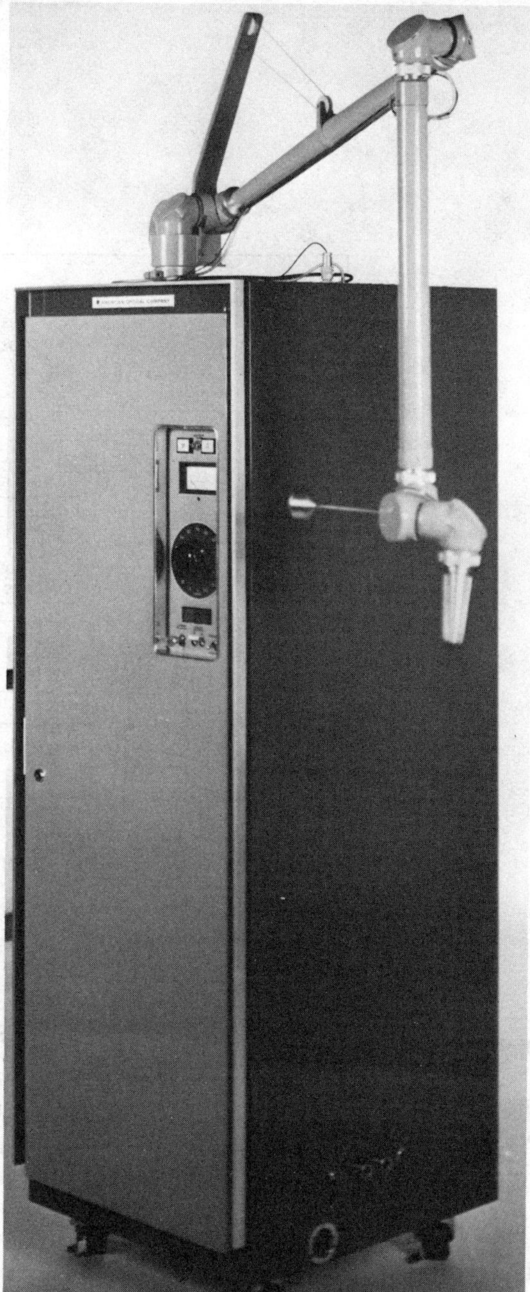

FIGURE 15-5A. Cabinet and arm.

surgery, chiefly the CO_2 and Nd-YAG lasers have been used. The argon laser has been used in dermatology patients, the ruby in dermatological surgery and cancer surgery. With high output ruby lasers there is some concern that the following reactions may block some of the effect, blocking plume from the target, "thermal blooming" of heated air in the beam path defocussing the laser, or "gas breakdown" of air. All these are important, of course, for high output laser weaponry efficiency.

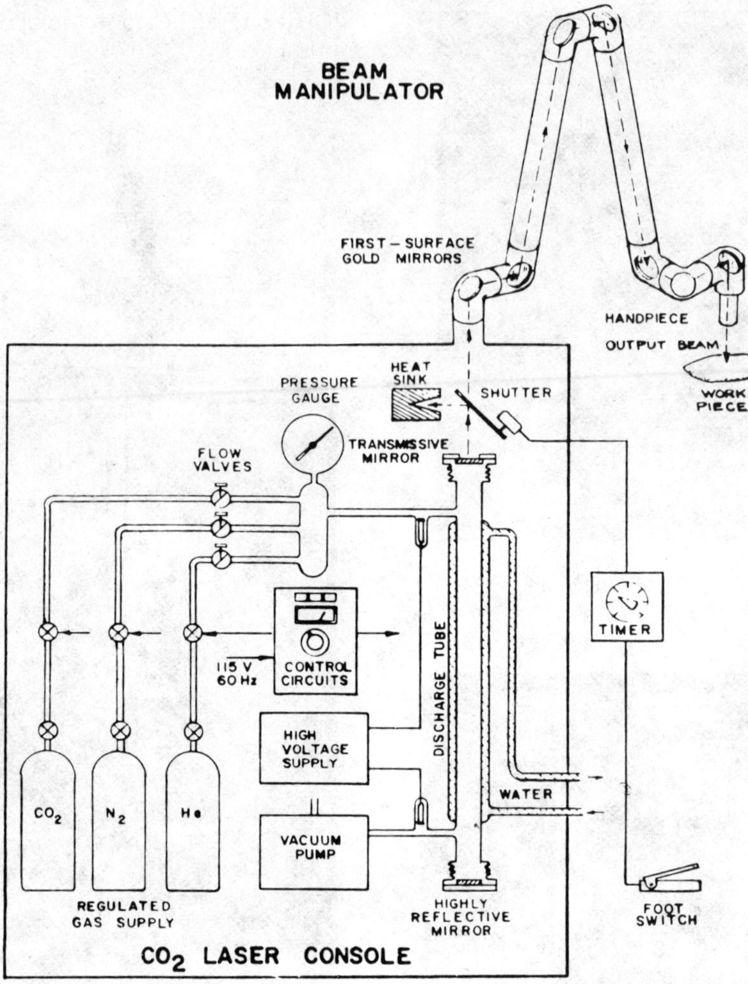

FIGURE 15-5B. Schematic diagram.

High Output CO_2 Laser Surgery

The ruby and argon lasers have been used chiefly in dermatological surgery. The high output CO_2 and Nd-YAG laser surgical instruments have been used in general surgery as well as in dermatological surgery.

At present, the most frequently used laser for surgery is the high output CO_2 laser. Two models have been used. The first model was the American Optical CO_2 laser developed by Polanyi and Bredmeier. This is a flexible system with a beam manipulator and an endoscopic system for transmission of the laser beam into cavities. One flexible type has an output about 45 to 50 W; the experimental model at the American Optical Company has 300 W. The beam manipulator is steel, heavy and thick, has relatively limited range of motion, and is awkward but not impossible to sterilize for the operating room. Improvements are underway for the beam manipulator. This will increase the use of this valuable instrument. This instrument was the first one to introduce and show the value of clinical laser surgery. An excellent operating microscope is available for this laser.

A recent flexible high output laser system is the one of Panengineering of Laser Industries Ltd. of Israel. This is supposed to be a single mode continuous carbon dioxide laser with a variable output horizontally mounted and from which projects an articulated arm of narrow bore lightweight metal tubing ending in a pen-shaped manipulator with various interchangeable end pieces. The whole apparatus is self-contained, portable

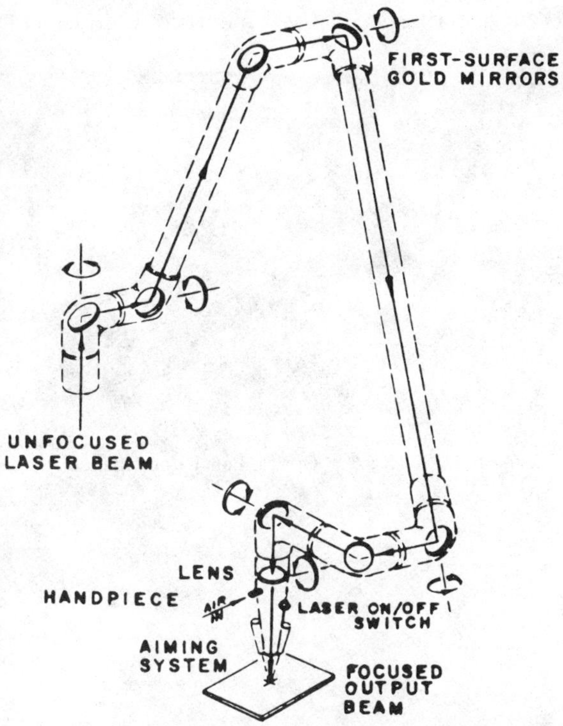

FIGURE 15-5C. Schematic diagram of beam manipulator.

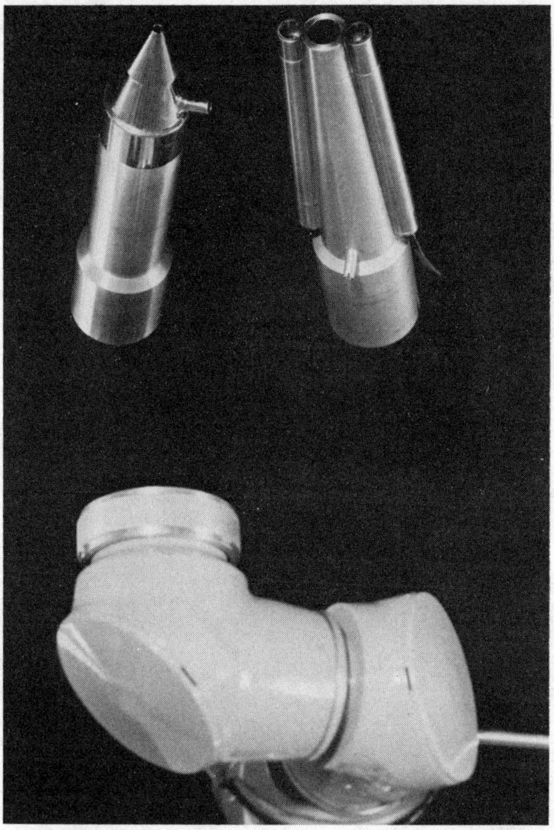

FIGURE 15-5D. Interchangeable focal length hand pieces for beam manipulator.

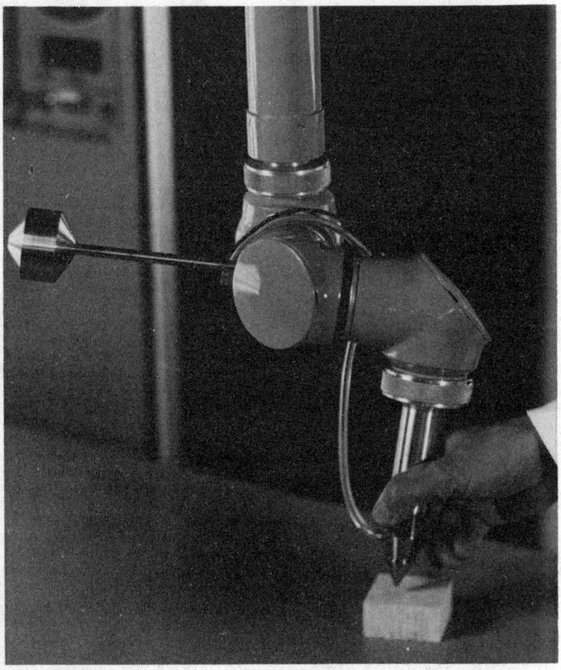

FIGURE 15-5E. Beam manipulator with 2 in. focal length lens with air hose to keep lens clean.

with a telescopic vertical beam, enabling the optical bench to be lowered and guided at will. A sterilizable remote control panel enables the surgeon to operate the apparatus independently. The spot size of this laser is smaller, with power density increased by a factor of 10. The hand piece is much more flexible and has a number of different adaptors, gas sterilized, to localize the beam on tissue. Less thermal-induced necrosis is reported with this system and safer techniques are possible for CO_2 surgery. Extensive major surgery in many patients has been done with this laser in Israel by Kaplan[26] and by Ger[9] at the Hospital of The Albert Einstein College of Medicine. This laser has been placed recently in the Laser Laboratory and also at Shriners Burns Institute of the Medical Center of the University of Cincinnati.

The American Optical CO_2 laser has been used in animal surgery at the Laser Laboratory for basic studies on tissue incisions, liver excisions, and bone surgery. At the Laser Laboratory, the CO_2 laser has been used in cancer surgery in patients

FIGURE 15-6. Single Mode High Output CO_2 Laser. (Panegineering Co. of Laser Industries Ltd., Israel.)

FIGURE 15-6A. The unit.

FIGURE 15-6B. Optical bench and balance.

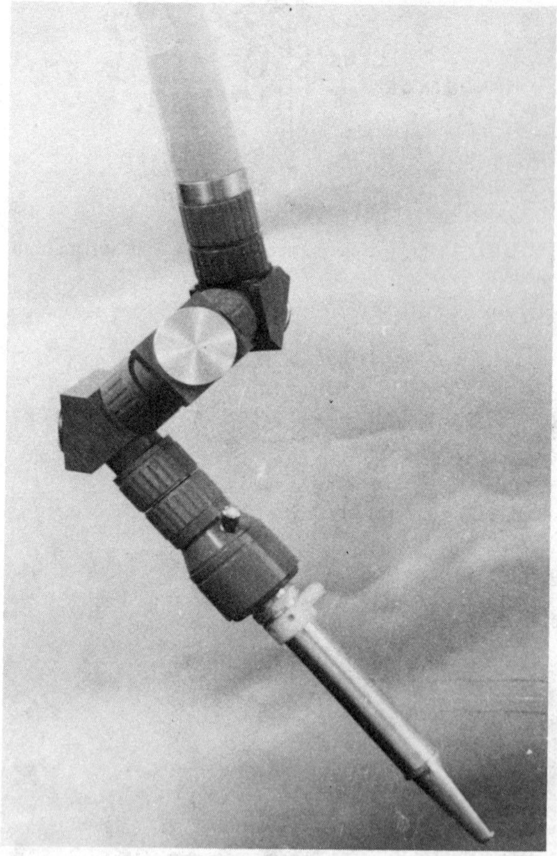

FIGURE 15-6C. Manipulator with long focal length lens.

and recently by Fidler at the Shriners Burns Institute in patients for laser excision of burn eschar with immediate graft replacement through relatively bloodless surgery. This has been done even on infants. Both Fidler and Kaplan[26] are investigating the possibility of developing a CO_2 laser dermotome. The CO_2 laser has been used also by Stellar for brain and burn surgery and bedsores, and by Goodale for the treatment of persistent bleeding stress ulcers of the stomach. Moore has done investigations on bone surgery. Additional clinical studies with the high output CO_2 lasers will be described in the section on laser cancer surgery.

High Output Nd-YAG Surgery

The flexible double tapered single fiber of Nath has removed the high output Nd-YAG laser from the optical bench and the industrial area into the surgical operating room. The experimental studies of dermatologic surgery and cancer surgery will be

FIGURE 15-6D. Manipulator with short focal length lens.

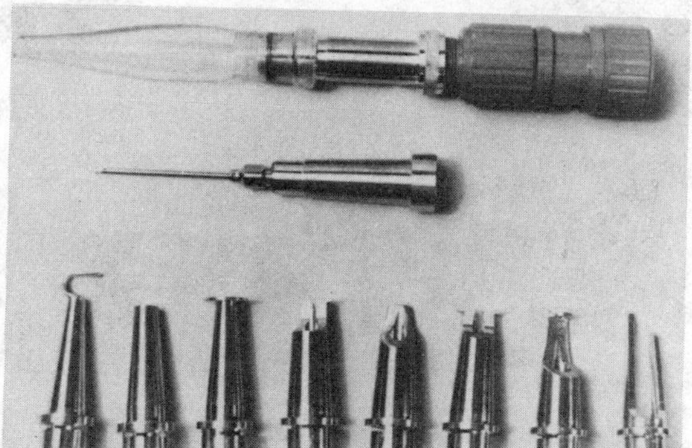

FIGURE 15-6E. End piece.

listed later. Fidler has used this laser for liver surgery in dogs. He found excellent control of hemorrhage, but deep tissue damage of the liver. Many more studies are needed since the property of blood coagulation, arrest of bleeding, is so important. Perhaps the CO laser with its more "practical" optics can be used as a substitute for CO_2 laser surgery.

FIGURE 15-6F. Sterile remote control box.

The special requirements for anesthesia for laser surgery have been shown by Birch in the Laser Laboratory. Currently, there is also research on the use of the laser in acupuncture for relief of pain and the role of the laser in Kerlian photography.

Laser Dermatology

Because of the accessibility of the skin and the definite need, hundreds of patients with skin disorders have been treated by lasers at the Laser Laboratory. The skin can be used as a test organ for initial clinical experiments even before eye surgery is done.[29] This is possible for the following reasons:

1. Accessible living tissue of man
2. Direct observation of acute and chronic reactions
   a. Gross
   b. Microscopic
3. Easy use of control studies

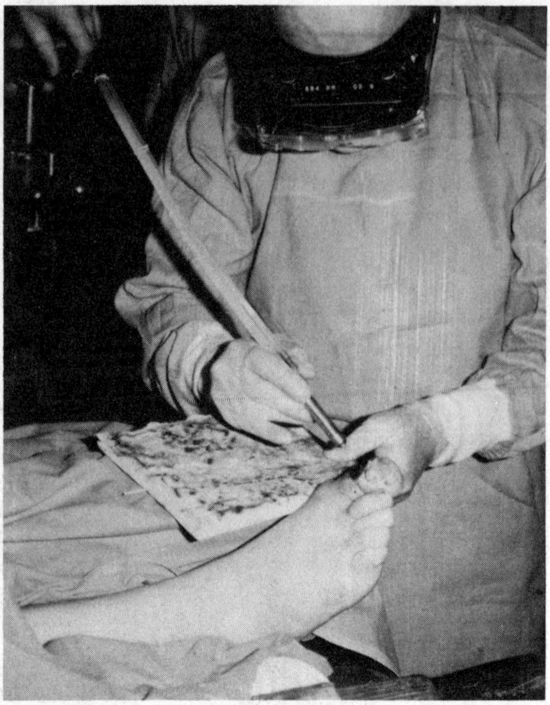

FIGURE 15-7. High output Nd-YAG laser with single fiber of Nath operating on squamous carcinoma of the toe.

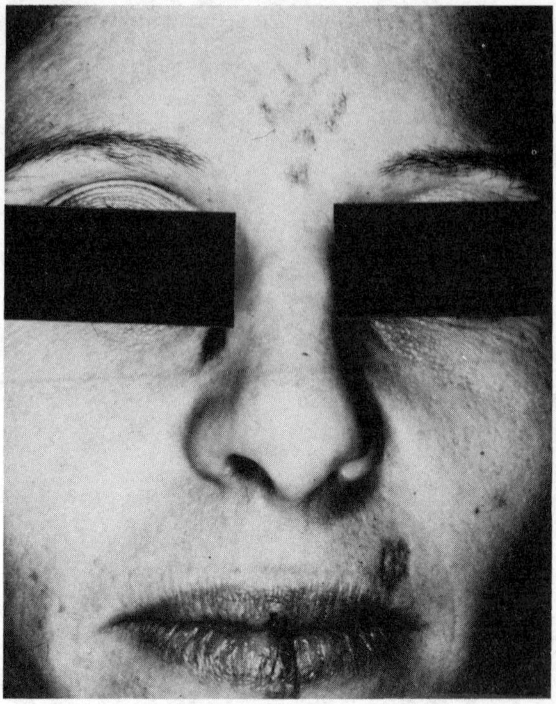

FIGURE 15-8A. Tribal tattoos.

4. Easy application of personal laser safety
5. Actual clinical applications

For more than 10 years the following clinical investigative studies have been done chiefly with normal mode ruby lasers and argon lasers. Some experiments have been done also with Q-switched ruby lasers. Oscillator amplifier with Q-switched ruby had been attempted to increase the value of the laser in the erasure of tattoo marks. Because of the difficulty of obtaining adequate instrumentation, unfortunately, these experiments were discontinued for the present.

Birthmarks
  Angiomas – cavernous
  Portwine – small
  Epidermal nevi – recurrent
Tattoos
Seborrheic warty growths (baso, squamous, acanthomas)
Epilation – removal of hair with small laser probes
Transillumination – for studies of soft tissue changes

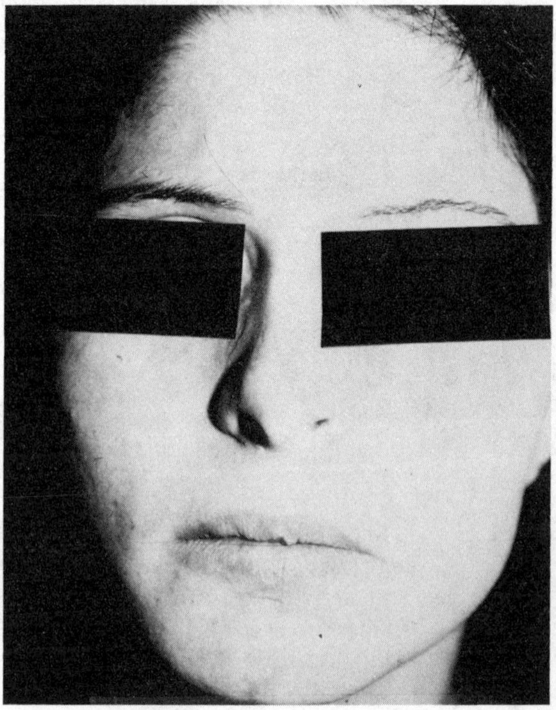

FIGURE 15-8B. After ruby laser treatments.

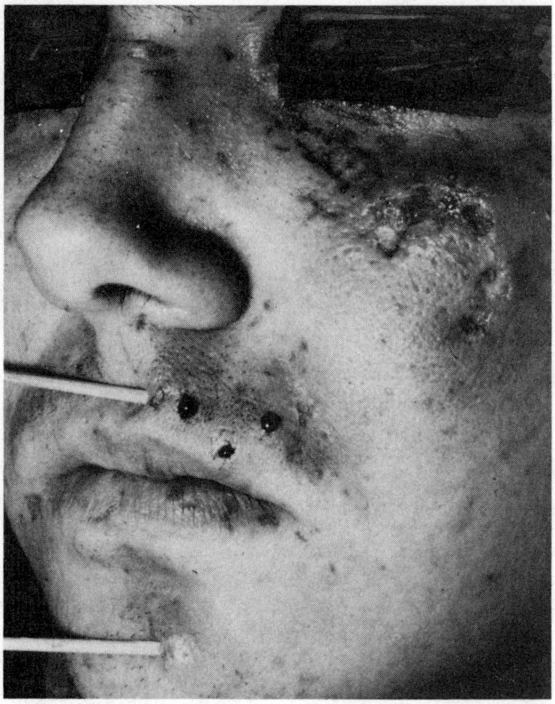

FIGURE 15-9. Small spot ruby laser treatment of debris tattoo from land mine explosion from Vietnam.

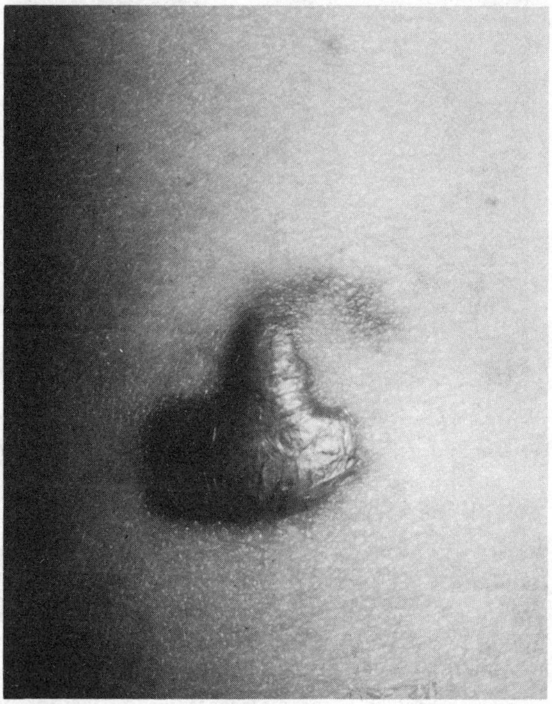

FIGURE 15-10. Hypertrophic scar following laser treatment of tattoo.

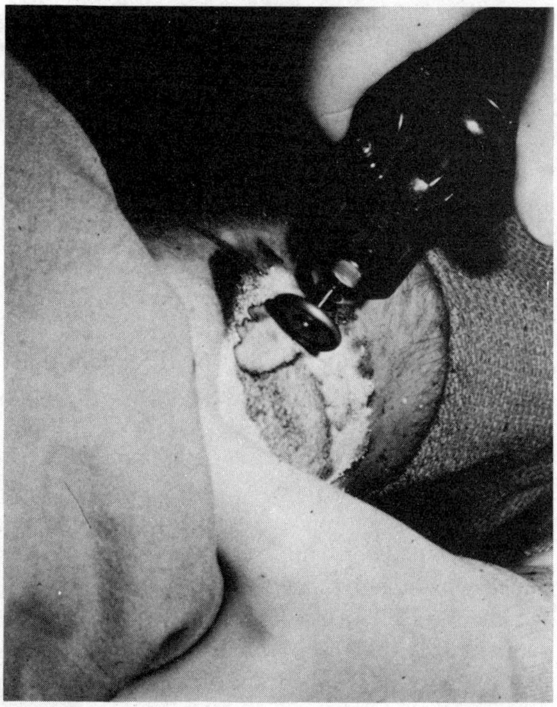

FIGURE 15-11. Adjuvant dermabrasion of tattoo after laser treatment.

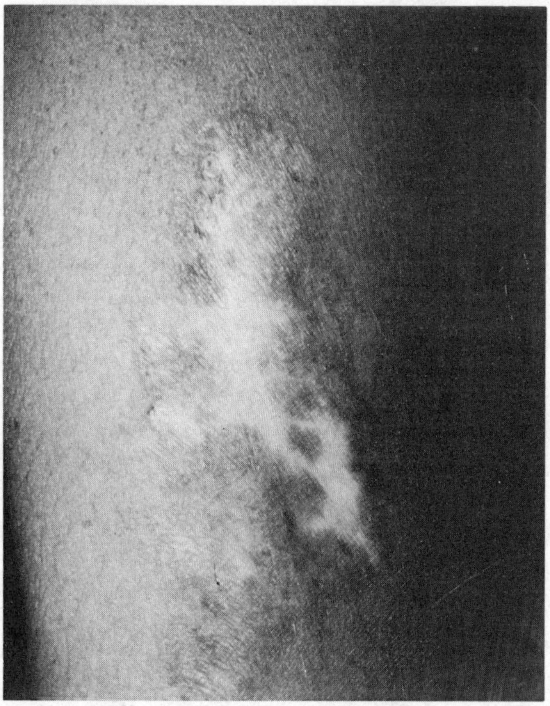

FIGURE 15-12. Repigmentation several years after elective laser treatment for tattoo on upper arm.

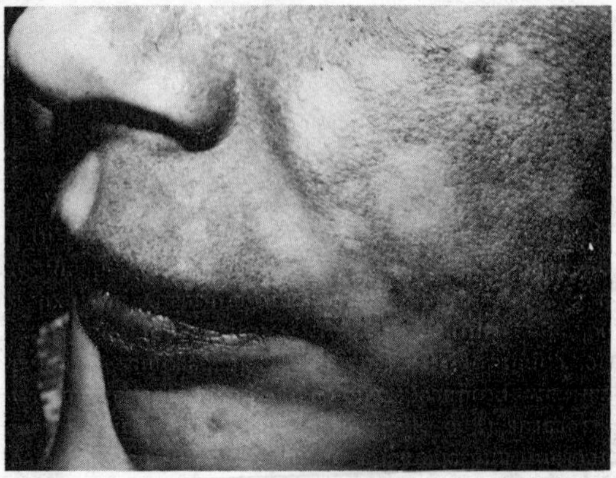

FIGURE 15-13. Test spots after ruby laser treatment of persistent portwine mark showing whitened areas.

With the ancient apparatus still available in the Laser Laboratory, called most appropriately by Schawlow as Wright Brothers Airplanes in a Jet Age, there were distinct limitations for clinical applications. Low priority at times because of lack of funding and at times by disinterest of some of the staff limited effective treatment to lesions of small size areas.

The basic tissue reaction in treating portwine lesions is a collagen tissue change with superficial fibrosis and decreased vascularity. The bleaching effects result from thrombogenesis and vessel rupture caused by the absorption of laser energy. These factors change the optical quality of the tissue.

One of the limitations of the laser therapy of these lesions includes a reticulated appearance resulting from the limited treatment areas now available with ruby, neodymium, or argon instrumentation. Hot spots on the target areas sometime produce more severe scarring in a few cases. Treatments with the argon laser are limited because of the small beam size; consequently, the ruby laser has been primarily used for clinical work. Since patient cooperation is necessary for adequate safety during treatments, therapy has been limited to children above the ages of 6 to 17 years. The light superficial type of portwine lesion has responded poorly to laser therapy. More recently, patients showing a good result for 2 months following the initial laser test dose have been anesthetized and large areas treated. One of the areas has been the eyelids. In the past, no other modality has been available for treating this area. Special protective shields allow laser treatments to be given in this area. Pantocaine is used in the eyes before the insertion of the shields.

Tattoos can be erased with the ruby laser. The Q-switched laser produced more acceptable scarring but the treated areas were too small. Poor quality ruby rods occasionally gave hot spots with linear scarring. Large tattoos required many periods of treatment. As an adjunct treatment, peeling with salt pads gave very little help; dermabrasion with diamond fraises was much more effective. Combined surgical excision of persistant areas or scarred areas can be done.

In the rehabilitation program for juvenile delinquency where gang tattoos have been made, the laser has been effective for the erasure of small linear scars on the face, forearms, and even the ankles. Hypertrophic scarring often develops on the dorsum of the hands, but usually not on the fingers.

The common seborrheic warty growth (senile warts, old lady brown spots, "flowers of the cemetery") is an excellent test model for the study of the laser on a controlled tissue growth. This spot is benign, darkly colored, and can be treated in a controlled fashion as compared to other treatment modalities.

Some experiments in wound healing have been done after incisional laser surgery. The experiments of Pratt and his co-workers[38] at M.I.T. offer the possibility of sterilizing wounds (and materials?) with the CO_2 laser. How selective this action is without any adjacent tissue destruction remains to be seen. The helium-neon laser has been used also to help wounds to heal.

In the treatment of many patients with skin disorders over a period of many years, no evidence of carcinogenic effect of the lasers has been found. It is necessary to repeat this statement again and again because of the anxiety of the uninformed patient. The skin has also been used to study picosecond impacts and TEA CO_2 laser impacts. Here too, benign nonspecific effects were observed over a period of two years.

With the development of reliable high output tunable dye lasers and ultraviolet lasers, the clinical investigative studies in dermatology can become more extensive. The comedone, blackhead pore, adjunct therapy of warts injected with dyes, vital or cytotoxic, extension of ferrofluid investigations, the selective treatment of skin pigmentations, and skin abrasion techniques all are areas of interest.

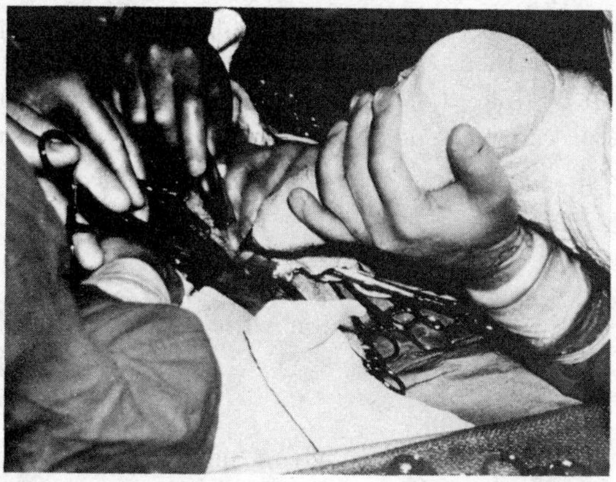

FIGURE 15-14. CO_2 laser excision of burn for immediate graft replacement.

Clinical Laser Cancer Treatment

From experiences in basic research in laser cancer studies, including animal cancer research, it is time now to continue controlled clinical treatment programs on a greater scale. As shown above, there are now effective high output laser systems, CO_2 and Nd-YAG and even ruby and hopefully soon the high output argon. There are now fiber optics such as the single fiber of Nath which can be used to transmit CW lasers Nd-YAG and argon now, and other systems later for CO_2 lasers. There are endoscopic instruments and operating microscopes available and more will be available shortly. Again, now is the time for properly funded, properly controlled laser cancer surgery by proper laser expertise even for primary cancer. All this should be done when the laser can do better than conventional therapy. Also cancer research has shown that the laser can be used as adjunct therapy to surgery, radiation, or cancer chemotherapy.

With these new flexible, effective, precise high output laser systems, Ketcham[27] may be able to change his initial pessimistic attitude about the role of lasers in clinical cancer treatment.

The initial experiments over the past 10 years have been done with skin cancers. Controlled studies can be done in patients with multiple primary cancers or with multiple metastatic lesions, here as a palliative measure.

Lee[34] has reported that the incidence of skin tumors, particularly melanomas, is rising in Caucasians in the U.S. Here, "substantial improvements in the rapidity of diagnosis and treatment of melanomas will be needed." The laser can assist in the early diagnosis by helping with skin microscopy in the living skin at moderate magnification by better illumination and transillumination procedures. The laser is very effective in the treatment of localized lesions. With the current interest in the immunobiology of melanomas, initial experiments have been done with the laser treatment of metastatic melanomas to attempt to affect the tumor antigen.

The other tumors treated on the skin were multiple tumors of basal and squamous cell types. Where there was no color to the cancer, the following materials were used to color tissues:

1. Vital dyes such as Evans blue
2. Tattoo pigments, by tattooing
   a. Dark tattoo pigments
   b. Cinnabar
3. Ferrofluids

The intensity of the laser reaction was increased by the pigment. The ferrofluids, as indicated previously, are of special interest, for these particles are not only affected by laser radiation, but also by magnetic fields and microwaves. The magnetic fields can move the particles as to regional glands, for example. Here, microwaves could be used to induce local hyperthermia, effective in the treatment of cancer. One patient has had ferrofluids injected into his basal cell cancer and laser radiation was used. His case is in press after an observation period of 2 years.

Vascular cancers and progressive angiomas that act like cancers have been treated effectively with high output CO_2 and Nd-YAG lasers if the lesions are small. The ruby laser has also been used for angiosarcoma and Kaposi's sarcoma.

Laser Maxillofacial Surgery

The initial animal experiments in this area by Hobeika, Jako, and Strong, and Stähle, Hogberg, and Engstrom,[41,55] have shown the value of such surgery for clinical applications. Hobeika has also done studies on pituitary ablation by the argon laser. These initial clinical studies were done by Goldman and Shumrick and recently by Strong and Jako.[54] With the CO_2 laser of the American Optical Co., Jako and Strong received the Hektoen Gold Medal Award at the American Medical Association meeting in San Francisco in June 1972. These were excellent studies on papillomas, cysts, and cancer of the vocal cords. More than

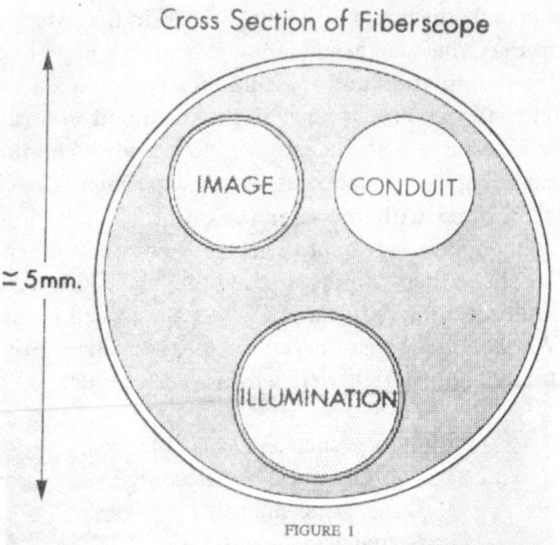

FIGURE 15-15. Basic structure of a flexible fiberoptic endoscope. (Michael Iseman, Resident and Staff Physician, February, 1973.) The operating laser fiber can go through the conduit for operative endoscopy.

100 patients have been treated at Boston University in the past 18 months.

Laser Colposcopy

In the cancer prevention program there is great interest in early detection of cancer, especially of preinvasive cancer. This can be done on the skin where such precancerous spots as actinic or solar keratosis are found on the lips and oral cavity where leukoplakic spots develop. These lesions have all been treated with the lasers, ruby with curved, tapered, special quartz rods, with total internal reflection, and with argon and Nd-YAG lasers. With lasers it may be possible also to affect the blood supply of cancers and consequently retard the growth of tumors.

A laser diagnostic and operating microscope has been developed for the cervix for examination of the tissue with deep transillumination for the study of cellular dysplasia with dyes under moderate magnification. Biopsies can be taken and laser treatment given. With the laser, it is possible to produce bloodless surgery. The instrument is under study now for the controlled clinical use not only for the cervix, but also for the oral cavity and the skin.

Laser Microsurgery

There are also areas such as in the ear, the eye, vascular patterns of the brain and other areas of neurosurgery where an operating laser microscope may be used. Since precise laser microirradiation can be done, the laser microscope is practical. The laser can be added to such operating microscope for:

1. Precision
2. Bloodless microsurgery
3. Color absorption with vital dye staining

Laser Endoscopy

With microscopy, where needed, and special fiber optics such as those of Nath, operative laser endoscopy may be done for the stomach, lungs, bladder, and other body cavities. It can be done even for viscera in the abdominal cavity. Nath has just developed his laser fiber optic gastric endoscope as an actual instrument. Laser laparoscopes have been developed for animal research. Laser hysteroscopes will soon be developed.

Controls For Laser Surgery

To answer the question of the value of laser surgery, it is necessary to have controls. It is often difficult to have controls of the proper power and energy density even for laser animal surgery. The basic factors are, of course, the value, healing, scarring, blood loss, secondary infection, and effect on the general body. The nearest rival to the laser is the high frequency electrosurgical unit. This causes, in brief, broader areas of thermal coagulation necrosis; it may stick to the necrotic tissue and hemorrhage loss is greater than with the laser.

The next rival, at present only in experimental animals, is the plasma scalpel.[18] The plasma is a gaseous state of cations, anions and electrons with neutral molecules, and light quanta. The hot gases at temperatures up to 60,000°F have a velocity up to approximately 10,000 mph. The Brayshaw model was used at the Laser Laboratory. Hishimoto[20-22] reported that this plasma scalpel was simple to use and produced good hemostasis with low trans-tissue heat conduction. Fatal argon gas embolization occurred during cardiac and liver surgery in the animals. Incropera,[25] who has done considerable work with the plasma scalpel, believes that argon depot in the tissues is not significant but that the argon is transmitted through the bloodstream. Control of the argon gas embolization is necessary before this excellent

FIGURE 15-16. LED spectacles as rival to laser radar spectacles for the blind.

control, especially for the CO_2 laser surgery, can be used even in animal surgery.

Physical Medicine and Rehabilitation

With computer technology, the laser can be used as a precise light scanning beam for muscle actintation and stimulation in a paraplegic patient. Laser transillumination may be used for fiber optic microscopes inserted in tissue to examine local muscle changes in living tissue. As shown, acoustical holography may be used to study movements of bones and soft tissues about joints without resort to X-ray films and fluoroscopy.

The lasers may be used for precise machining, drilling, and welding of prostheses. The CO_2 laser may be used also for cutting of materials.

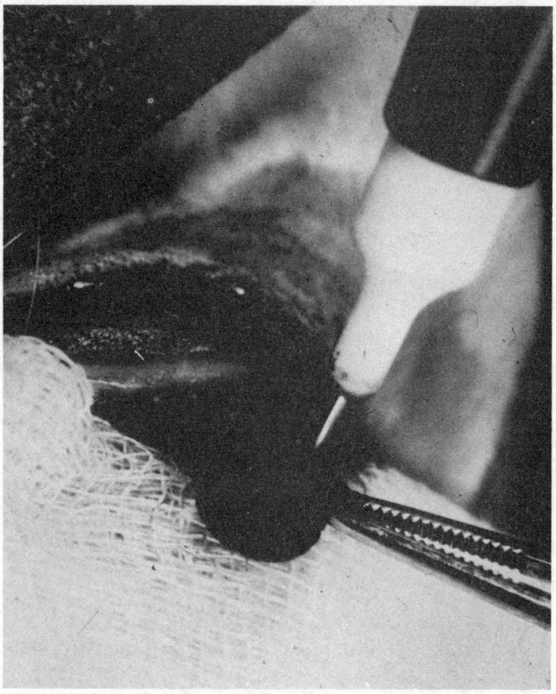

FIGURE 15-17. Plasma scalpel operation on liver with fatal argon gas embolization.

In the rehabilitation program, the crippled worker may be able to do precise laser work since heavy physical activity is not required. Also, closed circuit television control may also be used for laser safety in precision laser industrial work.

Laser Regulation for Medical Lasers

Regulation may be done for instrumentation,

FIGURE 15-18. Laser communication system to carry television signals between two hospital transmitting pictures of surgery and voice. (Laser Communications, Inc., Cleveland.)

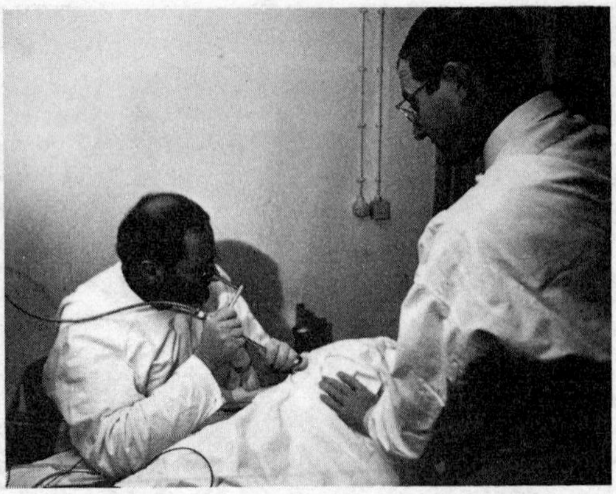

FIGURE 15-19. The argon laser laparoscope of Nath; yellow filter protects eye from argon stray light and transmits red color of blood. An 8-W CW argon laser with the single fiber of Nath connects the laparoscope to the laser. The laser operating gastroscope has similar arrangement.

standardization and safety, or for patient responsibility. As shown previously in the chapter, Safety, there will be standards for laser performance and safety. At present, the Laser Subcommittee of the A.A.M.I. (Association for the Advancement of Medical Instrumentation) is attempting to set up standards for laser medical instrumentation as regards value, need, reliability, and safety. Medical lasers are usually in the high-risk exposure situation and proper safety programs are required. Again, it is repeated, with planned programs laser surgery is safe for the operator and patient.

Since this is a new technology, reference has been made to the heavy responsibility of introducing this new agent in the treatment of patients. Special consent forms have been prepared and have been used for some years by the Laser Laboratory. A special form for developing the background for proper, as far as possible, informed consent for the patient has been prepared also. These examples are given.

CONSENT TO LASER OPERATION

DATE: ______________________

1. I authorize the performance upon ______________________________ of the procedure known as laser operation
(myself or name of patient)
to be performed under the direction of Dr. ______________________________.
2. I consent to the admittance of observers to the procedure and to the taking of photographs.
3. The nature and purpose of this procedure, possible alternative methods of treatment, the risks involved and the possibility of complication have been fully explained to me. No guarantee or assurance has been given by anyone as to the results that may be obtained.
4. I understand that in order to evaluate this procedure I will agree to tests of the treated areas. These tests may include surgical biopsies.
5. I understand that the laser treatment(s) I am about to receive involves treatment with an intense light. The purpose of this is to try to help the dental conditions which I have and which require treatment. The light is hazardous to the eyes and I understand that my eyes as well as the eyes of those who give the treatment will be protected; further, that there are no dangers to my eyes when they are protected. The laser treatment of the teeth inside the mouth is done with protection of the areas about the treatment. This will protect other places which do not require the laser treatment. I have been informed that it is not known whether this treatment will be helpful at all or not, but it has been used before and has helped in my type of trouble. The treatment is calculated to produce some tissue destruction with a minimal amount of scarring. The scars are not dangerous. There is to be no damage to tissues other than those which have been treated directly by the laser. The laser treatment is fast. There is occasionally some pain on the impact, but this pain varies with the individual and with the area which is treated. After treatment, sores, crusts and scabs may form. These must not be picked. I will be informed of what type of treatment, if any, the treated area requires. From time to time, tests may have to be made to determine whether the treatment is deep enough or adequate enough and whether additional treatments, or tests, including biopsies, will have to be done. I understand that should I desire any additional information about the laser treatment, the laboratory will be glad to give it.

Signed: ______________________________
(Patient or person authorized to sign for patient)

Relationship to patient: ______________________________

Witness: ______________________________

SPECIFIC CONSENT FOR LASER DENTISTRY

DATE: ____________________

1. I authorize the performance upon ______________________________ of the procedure known as laser dental
(myself or name of patient)
operation to be performed under the direction of Dr. ______________. The laser treatment is specifically for:
______________to help prevent tooth decay
______________for pulp cavity treatment
______________for treatment of caries
______________for the preparation of dentures
______________for the treatment of ______________in mouth
2. I consent to the admittance of observers to the procedure and to the taking of photographs.
3. The nature and purpose of this procedure, possible alternative methods of treatment, the risks involved and the possibility of complication have been fully explained to me. No guarantee or assurance has been given by anyone as to the results that may be obtained.
4. I understand that in order to evaluate this procedure I will agree to tests of the treated areas. These tests may include surgical biopsies.
5. I understand that the laser treatment(s) I am about to receive involves treatment with an intense light. The purpose of this is to try to help the dental conditions which I have and which require treatment. The light is hazardous to the eyes and I understand that my eyes as well as the eyes of those who give the treatment will be protected; further, that there are no dangers to my eyes when they are protected. The laser treatment of the teeth inside the mouth is done with protection of the areas about the treatment. This will protect other places which do not require the laser treatment. I have been informed that it is not known whether this treatment will be helpful at all or not, but it has been used before and has helped in my type of trouble. The treatment is calculated to produce some tissue destruction with a minimal amount of scarring. The scars are not dangerous. There is to be no damage to tissues other than those which have been treated directly by the laser. The laser treatment is fast. There may be occasionally some pain on the impact, but this pain varies with the individual and with the area which is treated. After treatment, sores, crusts and scabs may form. These must not be picked. I will be informed of what type of treatment, if any, the treated area requires. From time to time, tests may have to be made to determine whether the treatment is deep enough or adequate enough and whether additional treatments or tests, including biopsies, will have to be done. I understand that should I desire any additional information about the laser treatment, the laboratory will be glad to give it.

Signed: ______________________________
(Patient or person authorized to sign for patient)

Relationship to patient: ____________________

Witness: ____________________

THE LASER LABORATORY OF THE MEDICAL CENTER OF THE UNIVERSITY OF CINCINNATI

THE LASER TREATMENT

Laser treatment, often called laser surgery, is a form of treatment of the so-called non-ionizing type. In other words, it is not like x-ray. Laser beams destroy tissue, chiefly through heat. Some lasers have selective properties in that they affect those tissues which have color in them. As yet, no other form of surgery does have this selective property.

Laser beams have been used for laser surgery for more than ten years. Because this is a very bright light, the hazards are chiefly to the eyes. Therefore, eye protection must be done for all laser treatments. Patients as well as the laser operators must have eye protection.

The laser treatment may last for a tiny bit of a second, or with some lasers, the treatment may be continuous. Different conditions require different lasers. If the treatment period is very short, there may be little pain. This varies with the type of laser, the type of condition treated. Where many treatments have to be given at one time, general anesthesia may be required.

A selective burn of tissue is produced. Because of the precision and character of the laser, this burn is selective as regards type of tissue and depth of burn. As it heals, scars are produced. The character of the scar varies with the condition treated; unless it becomes infected or picked or burned, the scar is usually not excessive. For surface conditions, the scarring is usually not excessive; for deep conditions, the scarring may be deep. Spots which are deep may take a considerable period to heal.

The type of treatment to be used after laser surgery will be explained to you, as well as the follow-up program.

You are having the laser treatment because alternative methods of treatment, such as, plastic surgery, x-ray, chemical treatments or just covering the trouble, cannot be used.

If you wish any portion of this explained, or if you have any questions, please advise.

In addition to these forms, the proper program would include the Laser Faculty Advisory Committee to survey and direct the programs for this new technology.

Communication with patients is always important to show the need and value of this new technology.

Laser Information Handling in Medical Education

Case Western Reserve University has shown the value of a laser installation for two-way color television across the campus.[28] This can be expanded to include hospitals. One example of the value of this was the transmission of high quality color television over a half a mile, in a snowstorm, from Lakeside Hospital to the Veterans Administration Hospital. The cost of this system (1973) is $8,000 for a two-way channel 10-mi link as compared to $20,000 for a one channel microwave system, and $80,000 for a closed circuit one channel cable system.

Laser computer technology for medical information storage and retrieval, video cassettes, and the recent MC video disc technology are all new bits of educational aids to help medical education, both undergraduate and continuing education.

Laser Institute of Biology and Medicine

It is evident that more investigations are necessary for laser biology and medicine. From the author's experience, a multidisciplined approach is needed with brief effective narrow lines of communication. If the laser industry is to progress, more knowledge is necessary. If the laser is to be used in biology and for the treatment of patients, again, many disciplines need to be used and critical controlled investigations are needed.

With the experience with ionizing radiation, X-ray institutes were found to be necessary. If such institutes of radiation had been established early, much more effective use and much safer use could have been possible much earlier.

So, again, a plea is made now for the establishment of an Institute of Laser Biology and Medicine. It is needed and, for a change, cooperative funding without reduplication of needs would establish such an institute for the real good of all currently in laser technology.

Conclusions

It is evident that lasers have a great impact, as it were, for diagnosis and treatment of the patient. It is evident that there is a need often for such new applications, many of which have been investigated for over 10 years. Such programs should be continued in laser centers over the world in spite of official disinterest in some areas.

REFERENCES

1. **Allemand, C. D.,** Personal communication, 1971.
2. **Cameron, W. G.,** *Diagnosis by Transillumination,* Cameron Publishing Co., Chicago, 1923.
3. **Feleppa, E. J.,** Holography and medicine, *IEEE Trans. Bio. Med. Eng.,* 19, 194, 1972.
4. **Fidler, J.,** Personal communication, 1972.
5. **Flocks, M.,** *Laser Photocoagulation,* International Ophthalmology Clinics, Little Brown and Co., Boston, 1966.
6. **Gameleja, N. F.,** *Biological Effects of Laser Radiation,* Kiev, 1972.
7. **Geeraets, W., Jr.,** Observations of lens epithelium in cell cultures, *Med. Coll. Va. Quart.,* 8(4), 264, 1972.
8. **Geeraets, W., Jr. and Carroll, W. W.,** Some therapeutic considerations in diabetic retinopathy, *Med. Coll. Va. Quart.,* 8(4), 289, 1972.
9. **Ger, R.,** Personal communication, 1972.
10. **Gerson-Cohen, J., Hermel, M. B., and Murdock, M. G.,** Priorities in breast cancer detection, *N. Engl. J. Med.,* 283, 82, 1970.
11. **Goldman, L.,** *Biomedical Aspects of the Laser,* Springer-Verlag, New York, 1967.
12. **Goldman, L.,** Laser Applications Course, University of Cincinnati Medical Center, April 1-3, 1971.
13. **Goldman, L.,** *Laser Cancer Research,* Springer-Verlag, New York, 1966.
14. **Goldman, L.,** Snags in laser biomedicine, *Biomed. Eng.,* 6, 9, 1971.
15. **Goldman, L. and Rockwell, R. J., Jr.,** *Lasers in Medicine,* Gordon and Breach, New York, 1971.

16. **Goldman, L., Rockwell, R. J., Jr., and Richfield, D. F.,** Long-term laser exposure of a senile freckle, *Arch. Environ. Health,* 22, 1971.
17. **Goldman, L., Rockwell, R. J., Jr., Naprstek, Z., Siler, V. E., Hoefer, R., Hobeika, C., Hishimoto, K., Polanyi, T., and Bredemeier, H.,** Some parameters of high output CO_2 laser experimental surgery, *Nature,* 228, 5278, 1334, 1971.
18. **Henderson, M. R., Link, W. J., and Incropera, F. P.,** Gas Transport from Plasma Scalpel Surgery, Purdue University, Lafayette.
19. **Hillenkamp, F.,** Kaufmann Laser in der Medizin. Gesellschaft fur strahlen und Uniwetforschung, Munchen, June 1972.
20. **Hishimoto, K.,** Plasma scalpel and plasma surgery, Gekash in ryo, *Clin. Surg.,* 14(8), 1972.
21. **Hishimoto, K. and Goldman, L.,** Plasma Scalpel Surgery, Sanryo Co. Ltd., Japan.
22. **Hishimoto, K., Rockwell, R. J., Jr., Epstein, R. A., Yunis, W. G., Dreffer, R., Johnson, J. W., and Franzen, M. M.,** Plasma scalpel surgery, in press.
23. Holocamera, TRW Instruments, El Segundo, Calif., 1972.
24. Holoviewer, *Lasersphere,* May 31, 1971.
25. **Incropera, F. P.,** Personal communication, 1972.
26. **Kaplan, I.,** Personal communication, 1972.
27. **Ketcham, A.,** The laser – its role in cancer, *Ca-A Cancer J. Clin.,* 1971, 297.
28. **LaFleur, D. M.,** Laser Communications Inc., Cleveland, 1972.
29. **Lagunova, I. G., Savchenko, E. D., Likhovetskaya, L. L., Garvey, N. N., Shamayeva, G. G., and Klinan, A. D.,** The biological effect of lasers on the skin, *Med. Radiol.,* 16(9), 1971, *Laser Abstr.,* 1971-2253, 1971-20.
30. The Laser, Annals of The New York Academy of Science, May 28, 1965.
31. Laser Applications Course, Biology, Medicine, Dentistry, University of Cincinnati Medical Center, April 1-3, 1971.
32. The Laser Second Conference, New York Academy of Sciences, February 10, 1970.
33. Lasers bright future in medicine, *Business Week,* July 15, 1972.
34. **Lee, J. H.,** The trend in mortality from primary malignant tumors of the skin, *J. Invest. Dermatol.,* 59, 445, 1972.
35. **Leibowitz, H. M. and Luzzio, A. J.,** Laser induced cataract, *Arch. Ophthal.,* 83, 608, 1970.
36. **Levine, S. N.,** *Advances in Biomedical Engineering & Medical Physics,* Vol. 1, Interscience, New York, 1968.
37. **Mims, F. M.,** Personal communication, 1972.
38. **Pratt, G.,** Laser beam to cleanse wounds, Universal Science Service.
39. **Rockwell, R. J., Jr.,** The current status of laser, applications in medicine and biology, *CRC Crit. Rev. Bioengin.,* 1(1), 49, 1971.
40. **Shea, M., Maberly, A. L., Walters, J., and Freeman, R.,** Eye tapeworms, *J.A.M.A.,* 222, 891, 1972.
41. **Stahle, J., Hogberg, L., and Engstrom, B.,** The laser as a tool in inner ear surgery, *Acta Otolaryngol.,* 77, 27, 1972.
42. **Stark, A. M. and Way, S.,** Screening for breast cancer, *Lancet,* 407, 1970.
43. **Trubey, R. J.,** *Clinical Applications of Lasers,* Upjohn Co., Kalamazoo, Mich., 1972.
44. **Vahl, J.,** Applicability of scanning electron microscopy to the topography of the surface specimen after laser irradiation, *Ann. N. Y. Acad. Sci.,* 168, 664, 1970.
45. **Vahl, J.,** Personal communication, 1971.
46. Video disc technology, MCA, *Laser Weekly,* February 5, 1973.
47. **Wuerker, R.,** Personal communication, 1972.
48. **Yo-Han, P.,** Laser beam carries medical lectures, *Med. World News,* April 23, 1971.
49. **Zivi, S. M. and Humberstone, G. H.,** Chest motion visualized by holographic interferometry, *Med. Res. Eng.,* 9, 1970.
50. **Holbrooke, D. R., McCurry, E. M., Richards, V., and Shibita, H.,** Thorough Transmission Ultrasonic Imaging of Intrauteral and Fetal Structures Using Acoustical Holography, 2nd World Congress of Ultrasonics in Medicine, Rotterdam, June 6, 1973.
51. **Green, P. S.,** Ultrasonic camera images in real time, *Ind. Res.,* July 1973.
52. **Greguss, P.,** Non-electromagnetic holography and its impact on biomedical research and clinical practice, *Acta Biochim. Biophys.,* 7, 203, 1972.
53. **Greguss, P.,** Biomedical applications of holography, *Laser Unconvent. Opt. J.,* 39, 2, 1972.
54. **Strong, M. S. and Jako, G. J.,** Laser surgery in the larynx, *Ann. Otol.,* 81, 791, 1972.
55. **Stahle, J., Holberg, L., and Engstrom,** The laser as a tool in inner-ear surgery, *Acta Otolaryng.,* 73, 27, 1972.

Chapter 16

LASER DENTISTRY

Dental Caries

There are many actual and potential applications of the laser in dentistry. The greatest challenge is about one of the world's most common ailments, dental caries. In regard to this, Stern[2–4] believes that the greatest potential for the biomedical application of the laser is the use of the pulsed CO_2 laser or Nd-YAG laser on enamel for the prevention of caries. Whether this is the whole phase, or whether the time employed in caries removal is only a small fraction of the time required for the complete restorative procedure, is being discussed at the present time. The question of the effect of the CO_2 laser on the enamel with regard to the viability of the pulpal cavity is also of great concern. The effects on the pulpal cavity can be measured by temperatures of the pulp in vivo and also by detailed histological and electron microscopic studies. The following studies of Vahl are quoted in detail because of their importance:

> "The electron optical studies were undertaken to determine the possiblity of caries prophylaxis by laser irradiation. This is a continuation of previous studies. Structural changes of lasered enamel occurred to varying depths depending on the energy densities. Relatively high energy densities (> 500 J/cm^2) using a ruby laser produced secondary undesirable cracks. Stern, Vahl and Sognnaes (*J. Dent. Res.* 51:455, 1972) reported results employing a pulsed CO_2 laser with energy densities <50 J/cm^2. As an extension of this work, thin sections of lasered enamel did not show prism structures to a depth of 0.05 mm. Microspaces were filled with an electron optical amorphous substance. This could perhaps consist of numerous nuclei of crystallites, which were unable to grow into larger crystals during the process of recrystallization. 0.5 μm wide fissures were evident in the outer layer of enamel. Further structural changes were circularly formed micro-spaces of about 1.5 μm diameter, not in contact with the outer enamel surface. These spaces were randomly distributed and separated by relatively large distances and therefore the stability of the molten enamel is not reduced to any appreciable extent. By x-ray and electron diffraction of the lasered enamel it could be calculated that the crystal lattice of the hydroxyapatite was changed to a tricalciumphosphate."

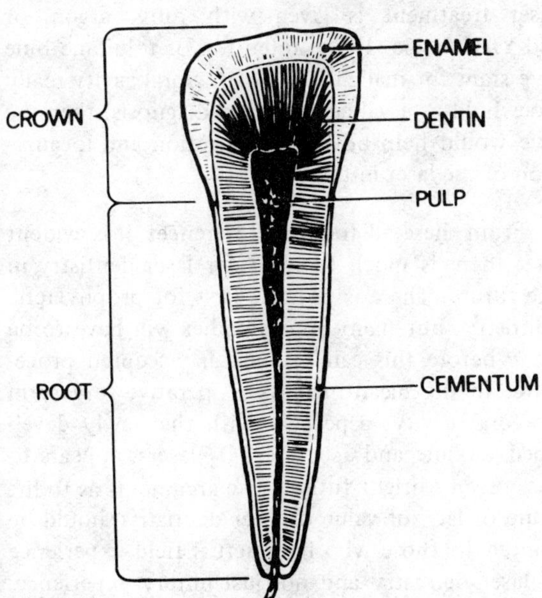

FIGURE 16-1. Basic structures of the tooth. (Ralph H. Stern.)

There is also the difficulty of uniform complete treatment of the entire tooth.

Pulpal Cavity Treatment

The next phase of the dental care program is the use of the argon laser for pulpal cavity treatment. These pulpal cavity treatments may be an alternative to extraction or root canal therapy. As yet, this technique has not been used on human subjects.

The real question of the cavity preparation by the laser has to be measured by controls of the cavity preparation by today's conventional dental drill that utilizes a high-speed air/water-cooled turbine. As long as these debates continue on theoretical means and not on patients, it is going to be difficult to answer some of these queries. With planned programs, again, the safety of the laser has been shown for dental work in man. Impacts of the teeth have been done in patients without hazards to the adjacent areas of the teeth or to the patient. Other parameters of cavity preparation include:

1. Concern for pulpal damage
2. The accuracy of the treatment
3. The expense
4. The effects as measured by controls

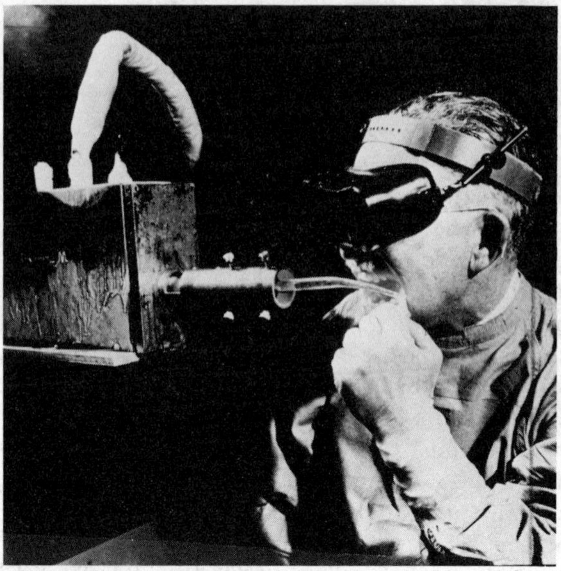

FIGURE 16-2. Pulsed ruby laser impacts 20 J/cm^2 on carious areas of the crown of a tooth prior to extraction. Transmission of the laser is through a curved tapered quartz rod. Temperature readings taken of adjacent areas during impact.

When all these factors are resolved, the true value of the laser in cavity preparations can be determined. Stern mentions a new filling material that is painted over the cavity and then fused with the laser to fill in space. The CO_2 laser is also used for fusing fillings.

Transillumination and Plethysmography

Another factor of dental application is laser transillumination which determines cracks, fissures in the tooth, and also the question of viability. There is also the use of transillumination with laser plethysmography as a test for circulation in the teeth. Whether this is more effective than the usual electrical testing apparatus remains to be seen from additional studies.

Dental Prosthesis

Another application of the laser is the use of the intra-oral dental prosthesis work with Q-switched research laser, single fiber of Nath. With the absence of heat in the use of this technique it may be possible to do prosthesis work in the mouth. Gordon has shown the value of the laser in the application of dental prosthesis. More investigative studies are needed to determine whether this laser metalworking technique can be applied to actual intra-oral work.

Oral Surgery

There is another important phase of laser dentistry, namely, laser oral surgery. In this program the laser is used as an operating tool for such conditions as precancerous leukoplakia of the mouth or gums, tumors, bleeding, etc. These lesions may be painted with vital dye before the laser treatment is given with ruby, argon, or Nd-YAG lasers. If the retention of toluidine blue dye stain for malignancies of the oral cavity really does hold as a valuable aid for diagnosis, then the dye would help both for absorption and localization of the laser in this area.

From these all too brief references, it is evident that there is much to offer for laser dentistry in the future. There is promise first for prophylactic dentistry, but many more studies will have to be done before this can become an accepted procedure. In the meantime, laser operative surgery in the oral cavity, especially with the newly developed flexible and accurate CO_2 lasers, appears to have a very bright future. The arguments as to the value or lack of value of laser dentistry should be limited to those who have actual field experience in laser dentistry and not just library experience. The dental groups must be interested in cooperative and controlled studies. Especially valuable in

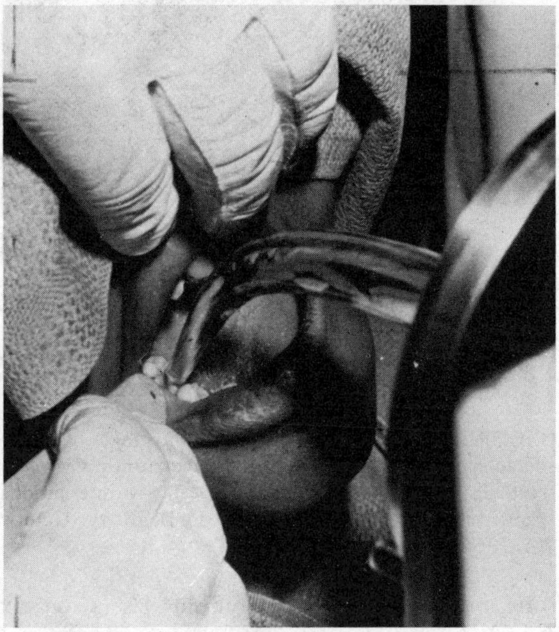

FIGURE 16-3. Recent high output pulsed ruby for impacts of actual carious areas with the same technique.

FIGURE 16-4. Single fiber of Nath used as a hand held welder to weld metal in studies of intra-oral denture technique, Nd-YAG, Q-switched, B picture of the weld x 42.

FIGURE 16-5. Welding of dental gold samples by Nd-YAG laser 8 J/pulse, 7 m/sec pulse length. He-Ne guide light with the single fiber of Nath. (Gunther Nath.)

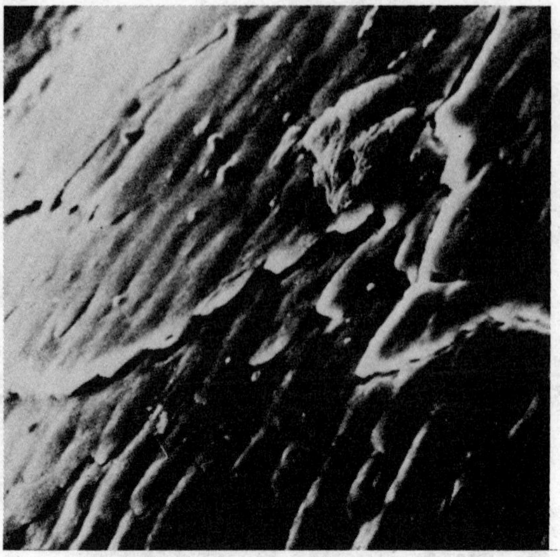

FIGURE 16-6. S.E.M. micrograph showing a beginning detachment of a lased enamel surface from a CO_2 laser exit energy 9 W. (Prof. Dr. J. Vahl – International Conference on Laser Application in Dentistry, October 15, 1969.)

the past and for the future are the scanning electron microscopic studies of Vahl[6–8] and similar detailed basic research investigations. Much more support for laser dentistry is provided for in Europe than in the U.S. Experiments with high output lasers on extracted teeth are scarcely proper test models for the actual treatment of the teeth in the patient. As in laser medicine and surgery, it is time now for properly controlled studies of laser dentistry in man.

REFERENCES

1. **Goldman, L. and Rockwell, R. J., Jr.,** *Lasers in Medicine,* Gordon and Breach, New York, 1971.
2. **Stern, R. H.,** The potential of lasers in caries prevention, Second Conference on the Laser, Annals of the New York Academy of Sciences, February 10, 1970, 642.
3. **Stern, R. H.,** Laser inhibition of dental caries suggested by first tests in vivo, *J.A.D.A.,* 85, 1087, 1972.
4. **Stern, R. H., Sognnaes, R. F., and Vahl, J.,** Lased enamel: ultrastructural observations of pulsed carbon dioxide, *J. Dent. Res.,* 51, 455, 1972.
5. **Vahl, J.,** Biocrystallographical studies of laser treated human enamel, in press.
6. **Vahl, J. and Ruter, W.,** Feinstrukturuntersuchungen an zapenförmig ausgebildeten menschlichen Zahnen, *Dtsch. Zahnarztl. Z.,* 28, 317, 1973.
7. **Vahl, J., Plackora, A., and Hohling, H. J.,** Mikroanalytische Untersuchungen der patholgisch bedingten anorganischen Ablagerungen in der menschlichen Pulpa (Elekronenstrahlen-Mikroanalyse), *Dtsch. Zahnarztl. Z.,* 28, 312, 1973.
8. **Vahl, J. and Mierau, H. D.,** Strukturelle und analytische Ergebnisse sekundärer Veränderungen am Dentin und Schmelz in Funder aus dem 11. und 12. Jahrhundert, *Dtsch. Zahnarztl. Z.,* 28, 322, 1973.

Chapter 17

TRAINING AND RETRAINING IN LASER TECHNOLOGY

It has been said repeatedly that information and knowledge increase progress and development. Since we have seen the recent birth of a new technology, the obligation is for us to teach laser technology, as it were, to our children and to our children's children. Oftentimes it is necessary to do such teaching without instruction in education or even without an understanding of our own process of learning. It is never too early for young children to be dragged, wandered, or herded through science museums. Short, entertaining visits may be made with participation by the children in the various activities. Then children begin early to understand and not fear science. Properly protected nursery school groups have visited the Laser Laboratory. Of course, it has always been a question as to which the children enjoyed most: the suckers, the staff talking with helium in their mouths, or the laser impacts on pennies. Anyway, they had their pictures taken with their safety glasses.

The Laser in High School Physics

The first real interest in lasers usually begins in high school physics classes. The Bureau of Radiological Health Laser Section of the U.S. Public Health Service has developed an excellent laser manual for the high school student. Even laser safety is included. As discussed in the chapter on Safety, this Bureau is disturbed because of the lack of safety in the use of the laser in high schools and colleges. The laser for high school physics is usually the low output helium-neon laser.

The Laser in College Physics

Here again, there is concern about the lack of laser safety. This is a pity, for here the laser scientist may be born. The use of the laser is more extensive in college. There is more use in college, and hopefully, not abuse of laser teaching sets. An excellent one is that of American Optical Co. Special projects may develop and the laser student goes into graduate laser physics with his social conscience and his ideals about laser technology.

The Laser Technician

In the new field of laser training for laser technology, perhaps more important from a practical standpoint than the laser biomedical engineer is the laser technician.[7] It is evident that as laser technology expands there must be a training program for technicians who combine training both in electrooptics and lasers. This is a new field of training which is fortunately already under way.

The Technical Education Research Center, Inc. (TERC),[8,9] a nonprofit public service organization in this country, is interested in technical education research. The Texas State Technical Institute of Waco has been selected to develop a pilot program for training laser electrooptics technicians. This has been successful. The curriculum was developed by 16 members of the electrooptics industry to serve as an Advisory Committee under the direction of Dr. Donald Pierson, Laser Electro-Optics Technology Department, Texas State Technical Institute, Waco. Rockwell[7] of the Laser Laboratory directs the laser phase of this excellent course.

Smith of the Laser Institute of America[5] suggested the various laser training guidelines for specific types of laser technicians interested in such fields as communications, computers, medicine, or metallurgy. Some specific types, as suggested by Smith, are

1. The laser operator to follow a specific place or operation procedure
2. The operating technician to guide operators to make mild adjustments to laser
3. The maintenance technician to maintain manufacturing or other functional equipment such as military lasers
4. The engineering technician to assist engineers or other technicians with a minimum of direction
5. The applications engineer to be responsible for developing laser applications

These different technicians require different training programs.

Rockwell[7] has indicated that the school is

FIGURE 17-1. Nursery School Group Visits the Laser Laboratory.

FIGURE 17-1A. Explaining eye protection.

FIGURE 17-1B. The class picture.

definitely expanding. It has been estimated that as many as 10,000 laser technicians will be needed over the next decade. Since the program for the development of laser technicians is so important, the Laser Advisory Board has developed modules of course materials for teaching. The following modules have been devised on such phases as:

1. Diffraction
2. Absorption
3. Light waves
4. Reflection and refraction
5. Energy levels
6. Emission of radiation
7. Operation of He-Ne, dye, ion, injection, and optical pump lasers

Rockwell[7] has used approximately 156 modules for this initial pilot training program to cover everything in technician training from absorption to the Mach-Zehnder interferometer. It is obvious then that there will be need for a number of technical schools to follow the initial pilot training program. For such a training program then as TERC (from the Southwest Center of Technical

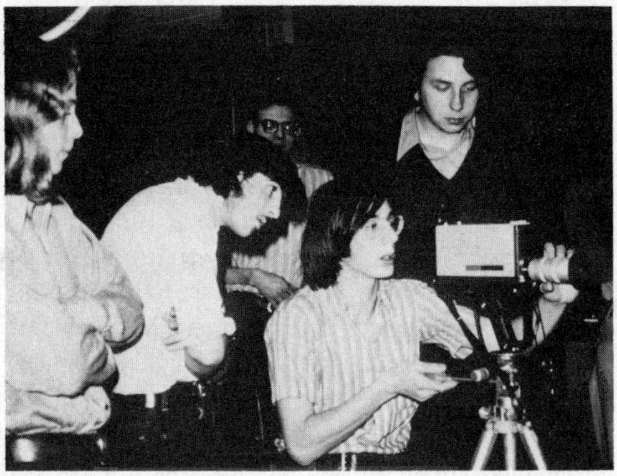

FIGURE 17-2. High school students producing T.V. program of laser technology at Laser Laboratory. Intensity of reflected beam burned Vidicon tube – phosphor through failure of protection of television camera.

Education Research Center), it is obvious that government, industry, professional societies, labor, and education share these objectives. The efforts of each of these segments of the economy have been commendable, but the burgeoning nature of the problems to man means an increasing effort on the part of all concerned. TERC welcomes the opportunity to develop the framework which will bring "cooperation and interaction among these groups." It is evident, then, that this program is necessary to provide for the increasing developments in laser technology to offer support to the disciplines in laser work, the physicist, the engineer, the electrooptics expert, the instrument engineer, and the specific operator. A two-year training program for laser technicians is under development by Rockwell at the Raymond Walters Branch of the University of Cincinnati.

Paramedical Laser Personnel

With the increasing use of the laser in the medical field, laser paramedical personnel will be a new aspect comparable in importance to paramedical personnel in other fields of medicine. After completion of the basic training in a technical center, additional training will be provided in the biomedical field. These additional courses can be in biology and in the established programs for paramedical training.

In biology, the laser technician will be able to service and monitor laser microscopy in addition to his work as a biology technician. For laser ophthalmology, the laser technician will assist in laser operations, maintenance and servicing, and research and development of laser ophthalmological equipment. In general surgery, the laser surgical technician will be involved in the applications of the laser in the operating room, its safety, its sterilizability, and its servicing, so important in the operating procedure. All this will be in addition to the valuable paramedical services of the surgical technician. Because of the complex training and background, the laser surgical technician will be in a different group and discipline than the ordinary, capable surgical technician. At first, laser surgery will be limited to the large medical center equipped with a laser surgical operating room. Later, this application will extend to the analytical uses of lasers for analyses in the laboratory, the lasers used in outpatient work, for cancer diagnosis, and treatment in outpatients for laser dentistry and the like.

It is possible in the future for the large medical center that there might be a separate Laser Biomedical Engineering Division to include the biomedical engineers and the laser technicians. This division will direct research and development and actual servicing and care of laser units in the large medical center. The time for planning these programs is now. As the Texas State Technical Institute has shown, the continued education programs should be developed now, initially with pilot programs for the various divisions within modern laser technology.

Laser Biomedical Engineering

It is obvious from the developments in biomedical engineering and in laser biology, medicine, and dentistry that new disciplines are required.[2–4] Specifically, these disciplines would relate to the so-called nonionizing radiation physicists, biomedical engineers including biologists, biochemists, physicians, and surgeons. This is the background of laser biomedical engineering. Actually, in the current training programs of biomedical engineering, optics has been considered very little if at all. Mechanical, electrical, and chemical engineering are usually the divisions included. It is now necessary to integrate many phases of modern basic and applied optics and the new fascinating field of optical electronics to develop this field of laser biomedical engineering.

First, is the program for the radiation physicist. It is pertinent at this time, at least, to call the radiation physicist a radiation physicist whether this refers to ionizing or nonionizing radiation. Now, the radiation physicist must also learn about lasers, microwaves, and ultrasonic and ultraviolet radiation. All of these should be included in his training program. Usually this is done with relatively few sessions unless he attends specific laser and optics instruction courses and laser safety courses offered throughout the country. As yet, for radiation physicists, there are no standardized or nationally adopted requirements for laser instruction and for laser safety and monitoring.

There are varied programs on biomedical engineering at undergraduate or graduate levels throughout the country. As yet, there are no formal offerings for laser biomedical engineering training except at the University of Cincinnati. To the already crowded curriculum at undergraduate level at least, optics technology must be added for all biomedical engineers irrespective of their future goals and plans. These may not be more than basic optics, including optical electronics. It is at the graduate school level that detailed programs on laser physics, measurements, safety, biology, and instrumentation should be developed and expanded. It is emphasized that such programs at undergraduate or graduate levels not be confused with the programs now currently in the planning stage for the development of laser technicians. These plans refer to brief training in laser technology to include instrumentation, monitoring, and servicing.

The challenge is there. The laser industry will continue to expand. Unfortunately, most of these industrial laser establishments have had no experience or interest (low financial yield) in possible biomedical application. Their activities in laser research, industrial and military, do not orient them for the exact, tedious safety performance standards and governmental control required for the development of instrumentation for the clinical treatment of patients. The training of the biomedical engineer with a background in biology and medicine will do much to help the development in this particular field. Biologists, physicians, and surgeons would do well to establish better lines of communication with laser biomedical engineers for research and development of laser biomedical instrumentation.

As laser medical instrumentation gains in popularity and importance, one should expect legislation as regards performance standards and safety. Even with laser eye instrumentation, the current market is meager compared to the fat laser contracts of industry and the military. The laser biomedical engineer has an important role in the performance and safety of laser medical devices. At present, the Association for the Advancement of Medical Instrumentation (A.A.M.I.) has a subcommittee on lasers.

Hopefully, financing will be arranged as laser technology continues to develop and the need for a separate Institute for Laser Biology and Medicine is finally recognized. This is an oft repeated appeal and will be realized when the low priority of the laser in biology and medicine is reversed. Laser biomedical engineers can function as laser safety officers and can coordinate planning for the protection of the workmen. Some specific problems challenging the biomedical engineer at the present time relate to calibration of instrumentation, area and personnel safety, and the development of flexible, reliable sterilizable transmission devices for high output laser systems so that these may be used in laser surgery. Also there are current studies on the applications of fiber optics and instrumentation design for the development of laser diagnostic and operative endoscopy. These are concerned with diagnostic and operative laser microscopy for the oral cavity and larynx, lips, skin, and cervix. The flexibility of laser tissue biopsy instrumentation and the degree of induced thermal coagulation necrosis produced are two important facets. Additional fields of investigation are the adaptation of the tunable dye lasers for biomedical applications, the grave concern for the development of high output ultraviolet systems, and the future concern with possible X-ray lasers, the grasers.

All these are but some of the problems that confront the laser biomedical engineer in the expanding field of laser technology. The concept of the development phases of lasers for the good-of-man, often more verbal than actual, by government means increased employment in physics, optical, electronic, and space research, and consequent development of trained capable and effective laser biomedical engineers.

Retraining

Retraining in laser technology may mean retraining from other areas into lasers or from lasers into other areas. Unfortunately, the most common way is the forced departure from lasers (for no one would voluntarily desert current laser technology). The reduction of the space technology program and some phases of the aircraft industry has been responsible for loss of laser personnel. Furthermore, the rather bleak immediate future for physicists adds another group for retraining. Most retraining has been on a personal basis.

One of the active programs for retraining is the one at Oregon State University funded from the Department of Labor.

It seems that laser expertise in the above areas could do well to stay in laser technology for work in other applications of the laser. The difficulty, of course, is the availability of risk private capital and federal funds.

Again, if the executive branch still disdains science, the funding could be for direct applied research for practical applications of lasers. It still appears better to do this than to develop a retraining program in another field such as construction or transportation. Some, of course, recently have been brought back into the fold by the increased demands of research and development for laser-induced thermonuclear fusion.

If applications of the laser greatly increase the need for personnel, then personnel from other fields such as electronics can be retrained as electrooptics specialists. By instruction and experience, biologists may become part of the new laser biology; chemists may become laser chemists. Other retraining programs can develop laser computer specialists, laser military specialists, laser radar personnel, holography and acoustical holography experts, and, of course, laser biomedical engineers and laser biophysicists, surgeons, and dentists.

Conclusions

There will be constant revisions of laser training, teaching, and learning programs not only for those who wish to go into laser technology but for those who are not in laser technology. The continued progress of this new technology requires refresher courses so that one remains young and learned in this new technology. The same principles apply, of course, to training and retraining in laser safety.

REFERENCES

1. **DuVal, M. K.,** Implications of advanced biomedical research and technology, *J.A.M.A.*, 220, 247, 1972.
2. **Goldman, L.,** Biomedical laser technology: a challenge to the engineer, *Biomed. Eng.*, 6, 22, 1971.
3. **Goldman, L.,** Snags in laser biomedicine, *Biomed. Eng.*, 6, 9, 1971.
4. **Goldman, L. and Rockwell, R. J., Jr.,** *Lasers in Medicine,* Gordon and Breach, New York, 1971.
5. **Gould, G.,** 100,000 laser technicians, *Lasersphere,* July 15, 1972.
6. **Rockwell, R. J., Jr.,** The laser electro-optics technician, *Lasersphere,* March 15, 1972.
7. **Rockwell, R. J., Jr.,** Growth in technical training, *Lasersphere,* September 15, 1972.
8. Technical Education Research Centers, 44-A Brattle St., Cambridge, Mass., Southwest Center, 1224 Austin Ave., Waco, Texas.
9. Texas State Technical Institute, James Connally Campus, Waco, Texas.

Chapter 18

LASER ART AND MUSIC

If the artist understands something of the mechanisms of light and its uses, then he can apply it to his work. The fascination of the great artists of the past in their early experimentation showed the wonderful uses of changing the patterns of light. Rembrandt, Vermeer, and Caravaggio were some of the great masters who experimented with light. In his preface to the first comprehensive laser art exhibit held at the Cincinnati Art Museum, November 1971,[1] Richard Boyle, Curator of Paintings of the Cincinnati Art Museum, showed the significance and importance in the development of light in art to its culmination in the introduction of the laser to modern art.

Fascinated by the atomic patterns of matter and the regularity and clarity of geometrical figures, the concepts of op-art gradually evolved. Static art forms were mostly characteristic of this op-art. Then the development of the laser came to excite and form a new dynamic type of modern art, the laser.[4,5] So, the modern artist must learn about lasers.

Safety Programs for the Artist

It has been shown that the laser is hazardous primarily through its effects on the eyes and next is the exposure of the skin. Finally, there is air pollution from the effect of the laser on material processed. How can the artist develop laser art so it is entirely safe? With the other applications of the laser, it has been demonstrated quite effectively that with safety programs the laser can be used. Laser safety for the artist is concerned not only for himself but also for the public who may view his material if lasers are used in the final design and arrangements.

With any sort of laser that the artist uses, advice and suggestions ought to be secured from a laser safety officer. These are people who are trained completely and effectively in laser safety. Any large laser installation has a laser safety officer who can instruct the artist. In general, the following program has been adopted. First, the proposal of the artist is submitted as regards laser safety. Then, if lasers are used in the final presentation, as for laser design, the entire setup is reviewed again for safety. This has been emphasized in the chapter on safety and is repeated here. Finally, before the exhibit opens to the public, the safety program is reviewed again.

If high output laser systems are used, for the present, the artist will work in laser installations where there will be adequate laser safety programs. In the future, the small output helium-neon lasers and, to some extent, the argon lasers may be purchased or rented by the artist. He then will have to take the sole responsibility for his own protection in the development of his laser piece. At all times, he must consider eye protection, never looking directly into the beam, watching the beam with care from its reflectance, especially, on fairly highly reflective or specular surfaces. Laser

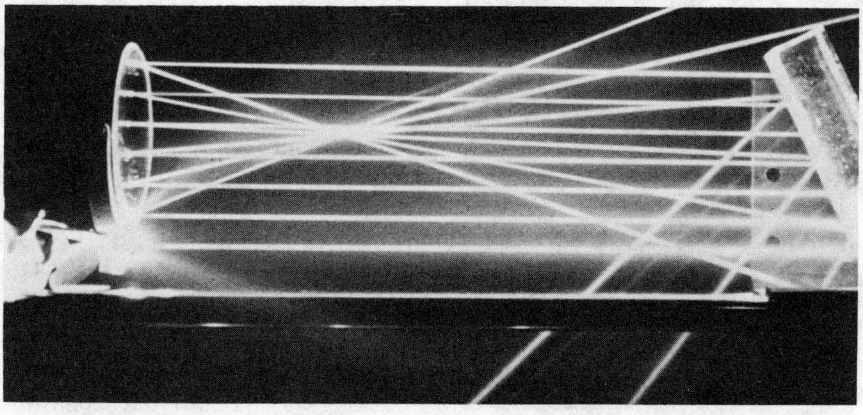

FIGURE 18-1. Laser optics as art. (American Optical Co. teaching kit.)

FIGURE 18-2. Color Absorption Studies.

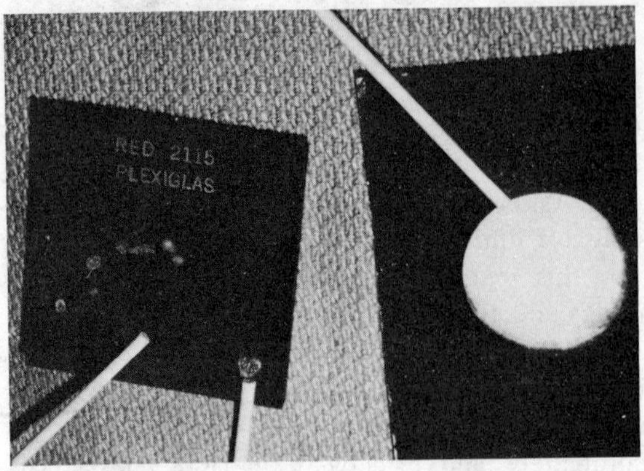

FIGURE 18-2A. On left – red Plexiglass – no effect ruby, bleaching by argon. On right – black paper – bleaching by ruby.

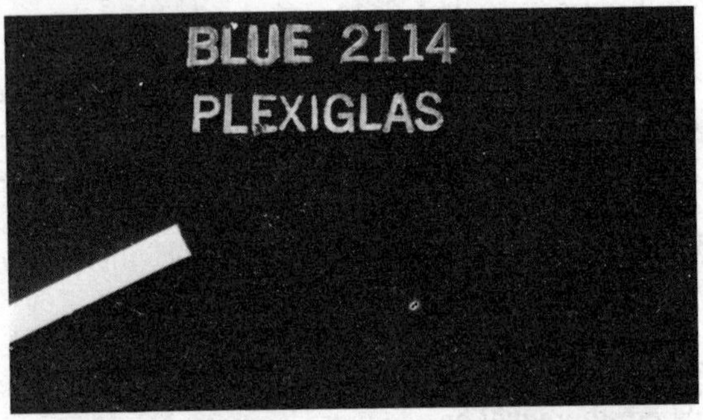

FIGURE 18-2B. Blue Plexiglass – no effect of argon.

FIGURE 18-2C. Red Plexiglass – design by argon.

safety glasses are appropriate for the specific laser, since there is not one laser safety glass suitable for all lasers. The protective glasses should be worn almost continuously by the artist as he works with the laser. Again, it is repeated that when the piece is completed, the laser safety officer reviews the project if there is a laser included in the piece, for safety especially for the viewing public. If no laser is used in the arrangement, such as for etching, painting, sculpture, or photography, then there is no concern about laser safety of the completed work. However, if the laser is an integral part of the design, such as may occur in laser design, holography, or even laser music, then the laser safety officer reviews the safety for the general public. The concern for the public viewing laser arrangements, is the continuous wave type of laser, such as the helium-neon or the argon. The audience must be kept away from any direct contact with the beam or, as indicated, reflections from the beam. This means that the design must be stable – the use of the laser must not be subject to vibration or loosening of any area that can reflect the beam in a different manner. There must be some type of railing or device that keeps the audience away from the laser system. This means protection for the adventuresome child or the inquisitive observer. Signs are not enough. There must be actual arrangements of physical restraints around the object so that no one can come in contact with the beam or its reflectance. The viewing of the object can be completely safe even in a semidarkened room if the laser beams are contained completely within the object. There have been instances of laser exhibits where the Board of Health Radiation Safety Division has withdrawn lasers from the exhibit because of the potential or actual hazards to the viewing public. So, the artist and the laser safety officer must consider from the planning stage on, the worst possible hazard if the laser is to be shown on exhibit, especially in a crowded area. Most often, the difficulty with the lasers in a laser art exhibit is the laser burning out or exhaustion of the laser system. Accidents may develop in changing the reflective mirrors in that the beam is directed other than the place for which it is intended. This holds also for laser beam in the reconstruction of the hologram. People have been observed at unprotected exhibits, especially of holography, looking at and into the reconstructing beam. If these precautions of safety are taken, these colorful and fantastic laser designs and exhibits may be shown with complete safety to the general public.

If the artist considers the sale of his composition to an individual, and this composition has a laser contained in the unit, it is the responsibility of the artist to provide complete instructions in laser safety and maintenance to the individual. The individual in his home or private viewing area must then assume the responsibility for the protection of any viewers to whom he proudly shows his exhibit. In some states there is registration of even a very weak output laser system. For the present, there must be this registration; both the artist and the individual purchaser must be made aware of this.

A hammer and a chisel are dangerous to the artist. Welding is hazardous. The use of lead in art work is also of considerable hazard. All things must be used safely and the laser is no exception. Because of its powerful bright light, the artist must be reminded continuously, no matter how individualistic he is, that he is not a superior being and must maintain complete understanding of laser safety. In the research laboratory, unfortunately, it is not uncommon to see the research physicists, highly individualistic and enthusiastic, forget about the safety glasses and leave them hanging on the wall. It is very easy to protect one's eyes from the hammer, mallet, and chisel in sculpture and from high output electrical drills. It must be routine to have the artist put on his safety glasses whenever he starts to work on his laser objects.

Very little has been said about air pollution from these very sophisticated, expensive, and high output laser systems which could be used in sculpture. Fortunately, at present, their use would be limited to laser installations where the artist would either grudgingly be allowed to work, or custom work would be done at a certain and usually expensive fee. The creative developments of laser sculpture are actually unlimited. The hazards from laser processing of plastics have been listed.

If the laser is used with ultrasonics, as it may be by the artist of the future, the hazards of ultrasonics must also be considered. If the laser is an adjuvant to other types of art techniques, the artist must know that the processing by the laser produces no harmful products for him to breathe. Again, in working with plastics, toxic materials may be vaporized sometimes in the development of laser sculpture and adequate ventilation must be arranged for. All of this, again, will be taken care of by the laser safety officer. The artist will scarcely have to consider atmospheric propagation of lasers at the present time, although, in the future, such techniques may be used to develop three-dimensional color imagery in space, such as in designs, figures, and displays. There would have to be laser safety programs in regard to atmospheric transmission of the laser. In other techniques that the artist may use in the future, such as computer technology, video tape cassettes, or three-dimensional laser imagery, some of these systems may not be enclosed systems and, therefore, completely safe. Thus, by knowing about laser safety, the informed artist will have greater potential application in many areas of modern art.

Laser Designs

The bright lights of the laser may be used in design motifs, for this is a very powerful light. A small helium-neon red laser, for example, can transmit light at a tremendous distance in a large art display and still emit a gorgeous red beam at the final target. Some familiarity with optics transmission, as indicated, will provide the mirrors, the prisms, the quartz tubes, fibers, and the lenses which make for the transmission of the laser to different respective areas.

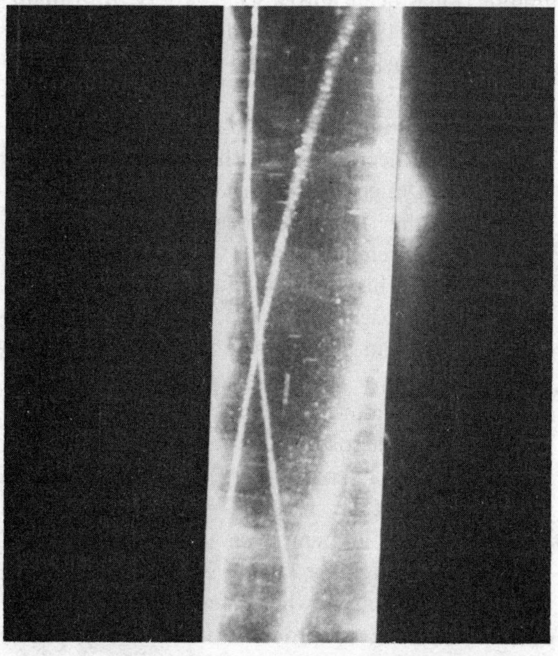

FIGURE 18-3. Laser design – Experiment No. 7, John Peterson, helium neon and Plexiglass. (Cincinnati Art Museum – photo, Michael Campbell.)

One of the interesting materials used for laser design is plexiglass with its full transmission power. Intricate designs can be worked into the plexiglass. John Peterson, of the College of Design, Architecture, and Art, at the University of Cincinnati, has done dynamic studies with the helium-neon laser about plexiglass rods. His other designs have included argon laser impacts of plexiglass. Baron Kody has also done some interesting design patterns with plexiglass on exhibit at the Cincinnati Art Museum. The design patterns of laser outputs have also been photographed and have used designs of textiles and papers. When the plexiglass is colored red or black, the color may be vaporized without splitting the plexiglass by certain laser systems, such as, the pulsed ruby laser, Q-switched ruby laser, and neodymium-YAG. Fascinating design patterns can be developed inside the colored plexiglass. This also has been shown adequately by John Peterson. Thus, the laser in plexiglass produces unlimited opportunities for laser design. With the CO_2 laser, etchings can be developed in sheets of plexiglass. This will be described in the portion on laser etchings.

There are other types of laser materials, such as diffraction gratings, special Fresnel lenses, and even colored transparencies that can affect the transmission of the laser to turn the single, cold, unidirectional beam into a wide variety of brilliant patterns. With the neodymium-YAG lasers and the argon, the plastic may have fascinating designs inside or may be heated and melted and the diffused reflecting beam enfold dynamic patterns as the reflectance changes with the melting in the target area. Designs of this have been shown in art shows by the Laser Laboratory. This design pattern has been correlated with production of laser music. This will be described later. The designs in plexiglass can be made safely and colorfully. Helium-neon beam designs in plexiglass have been shown also in France and Italy. In Scotland, a helium-neon laser has been used to direct a beam across the church to impact glass fragments on a cross, adding to the diffuse reflections about the cross.

Further developments in laser designs may include actual etchings, as it were, by high power laser output beams of the surface of brick or any colored material to produce distinctive patterns for decoration. Again, the manipulation of various types of transmission systems will increase the developments of these patterns in laser design. Laser design may also be used to produce patterns on black film with the argon or ruby laser. When these are projected, fascinating patterns are seen on the screen.

CW lasers continually lend themselves to all types of design patterns which vary with the dreams, imaginations, and desires of the laser artist. The Laser Laboratory has shown this in numerous popular art exhibits and is continuing to develop new exhibits.

Laser Etchings

With the CW laser systems and the color absorptions of the argon and the neodymium-YAG, it is possible to provide a suitable medium for the development of laser etchings. The etching in such new medium will vary according to the output and its transmission; the duration of use can vary the character and the quality of the etching produced. This has been shown by the excellent etchings of Robert Fabe and Edith Smilack, among other artists working at the Laser Laboratory. Robert Fabe, of the College of Design, Architecture, and Art, of the University of Cincinnati, was the pioneer in the use of the laser for the technique of etching in various media. He has shown how an artist can use this new technology in forms which were not possible previously. The lasers used for laser etching have been the argon laser, 1 to 5 W output, the neodymium-YAG, 10 to 12 W output, and recently the CO_2 laser with paper, wood and plexiglass.

The etchings provide contrasts of light, dark, flat, and elevated patterns, according to the artist and his laser. The raised lines in many areas lend a distinctive form, a three-dimensional phase. When these etchings are done on the special heat-sensitive paper of 3M, the etchings are visible on both sides of the paper. In the exhibit, such etchings are used with glass or plexiglass coverings so that both images can be seen.

The CO_2 laser has been used to etch plexiglass. This provides an incised design through absorption and thermal decomposition of the plastic. Combinations of focused ruby impacts, patterns of incised designs with diamond pencils, have been added to the patterns of the etching. Various types of lighting techniques and colored filters can bring out qualities of this design pattern, especially in the semidarkened room.

Sculpture

Another interesting phase of laser art is sculpture. One technique is the use of focused impacts

FIGURE 18-4. Laser Etching – Paper – Argon Ion Laser.

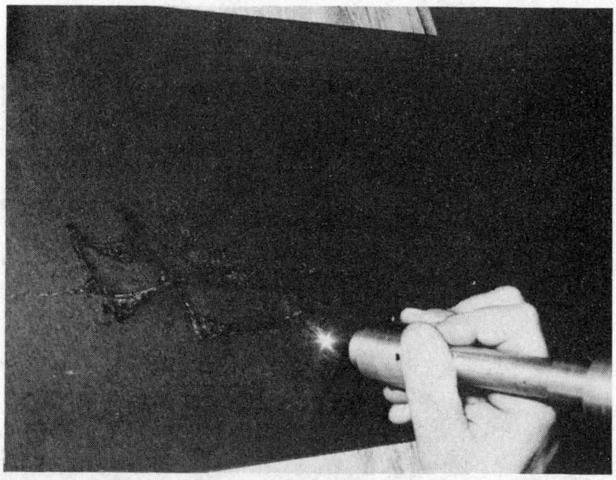

FIGURE 18-4A. Technique – Robert Fabe.

FIGURE 18-4B. Nazare fishermen – Robert Fabe.

FIGURE 18-4C. Faces of Eve – Edith Smilack.

FIGURE 18-4D. Reflections – Edith Smilack.

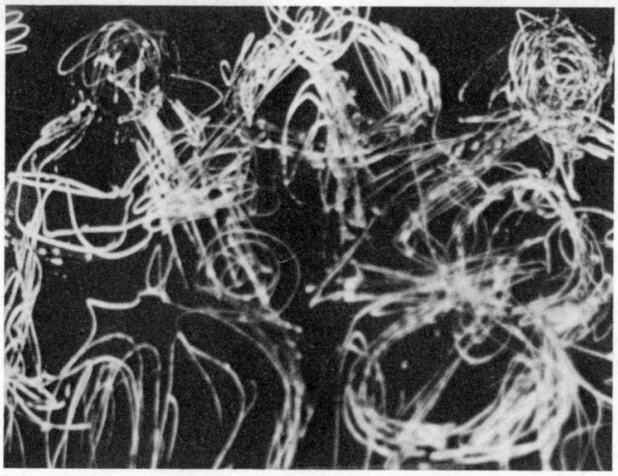

FIGURE 18-4E. Laser symphony – Edith Smilack.

FIGURE 18-5. Laser Etching – Paper Nd-YAG With Nath Fiber.

FIGURE 18-5A. Technique – Robert Fabe.

FIGURE 18-5B. Fisherman – Robert Fabe.

FIGURE 18-6. Laser Etching – Plexiglass – CO_2 Laser.

FIGURE 18-6A. Portuguese dorries – Robert Fabe.

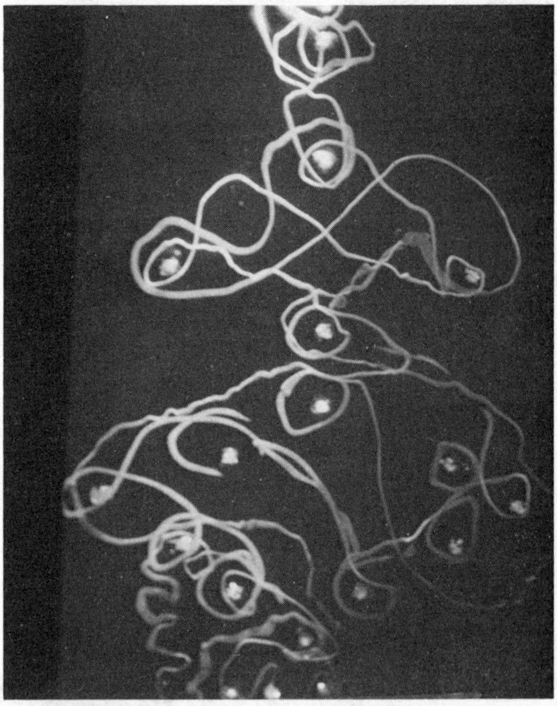

FIGURE 18-6B. Experiment No. 4 with ruby laser – Leon Goldman.

FIGURE 18-7. Laser Design – on Cross. (From Koebner, H., *Laser Elektro-Optik.* With Permission.)

FIGURE 18-7A. Helium-neon source.

FIGURE 18-7B. Impact.

FIGURE 18-8. Ruby.

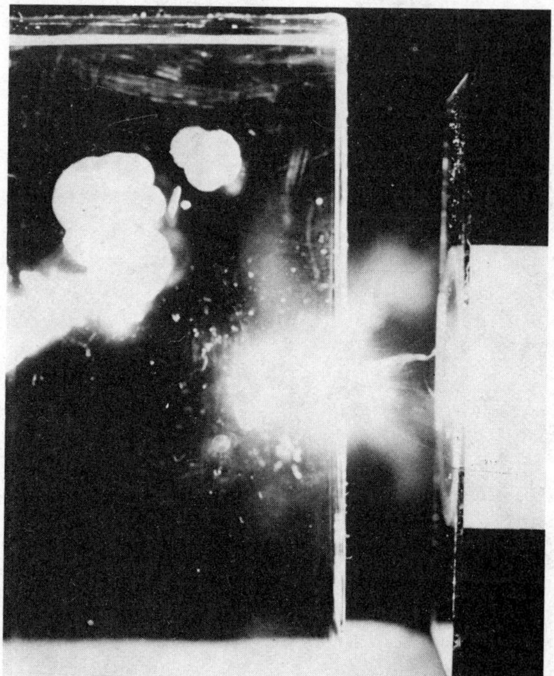

FIGURE 18-8A. Impact.

of pulsed ruby laser in plexiglass. The early models were from Martin-Marietta Company and were shown at the First Laser Safety Conference by the Laser Laboratory. The challenge of these pieces was met and the following types of sculpture patterns in plexiglass were done by members of the staff at the Laser Laboratory:

1. Impacts of clear blocks of plexiglass[4,5]
2. Cast blocks with many types of inclusions to absorb and reflect the laser
3. Laser impacts on commercial products embedded in plexiglass

The thermal decomposition patterns produced by focused beams in plexiglass varied with the type of plastic, the energy density, and the focusing. With predetermined designs it was possible to do dynamic patterns in the plexiglass.

The forms of inclusions embedded in the plexiglass included colored stones, colored plastics, and black films. When these were impacted with

FIGURE 18-8B. Martin-Marietta Co.

FIGURE 18-8C. Prisms – Ron Dreffer.

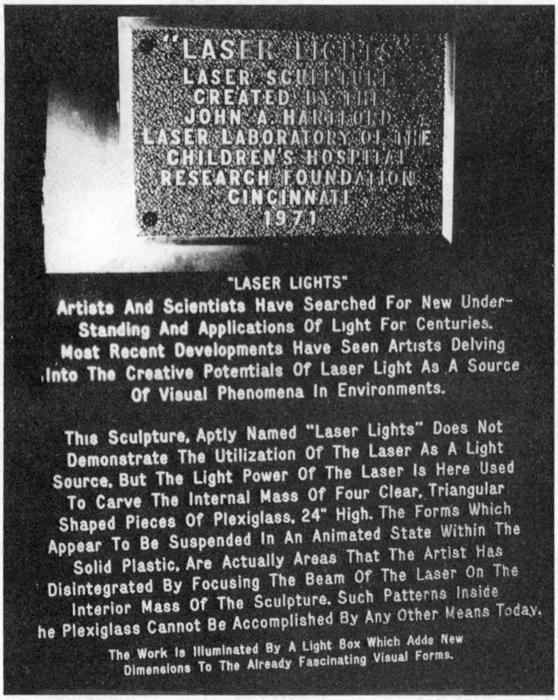

FIGURE 18-8D. Inscription to laser sculpture – Children's Hospital of Cincinnati.

FIGURE 18-8E. Mish mash – Leon Goldman.

FIGURE 18-8F. Laser lights – Leon Goldman. (From *Ohio State Med. J.* With permission.)

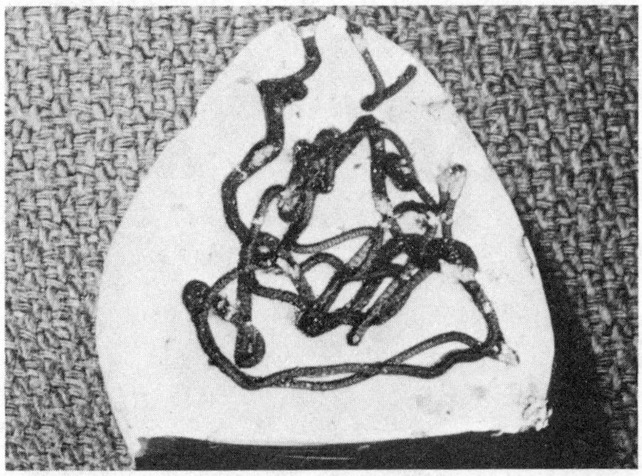

FIGURE 18-9. Carbon dioxide laser in turquoise.

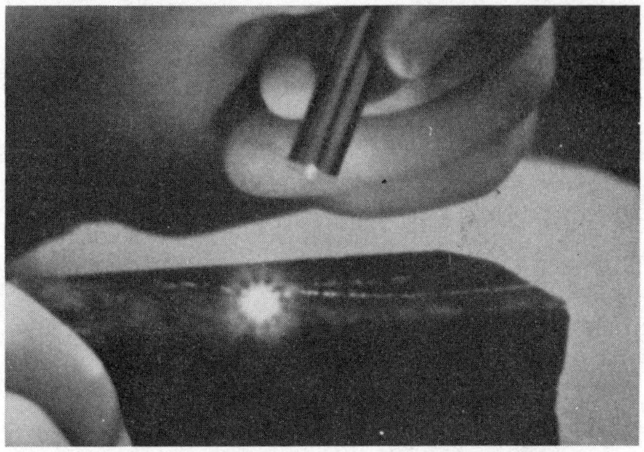

FIGURE 18-10. High output Nd-YAG laser carving in Alaskan jade.

argon, neodymium, and ruby lasers, colorful patterns developed inside the plexiglass. With regulated outputs, there were no fractures of the plexiglass. Commercial pieces of plexiglass, furnished by International Plastics of England, were impacted and interesting designs were made with such blocks containing screws, watches, spark plugs, silicon wafers, and the like.

Special types of lighting devices were developed for transmission of light through various colored filters and color wheels which formed ever changing patterns.

One interesting exhibit of the laser art show at the Cincinnati Art Museum was the revolving cylinders, with patterns cut with the CO_2 laser by Mack Smith.[1] Light transmission was from the sides. CO_2 lasers also were used to incise plexiglass by Robert Fabe and the staff at the Laser Laboratory. Several prize-winning sculptures in plexiglass were made by the staff.[4,5] Adequate ventilation and eye protection, other than the usual glasses for CO_2 laser work, were necessary because of the intense brilliance of these impacts.

Experiments were done with impacts on a number of semiprecious stones with argon, CO_2, and ruby lasers. With jade, it was possible to drill through green jade, producing bleached areas from the focused ruby laser impacts. It was possible to produce interesting designs through bleaching of black jade with the CO_2 laser. CO_2 laser designs were used on soft stones, such as meerschaum and alabaster. With harder stones, such as agate, the patterns were not as interesting or distinctive as those, for example, on softer stone like turquoise and chrysocolla.

With Ervin Pauly of Idar-Oberstein, a famed lapidarist and one of Germany's foremost workers in lapidary art, impacts were done on black agate with the high output neodymium-YAG lasers and small areas of bleachings were produced, occasionally with the fracture of the stone. An effort was made to utilize some of the techniques of hard stone mining to determine the value of sculpture in hard stones. The high cost of the loan of these lasers was an obstacle to any investigative studies.

FIGURE 18-11. Laser Art Ceramics – Carbon Dioxide Laser on Dry Clay.

FIGURE 18-11A. Designs – Leon Goldman.

Laser Ceramics

Some of the most interesting findings in art were produced on dry rough clay by the high output argon and the CO_2 lasers. The lasers produced immediate firing in the areas of the laser impact. In brown clays, destructive bleaching was found. To preserve the natural beauty, as it were, of these impacts, epoxy resins were used to cover the pieces. A number of interesting pieces were made at the Laser Laboratory by the staff and Edith Smilack.

Paintings

The effect of the laser on paintings was studied by impacts of the argon, ruby, and low output CO_2 lasers on acrylic and oil paintings. Black markings, especially in the acrylics, were noted with the lasers. These were developed to produce interesting designs. One such painting of free design lines made at the Laser Laboratory was called "Explosion in Jelly Bean Factory."

John Tcheng attempted to do some experiments with the laser in the field of Chinese calligraphy in which he is a master. The narrow beams of laser systems were an undesirable obstacle to his studies. These experiments are continuing with higher output systems and diffusing lenses to try to increase the width of the CW laser beam to meet the formidable challenges of the brushes. Calligraphy designs in English script were also done.

Holography in Art

A number of art exhibits have been held to illustrate the tremendous interest of holography to artists. Unfortunately, many of the subjects selected were more of the still life type than real possibilities which are offered, like statues and other art pieces. For the laser art exhibit at the Cincinnati Art Museum, a 14th Century Madonna and Child was selected.[1] The added art effect of the hologram was achieved by blowing cigarette smoke around the statue. The interesting holo-

FIGURE 18-11B. Forms #1 – Edith Smilack.

grams of this artistic piece were developed by Ralph Wuerker and his associates at TRW. Striking holograms were proposed by Salvador Dali and were executed by laser experts into interesting designs in which the experts carried out, themselves, the excellent draughtmanship of Salvador Dali.

The dynamic examples of acoustical holography are another field of great interest to the young artist. Some of the famed pictures shown of the acoustical holography, for example, the fish studies, are similar to ancient Japanese prints. Both holography and acoustical holography, then, open up wonderful areas of investigation by the architect, designer, and artist.

FIGURE 18-12. Laser Art as Holography.

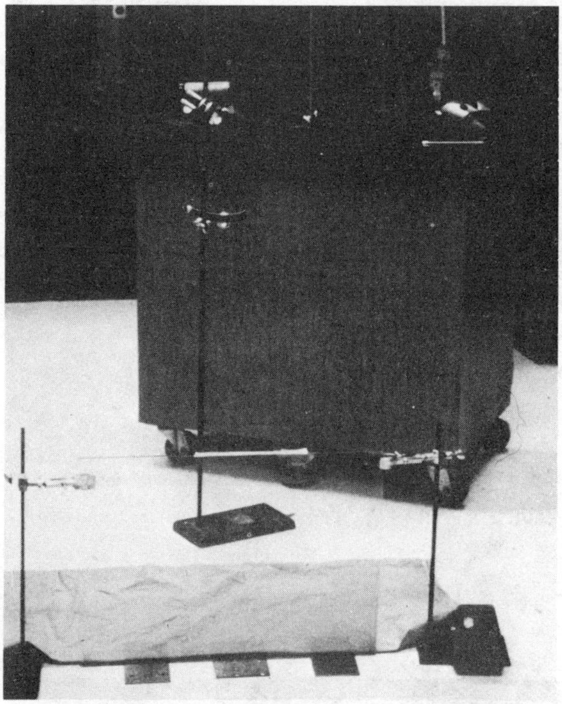

FIGURE 18-12A. Large glass plate hologram. (Cincinnati Milacron.)

FIGURE 18-12C. Circular reflection hologram. (International Holographics.)

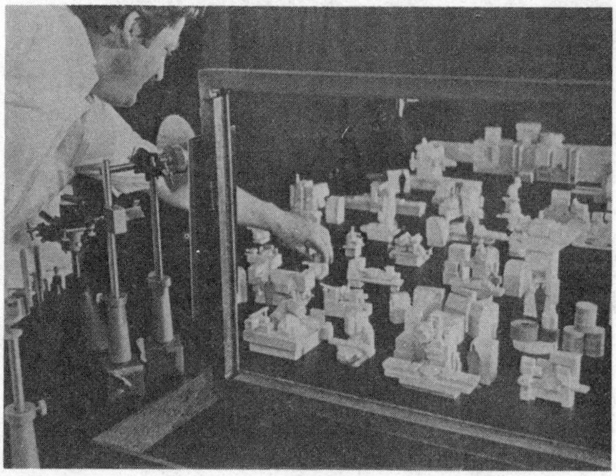

FIGURE 18-12B. Six-foot reconstruction. (Cincinnati Milacron.)

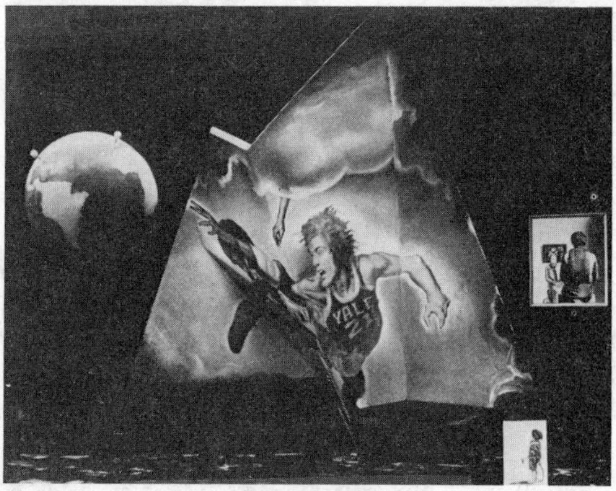

FIGURE 18-12D. Polyhedron – Salvador Dali. (M. Knoedler & Co. Inc., New York.)

Laser Art Photography

Laser photography may be divided into the following types of subjects:

1. The routine use of photography as related to the laser instrument and its transmission and application
2. The high-speed photography
3. Special film techniques with holography
4. Special film procedures for technical infrared laser systems
5. Photomicrography of laser impacts
6. The photography of diffraction grating interference patterns
7. Purely artistic designs of laser impacts
8. The use of wide-angle photography for study of reflectance patterns in the development of laser safety programs
9. Kirlian electrophotography

All of these show that there is considerable interest and concern about the role of photography in laser. The detailed techniques will be presented in Chapter 19, Photography.

Laser Art Exhibits

It appears that certain principles are necessary in the development of a laser art exhibit:

1. The exhibit may be developed to show the results of laser art and the possible applications for the present and future
2. The exhibit may require lasers to show laser art

The following program on laser art exhibits has been worked out as indicated by the Laser Laboratory. Since this is an important area of laser safety and may affect laser art of the future, it is necessary to repeat. Redundancy and repetition are always necessary for discussions of laser safety.

1. Laser expertise to survey the initial plans of laser artist as regards the safety programs
2. The survey of the finished material usually in the studio of the artist again to study the safety programs
3. The setup in the exhibit areas with monitoring by federal, state, and city authorities for approval of the general setup

Then follows the constant attention to this

FIGURE 18-13. Diffraction Photography – Edward Perry.

FIGURE 18-13A. Person with sunglasses (0.6328b).

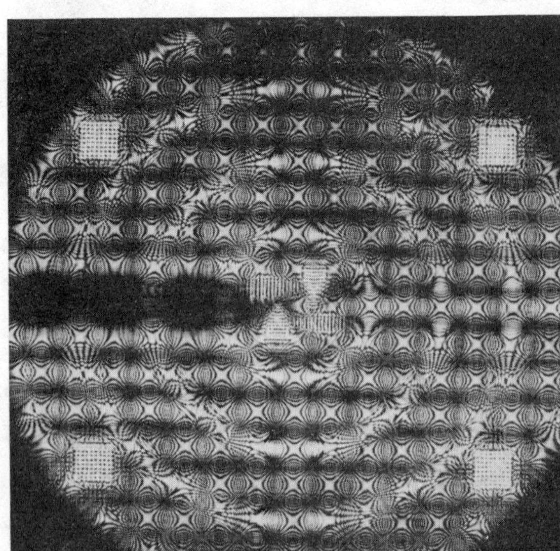

FIGURE 18-13B. Pattern test Air Force.

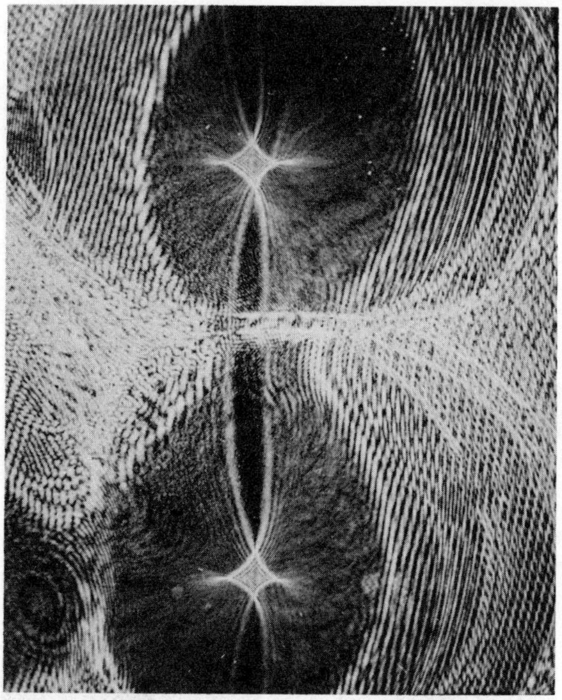

FIGURE 18-13C. Multisurface reflection stress in plastic.

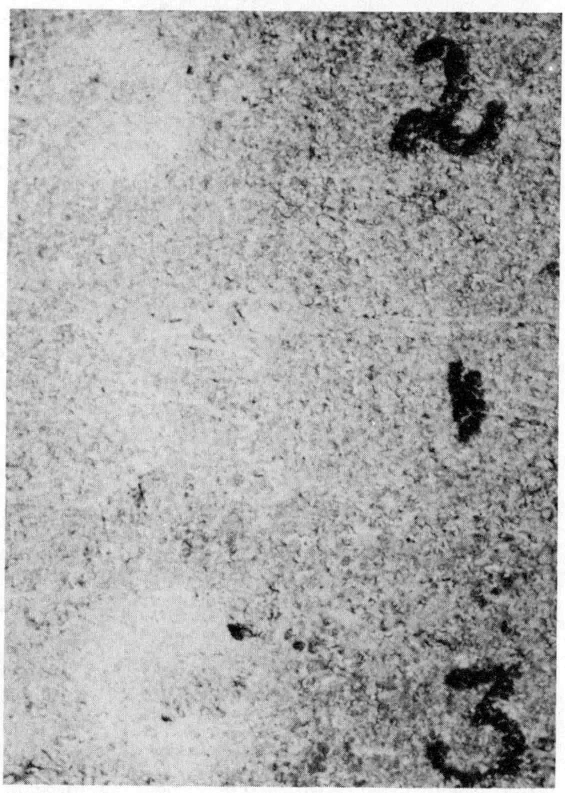

FIGURE 18-14. Ruby laser impacts on limestones for studies on effect of environmental pollution statistics.

during the crowded art shows with the adventuresome children and adults milling around art materials using actual laser systems. For in extensive laser systems, definite fail-safe devices must be included.

The first permanent laser art exhibit has been set up by the Art Museum of Cincinnati. Here, also, the first permanent laser art gallery was developed. One advantage of such a gallery is that one can show off a constant collection of distinctive laser art for the interest of the young artist and the general public. The disadvantages are the constant monitoring of such a system to be sure that the laser beam is in alignment and cannot strike the general public and the continued upkeep, especially of the laser systems that require constant supervision.

Again, when there is an exhibit in the home, it is the artist's responsibility to explain the safety program for any lasers that may be included in the art material that he has sent. When lasers are registered by the specific federal and state laws, more definite and more vigorous control programs can be done.

Laser Facilities for the Artist

The artist needs to have the opportunity not only to see, but to actually work with the new laser systems. This produces difficulties when there are no facilities for easy accessibility of the artist even on a university campus. There should be tolerance in the laser industry to develop facilities to allow the artist to do pilot studies of interest and value to laser art. Unfortunately, such tolerance for laser art is uncommon except for a few groups in the U.S. It is hoped that there will be adequate funding for artists to be in residence at a laser facility so that they could do actual experiments in that particular field to determine the current values and limitations of various laser systems. Unfortunately, such proposals made by the University of Cincinnati were not successful, but efforts are continuing because of the interest of the young artists and the great potential of laser art.

Laser Music

There has been considerable interest in the relationship of laser to music. The following associations can be drawn:

1. Dynamic laser designs as a background for a musical performance

2. The laser designs as an integral part of the actual performance

3. The use of laser in the opera or in the ballet

It was found early that the use of fine, small mirrors attached to speakers served to reflect the laser beams onto a screen. The patterns of the laser beam would vary according to the patterns of the music. This produced beautiful pictures, for example, Mike Campbell's "Fantasia" at the Laser Art Exhibit at the Cincinnati Art Museum and at the Smithsonian Institute. The use of different lasers, more commonly the low output argon, helium-neon, and helium-cadmium, would produce variations in beam movement. To relieve the monotony of the continual patterns of beams, a number of experiments can be done with the single reflected beam. Fresnel lenses, diffraction gratings, filters, etc. can produce myriad patterns to vary with the music.

Eric Kunzel, Resident Conductor of the Cincinnati Symphony Orchestra, has been interested and is developing a laser symphony with the laser accompaniment at intervals rather than throughout the entire course of the symphony.

FIGURE 18-15. The Mona Laser. (Martin-Marietta Co.)

Safety Program for Music

It is evident that specular reflection is a part of so-called laser music. Low output systems must, therefore, be used or the beam must be attenuated so that the reflections from the screens or from similar types of reflecting media produce reflections many orders of magnitude below the permissible exposure levels. Hazards from low output helium-neon laser, with planning, should be minimal. This is possible with the reflected beam transmitted through many different kinds of materials. Arrangements must be made, also for fail-safe devices to avoid hazardous unplanned reflections.

The new laser disc recorders, with videodiscs, and low output helium-neon lasers should be safe. These and laser-made video cassettes for the future will present no hazards to the consumer.

Again, the use of adapters in transmission systems that may cut down the actual speed of the music to produce slower evolving languid patterns rather than flashing patterns may also help considerably in this area. It is possible that this laser symphony can be done and be agreeable and entertaining, similar to the patterns of most op-art.

The laser has been used as a backdrop setting for the Royal Ballet in Copenhagen and at the Bayerischen Staatoper in Munich. Walter Haupt presented "Die Puppe," a laser ballet. With the music and the ballet, the laser had a tremendous dominating effect.[3] The initial performances indicated hazardous reflections. Safety measures were instituted by Hillenkamp later to insure performances safe for the dancer and for the audience.

It is evident that laser music is still in infancy or perhaps in its fetal phase so that its potentials are really unknown. The few examples given, where the laser is directly or indirectly part of the musical presentation and interpretation, suggest, of course, the application should be colorful and tuneful, and not boring. Hopefully, this will be the goal of the first written laser symphony.

Conclusions

Laser art and music are now part of the New World. These applications also have low priorities in the present order or disorder of the universe, but this will not continue. The appeal, again, as in many phases of the new technology, is to youth. Presently, the feeble attempts are to show the great potential and the planned safety programs of such goals. The laser has been used to detect fakery in art, to produce art, and hopefully even to protect the surface of art objects affected by environmental pollution. All these applications develop from laser research and fields of laser chemistry, biology, medicine, surgery, and dentistry.

REFERENCES

1. Laser, A New Visual Art, Cincinnati Art Museum, Cincinnati, November 1970.
2. Laser Photography. A Comprehensive Exhibit, Taft Museum, Cincinnati, March 19, 1971.
3. Erstes Buhnenstuck gennant, Laser, *Laser + Electro-optik.,* Nr. 2/1972.
4. **Goldman, L.,** Laser – the light fantastic, *New Sci.,* January 1971.
5. **Goldman, L.,** Laser art, *Abbott-tempo,* 4, 2, 1972.
6. **Goldman, L.,** The laser as an art, *Ohio State Med. J.,* 68, 891, 1972.
7. Laser Laboratory – The Art of the Laser, Augusta Richmond Museum, Augusta, Ga., March 2, 1973.
8. **Goldman, L.,** Progress of laser art in lapidary work, *Lapidary J.,* 25, 328, 1971.

Chapter 19

LASER PHOTOGRAPHY

Laser photography is photography of, by, and about lasers. Laser photography presents many challenges for dynamic photography for today and for the future. Laser photography may be used to

1. Record the output of the laser
2. Record the impact on the target
3. Monitor safety hazards from reflected light
4. High-speed photography
5. Holography
6. Show the effect of the laser
7. Purely artistic forms

In the area of biomedical applications, medical photography is important for record purposes in all phases of the work. The photographers in the Laser Laboratory who have developed excellent photographic techniques including artistic laser photography have included Patty Smith, Michael Campbell, Janet Borden, Marilyn Franzen, Scott Meyer, and Mark Wilson.

Unlike the light from photofloods and electronic flashes, laser light is coherent and can be used in very precise techniques. It is also stronger than any light previously used in photography. There is, of course, not one laser but many with frequencies from the ultraviolet far into the invisible. With wide-angle lenses and colored film, especially colored infrared film, the paths of the beam patterns in a laser installation may be seen and precautions taken accordingly.

First, color pictures are taken of the laser instrument usually in action. An oscilloscope is used to study the detailed characteristics. To film laser impacts with a duration of 0.5 to 5 msec and continuous wave lasers, a Tektronic® oscilloscope Type 585A is used. This is fitted with a Polaroid® camera with a series 125 F 1.3 Ilex® lens. Using Polaroid B/W roll film rated at 3,000, pictures of these impacts can be produced. For extremely fast pulses, the same camera is fitted to a Tektronic 519 oscilloscope, using Polaroid film rated at 10,000. In this manner, laser pulses in the nanosecond region can be recorded. Special techniques are used to film impacts in the picosecond range.

With the flashing lights of the various laser beams and their developing in an amazing technology, it is not surprising that photography should be an exciting event from an artistic viewpoint. This is so not only because there is photography with brilliant amazingly precise beams of many colors, but, also, because it is now possible that pictures can be made, even in color, without camera or lenses and even in the third dimension. The technology is still new, colorful, and requires much research and development.

Laser Microphotography

With the laser attached to the microscope, brilliant photomicrographic pictures may be taken with the laser as illumination of the impact of the area. In this instance the laser beam is transmitted through the optical systems of the microscope. Pictures may be taken, especially with illumination by the red helium-neon laser beam. Microphotography, however, is most often taken of the effect of the laser, through the microscope. These pictures refer to impacts often on metals, liquid crystals (changing color with changes in temperature), and living cells. For art, the most colorful pictures taken in our laboratory have been impacts on dyes, crystals, etc. Those lasers that are absorbed by color are the ruby, the neodymium, the neodymium-YAG infrared, helium-neon, helium-cadmium, and helium-selenium. The best film for this type of photography is Kodachrome II A®, using an 80D wrattan filter on a Zeiss photomicroscope. In art pictures, a variety of effects can be made by changing and combining color filters. Also, the scope prism is such that impact pictures can be taken.

Art Photography

The energy and power of the laser beams enable these to be used with mirrors, lenses, plexiglass, etc. for colorful patterns not possible with ordinary colored light. Pictures may be taken of the effect of pulsed lasers, especially in darkrooms where the reflections of the laser may be seen especially through a wide-angle lens. The technique is to vary the exposures for a variety of light intensities and patterns. Long exposures coupled with beam movement produce trailing effects. Of course, it is necessary for the long exposures to use a tripod for stability and smaller apertures to achieve clarity and depth in the photographs. Tests for exposures should be taken with each new film and each new laser. For pulsed ruby lasers with Kodachrome film, the shutter is opened in the darkroom with a countdown from a person firing the laser and closing approximately f-11 or 16. That way only laser light is exposing film.

With the help of Eastman Kodak Company, exhibits of the artistic value of lasers were held at the Taft Museum in Cincinnati,[6] at the Ontario Museum of Science and Technology, and at the Augusta Richmond Museum.

The technique for attempting to photograph a picture from the hologram is as follows: Since holograms, as yet, are somewhat dimly lit and hazy it may be difficult to focus, but not impossible. Long exposures will be necessary with smaller apertures (approximately f-11 or 16; 1 sec Ektachrome®) to insure enough depth of field to show the hologram. Although the picture of the hologram will never do it justice, it will convey the third dimension.

For selecting film for the hologram, Robert D. Anwyl[2] of Eastman Kodak recommends that holography and several other photographic procedures involving laser sources require recording materials characterized by extremely high resolving power and correspondingly low speed. However, there are several applications involving laser exposures where the speed and resolving power of films used for conventional photography are clearly indicated. Kodak recommends KODAK Spectroscopic Plate or Film®, Type 649-F, which requires an exposure of 800 to 1,000 ergs/cm^2 for helium-neon or argon lasers. For the typical side-referenced 3-D holograms, the author suggests a reference-beam to object-beam intensity ratio in the range of 5:1 to 10:1. The plate or film should be processed in KODAK HRP Developer® (1:4) or KODAK Developer D-19® for 6 to 8 min at 68°F (see "Processing"). Exposures should be adjusted to yield an average diffuse density on the processed hologram of about 0.6. KODAK Spectroscopic Type 649-F® materials and several other Kodak products which have been used effectively by holographers are described briefly in the following paragraphs.

"KODAK Spectroscopic Plate and Film, Type 649-F

These materials have an extremely high contrast, panchromatic emulsion. Their useful sensitivity extends almost to 700 nm, so that they must be handled in total darkness during handling and processing. Their useful sensitivity (speed) is characteristic of an emulsion with extremely fine granularity and high resolution (over 2000 lines/mm).

The rigidity and dimensional stability of glass plates are important considerations for most holographic applications, and Type 649-F Plates are not significantly more costly than their companion film products. Except for those situations requiring the flexibility of film supports, we therefore recommend the use of glass plates and attempt to maintain this otherwise special-order product in stock in Rochester in several commonly specified sizes. The emulsion thickness of Type 649-F Plates is 17 microns before processing, while on tri-acetate film base, the coating thickness is 6 microns.

KODAK High Resolution Plate and Kodak High Resolution Film (ESTAR Thick Base) SO-343

These products have an orthochromatic-sensitive emulsion suitable for recording laser emission lines out to 560 nm. These high-resolution materials are otherwise identical in performance characteristscs with the Type 649-F products described above and should be processed in the same manner. They may be handled in the darkroom, however, using a KODAK Safelight Filter No. 1A or OC. Both the film and plate versions have an emulsion thickness (unprocessed) of 6 microns. High Resolution Plates are widely distributed in a number of standard sizes. SO-343 Film is factory stocked in two commonly used sheet sizes.

KODAK Spectroscopic Plate and Film, Type I-Z

These materials are included because they are the only commercially available products with useful sensitivity for wavelengths of 1060 nm (neodymium-doped laser) and 1150 nm (helium-neon laser). In fact, the type I-Z emulsion has its peak sensitivity at 1060 nm. However, they do not have the resolving power of the materials described above. Use a KODAK Safelight Filter No. 7 (Green). KODAK Developer D-19 is recommended (5 min at 68°F).

Processing

Careful attention to proper processing techniques is strongly recommended. In many holographic experiments, results are observed or measured on plates or films that have been exposed singly and processed at different times. In such cases, it is especially important that processing conditions – developer conditions, time, temperature and agitation – be standardized in order to minimize the effect of processing variability.

Proper agitation is especially important. For a complete description of agitation techniques for processing plates and films, see Processing Chemicals and Formulas, Kodak Publication No. J-1 (available at Kodak dealers).

Kodak Developers

Two developers are recommended for processing the Kodak materials mentioned in this publication. KODAK Developer D-19 is a high-contrast, clean-working developer that produces high-quality holograms with short development times. It is available in prepared packages to make 1 gallon and 3½ gallons of working solution. KODAK HRP Developer, originally designed to yield high-quality micro-images on Kodak High Resolution Plates, is a liquid concentrate available in sizes to make 1 and 5 gallons at a 1:4 dilution. Holograms processed in KODAK HRP Developer will generally have slightly higher contrast than those processed in KODAK D-19.

Develop the film or plate with continuous agitation for 6 to 8 minutes at a developer-solution temperature of 68°F (20°C).

Rinse with continuous agitation for 30 seconds at 65-70°F (18-21°C) in KODAK Indicator Stop Bath or KODAK Stop Bath SB-5.

Fix with frequent agitation at 65-70°F (18-21°C) as follows:

KODAK Fixer or KODAK Fixing Bath F-5	KODAK Rapid Fixer
Plates: 5 to 10 minutes	Plates: 2 to 4 minutes
Films: ½ to 1 minute	Films: ½ to 1 minute

NOTE: Fixing times are based on 2 to 3 times the time required for the plate or film to clear.

Wash with moderate agitation for 20-30 minutes at 65-70°F (18-21°C). A more rapid wash can be achieved by using KODAK Hypo-Clearing Agent. Used after recommended fixing, Hypo-Clearing solution reduces washing time and conserves water. First, remove excess hypo by rinsing the film or plate in water at 65-70°F (18-21°C) for 10 seconds. Then bathe the film or plate in Hypo-Clearing Agent solution for 15 seconds with moderate agitation. Follow this with a 30-second wash at 65-70°F; allow at least one change of fresh running water during this time. NOTE: For highest quality holograms, keep the temperature of all processing solutions, including the wash water, the same. In this way, image movement due to random shifts in the emulsion layer – as the gelatin alternately swells and shrinks during processing – will be minimized.

Dry in a dust-free atmosphere. Drying marks can be minimized by treating the plates or film in diluted KODAK PHOTO-FLO Solution after washing, or by wiping the surfaces carefully with a damp KODAK Photo Chamois or a soft viscous sponge. The use of PHOTO-FLO Solution will also aid in drying plate and film surfaces uniformly. For best results, dry plates or films slowly at room temperature."

High-speed Photography

High-speed photography has been used not only to record the ultrashort pulses of the laser but has also been used with the laser as the light source!

Hyzer[3] has critically reviewed what is called, or sometimes miscalled, "high-speed photography."

"A quantitative definition of the term "high-speed photography" has never really stabilized; with continuing advances in technology, the temporal boundaries of the field keep changing. The broad subject of high-speed

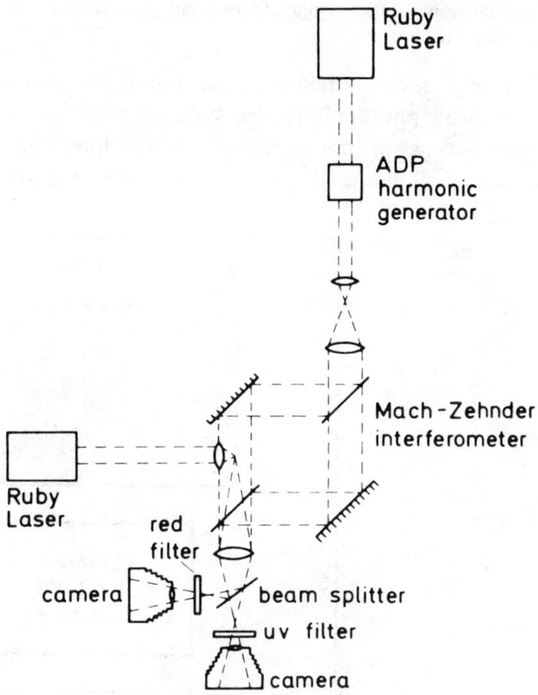

FIGURE 19-1. Mach-Zender interferometer for two wavelengths. (From Büchl, K., *Laser+Elektro-Optik.*, 17, 23, 1972. With permission.)

photography today includes exposure times extending from 1 millisecond down to the order of 10 picoseconds or less, exposure rates from 10^2 to 10^8 sec^{-1} or more, and time resolutions that span the temporal band from the limit of human perception (1/15 sec) to the order of 10^{-11} sec or less. Each of these time and rate spectrums have been traditionally subdivided into three spectral bands referred to as high-speed, very-high-speed and ultra-high-speed photography. The current state of the art is reflected in the time and rate regimes shown in the drawing at the top of this page. Ultra-high-speed photography of a few years ago was measured in nanoseconds, but these minimum exposure times have now been compressed by two orders of magnitude to durations of the order of 10 psec 10^{-11} sec).

The extreme brevity of exposure at this far end of the time spectrum is not easily comprehended. It has been brought somewhat closer to reality through an experiment performed by Duguay and Hansen of Bell Telephone Laboratories, who succeeded in photographing the world's fastest moving subject – a pulse of light – with a movement blur of only 1.7 millimeters during the 10-picosecond exposure time! The 1½ millimeter long pulse of light was produced as a second harmonic in a mode-locked Nd:glass laser. The light bundle was photographed by means of a Kerr cell shutter (pulsed open and closed in 10 psec) as it passed through a vial of milky fluid. The electronic shutter utilized by Duguay and Hansen is unique: it requires no electrodes and the conventional nitrobenzene has been replaced by CS_2. The first-harmonic pulse of light from a Nd:glass laser is used to drive the shutter."

Büchl[1] recently reviewed the role of the laser in high-speed photography. He believes that for this special type of photography, conventional light sources will be used less and less. Tunable dye lasers, picosecond pulses, and infrared laser systems will be used in this important photographic technique.

Safety Program for Laser Photography

In general, the same principles hold as discussed

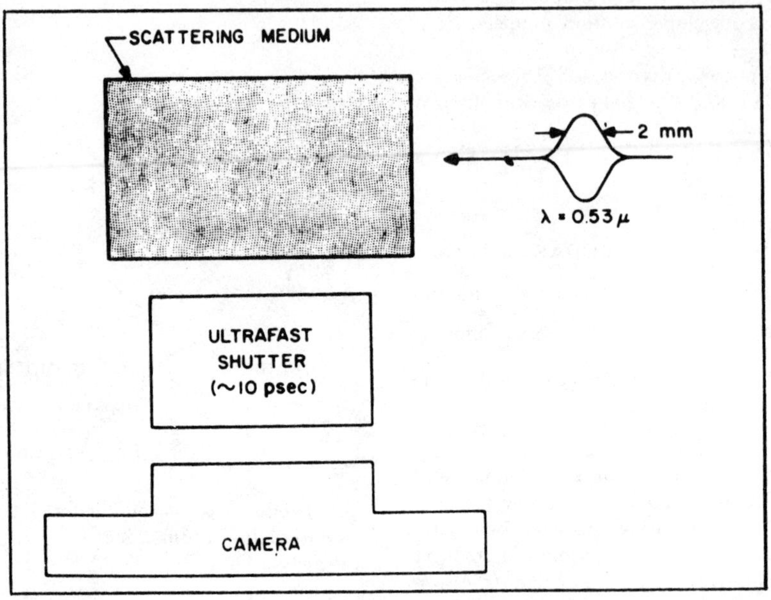

FIGURE 19-2. Experimental setup of photography of light in pulses with an ultrafast shutter. (From Büchl, K., *Laser+Elektro-Optik.*, 17, 24, 1972. With permission.)

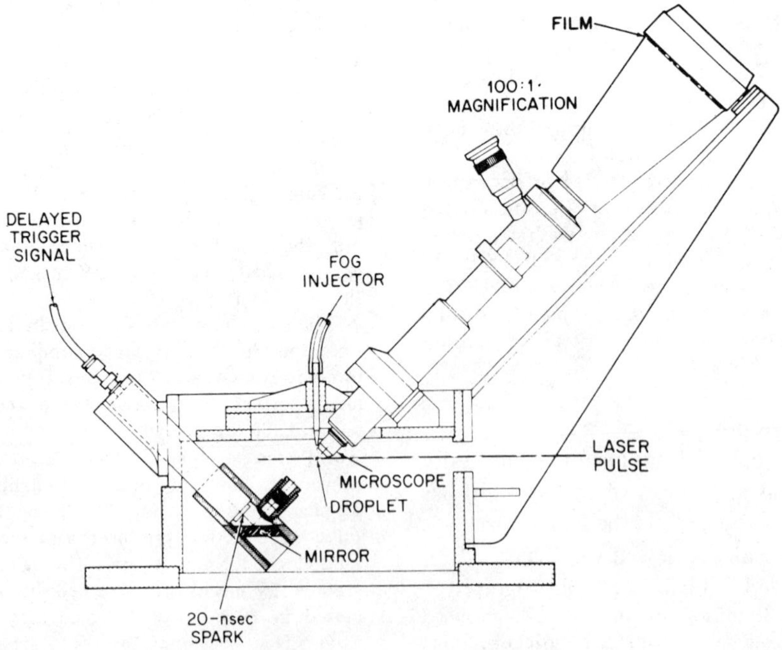

FIGURE 19-3. Partial diagram of high speed photomicrography system. (From *Applied Optics.* With permission.)

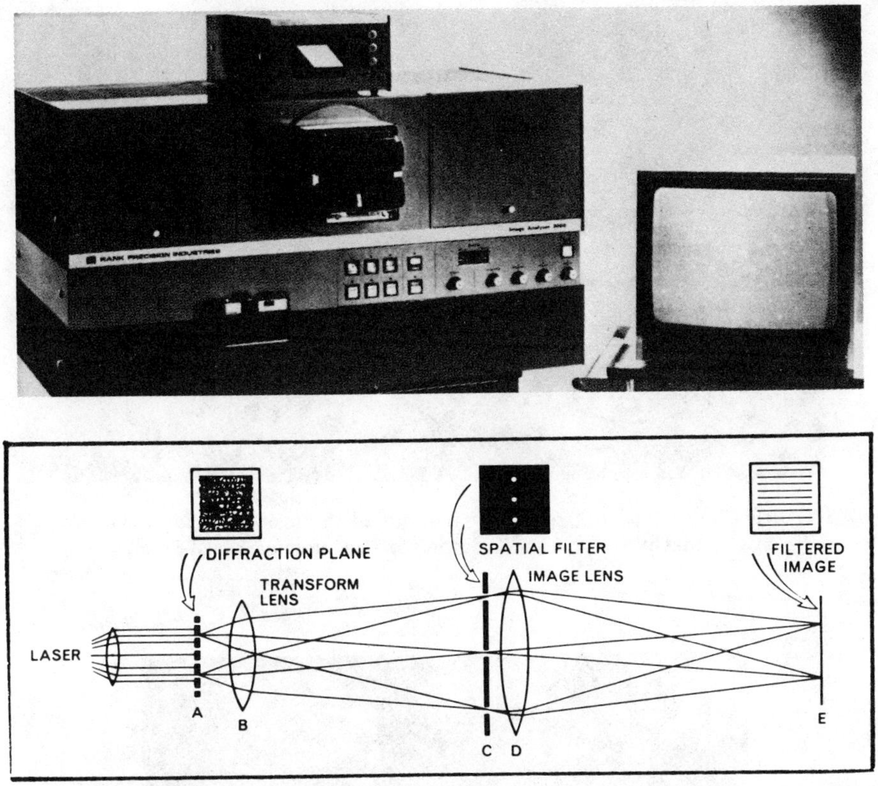

FIGURE 19-4. Image analyzer of Rank Precision Industries, Metrology Division. An optical diffractometer for detailed analysis of electron micrographs and other photographic transparencies. A. Apparatus. B. Principle of operation. The coherent collimated light illuminating transparency A is diffracted by the two-dimensional structure of the image. This light is transformed by Fourier transform lens B to yield a series of complex integrals at plane C. If a spatial filter mask is inserted at C, then the light through lens D will form an image at E consisting of only those spatial characteristics that are passed by the mask.

in the safety program for holography. The concern is attempt for focusing close-up pictures. The laser is fired only after the focusing is done and protective glasses are worn. For impact photography in the dark, the shutter, as indicated, is open in the darkroom, protective glasses down, and the released cable is used to open and close the shutter. With high output lasers there is concern about the reflectant beam on the lens. This is a definite hazard, of course, when the television camera is used for close-up studies. Spots have been burned in the phosphor layer of the vidicon tube. This is equivalent to the development of the retinal burn in man. Black unexposed films and liquid crystals may be used to monitor impact about the target area and, also, to produce art pictures, especially with the argon and ruby lasers.

Kirlian Electrophotography

In Kirlian chemiluminescent electrophotography, the film is exposed to a high frequency electrostatic AC field. The film records a striking pattern of some type of excitation from matter. According to Moss and Johnson,[4] it is not certain whether the phenomenon is a so-called bioplasma or a corona discharge, a commonplace electrical phenomenon.

The Laser Laboratory is investigating Kirlian electrophotography from laser impacts. With proper controls, impacts are given on the fingers in patients, normal, with various emotional and physiological states, and with a variety of pathological lesions. First, it is essential to establish standard techniques in all areas, including electrical and photographic parameters for the normal subject. Monteith, an expert in this new photography, directs the studies at the Laser Laboratory.

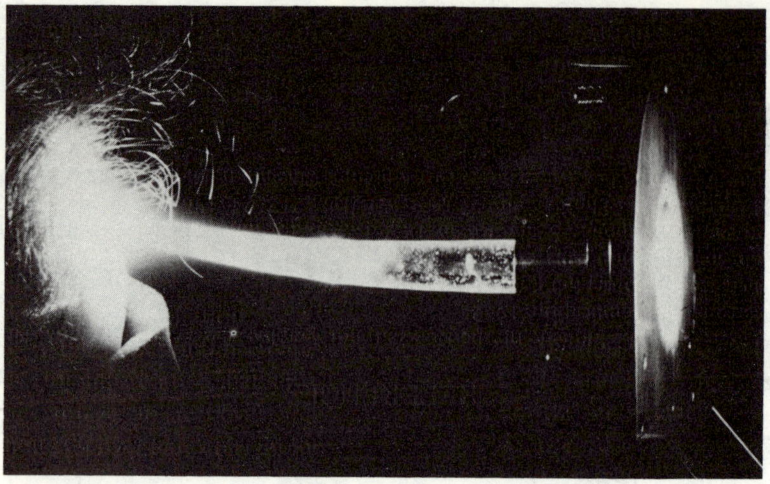

FIGURE 19-5. Impact of ruby laser transmitted through curved tapered quartz rod – black-and-white copy from Kodachrome – taken in darkened room.

FIGURE 19-6. Impact of pulsed ruby laser on unpolished sapphire window – black-and-white copy of Kodachrome – taken in darkened room.

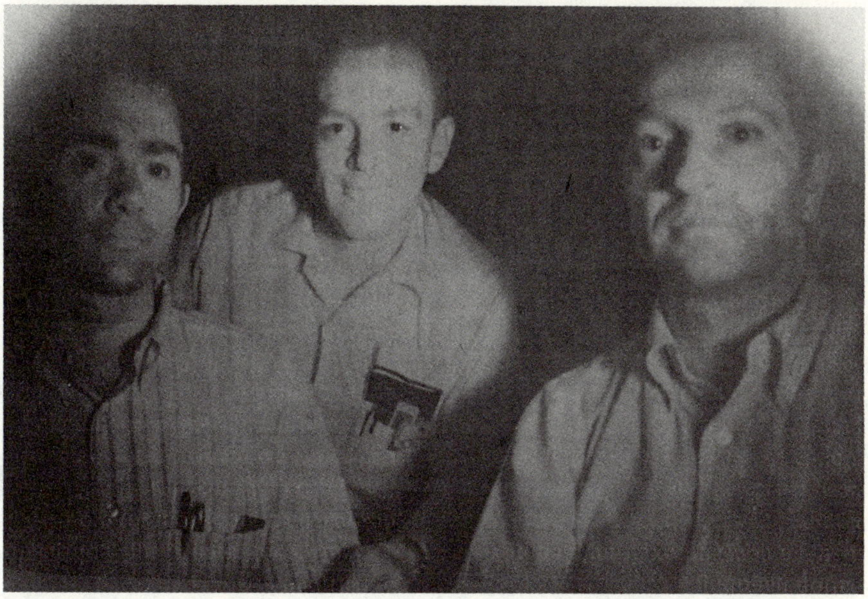

FIGURE 19-7. Holographic portrait – laser hologram – no camera. (Hughes Research Laboratory.)

Conclusion

The laser photography is an important technique for recording data including the laser, its output, the oscilloscope, the target area, and reflectance patterns. Photography actually belongs to every field of applications of the laser. The attractive artistic aspects of laser photography will be reviewed in the chapter on Laser Art.

REFERENCES

1. **Büchl, K.,** Der Laser – eine Lichtquelle fur die Kurzzeitphotographie, *Laser+Elektro-Optik,* Nr. 217, 24.
2. **Anwyl, R. D.,** personal communication, 1972.
3. **Hyzer, W. G.,** High speed photography, *Res. Dev.,* 23, 52, 1972.
4. **Moss, T. and Johnson, K.,** Bioplasma or corona discharge, in *Galaxies of Life – The Human Aura in Acupuncture and Kirlian Photography,* Krippner, S. and Rubin, D., Eds., Gordon and Breach, New York, 1973, 25.
5. **Leith, E. N. and Upatnieks, J.,** Photography by laser, *Sci. Am.,* 212, 24, 1965.
6. Laser Photography, A Comprehensive Exhibit, Taft Museum, Cincinnati, March 19–April 11, 1971.

Chapter 20

SOCIAL CONSCIENCE OF THE LASER OR THE LASER ALSO GOES INTO BATTLE FOR THE GOOD-OF-MAN

At present, at least in the U.S., there are high energy and low morale among laser scientists. Hopefully, this dichotomy (?) does not exist in laser laboratories in other countries.

The recent publicity about laser guided bombs (LGBS) in Vietnam and laser weaponry has awakened the old flashing fears of the laser as a death ray. It is not as well known that the laser is also smart and capable enough to battle for the good-of-man. There are some obvious factors why too little is known about how much good the laser can do and is doing. In the poverty stricken area of basic laser research where too little funds are still available, except for so-called in-house research (extensive in the military and industry), there are still too few lines of communication available or used by the multiple disciplines interested for laser research.

The current state of the social conscience of the research physicist is also a factor. This would seem a good time to show that there is something besides just a few gadgets to make life better. The physicist should also remember that exploitation of the wonderful laser discoveries is done not by the physicist, but by man himself. It will not be very long before people will praise the laser physicist and not continue to condemn him for some of the "fashionable ills" in this world, ills that are coming from the discoveries of science.

Regarding the uneasiness among physicists, Stubbs[3] states:

> "Physicists have consistently underplayed the major extent to which their subject is responsible for the modern horrors of war. The post-bomb generations have a dramatically changed attitude to the world in direct consequence of what, even 27 years after, remains a terrifying Sword of Damocles. A threat, moreover, whose fearful reality is in no sense diminished by lesser technological evils. It was physicists who produced laser bombing; it was physicists who invented the electronic battlefield; it was physicists who devised plastic antipersonnel bombs (incidentally, acoustical holography can pick up these transparent foreign bodies in tissues and x-rays cannot), "gravel," "spider mines," "daisy cutters" and a plethora of other perversions. 'Why shouldn't the public distrust them as a race? They do little to purge their own ranks of the monsters who contrive such appalling inhumanities.'"

There is, of course, the obvious need to be strong in this best of all possible new worlds. So we need advances in laser technology to be able to communicate, through strength, with other people across too-large round tables. Sackler[2] claims that:

> "To a considerable but not accurately known extent, the people, the Congress and the executive have lost confidence in the scientific establishment and its governing elite. The long series of reports from the National Academy of Sciences on the present state and foreseeable needs of science, painstaking and thorough though they were, may actually have been counterproductive, with their undeniable odor of special pleading for more of the same.
>
> "It's a new ball game, and the scientific world might well give more thought to how to play it."

The concern for us now in laser technology is the disillusion of the youth, both those who may venture into science and those already in science. Perhaps to some countries it seems strange that we listen to youth who feel this disillusion and that we are ready to communicate with youth and to attempt to understand their concerns.

As youth surveys the field of laser technology, they should see many challenges all the way from the physics laboratory to the moon and to the bedside of the patient.

These are but a few of the many challenges now:

1. For the physicist
   - A. New lasers
   - B. The confusion of the true current existence of X-ray lasers
   - C. Increasing the efficiency of output
   - D. Decreasing the cost
2. For the chemist
   - A. Analytical techniques
   - B. Chemical lasers

3. For the botanist
   A. Photochemistry
   B. Growth stimulation
   C. Control of disease – plant pathologist
4. For the zoologist
   A. Infrared receptors
   B. Semicircular canals of birds
   C. Neurosurgery as it were, of spiders, as portrayed by changes in web patterns
5. For the biologist
   A. Cytology – including microsurgery of chromosomes
   B. Cytogenetics and genetic engineering
   C. Experimental embryology and teratology
6. For the biophysicist
   A. Test models for studies of impacts of pulsed and CW lasers
   B. Doppler techniques
   C. Optics of tissue transillumination
7. For the engineer, the new multidiscipline of electrooptics, for information handling from computer storage to retrieval and transmission, holography, acoustical holography, thermonuclear fusion, etc.
8. Biomedical engineer
   A. Introduction of electrooptics for instrumentation research and design
   B. Laser biomedical servicing
   C. Laser safety
9. For the field of education
   A. Information handling
   B. Visual education aids
10. For the social scientist
    A. Law and order programs
    B. Establishing better and shorter lines of communication among people
    C. Rehabilitation program
11. For the librarian
    A. Information handling
12. For the artist – a new and exciting medium
    A. Design
    B. Etching
    C. Sculpture
    D. Photography
    E. Holography
    F. Analyses
13. For the musical student
    A. Visual expression of music
    B. Laser ballets
14. For environmental protection
    A. Monitoring of pollution
    B. Treatment of pollution
15. For the philosopher
    A. A current example of the need for developing social responsibility
16. For the military scientist
    A. Ranging
    B. Communications
    C. Weaponry
17. For the business major
    A. Information handling to include inventory scanning
18. For the lawyer
    A. Computer technology
    B. The law and order program
19. For the astronomer
    A. Space technology
20. For power supplier
    A. Laser-induced thermonuclear fusion
    B. Laser transmission of power
21. For the dentist
    A. Diagnosis
    B. Treatment
22. For the physician
    A. Diagnosis
    B. Treatment
23. For the cancer specialist
    A. Cytology
    B. Analytical techniques
    C. Animal surgery
    D. Immunobiology
    E. Cancer surgery of man

There is much to be done to turn youth from the current popularity of the social sciences to the real sciences of this new world.

Even medical science is not above concern and anxiety for the good of man. Some problems here relate to birth control, genetic engineering, organ transplants, selection for artificial kidneys, human experiments, etc.

Will governments and politicians support these desires and ambitions of youth? That is perhaps the problem only for the U.S., perhaps not for other governments, or maybe, all establishments are interested in the immediate applications, applied research, to include we hope, applications also for the good-of-man.

With the so-called antiscience backlash now, will there be funds for the youth who may wish to venture into laser technology or to continue with science?

The care and feeding of the scientist-to-be starts

in early schooling where science is presented by capable and interested teachers, not just assigned teachers. Visiting lecturers from the community, from the dutiful P.T.A., etc. should supplement this interest. The young student must know that a true scientist, unlike a technician, is one who understands his particular branch of science in the order of the universe, not just as a specific type of impersonal and highly efficient technology.

In the university, adequate facilities in the basic sciences with adequate funding for special projects should be available. Federal, state, and local support must be available even if it means taking funds from the ever popular and ever needy medical sciences.

Industry must also support fundings with no attached application, limited strings, or directed projects. For the graduate scientist, there must be an adequate future visible either in professional fields or in the academic life. Planning boards and commissions must be available to develop long-term planning so crises in job unemployment or job needs will be prevented as much as possible. For science graduates unable to find employment, as the physicists and space experts are experiencing at the present time, job redevelopment and reorientation must be done so that their considerable talents are not wasted in research and development of plush pizza parlors. Thus, science should again become popular and feel needed and give TLC (tender loving care) without the crown of a feared and mighty emperor. There may be signs that training grants are being reinstated.

Conclusions

The new and important laser technology cannot afford to be neglected by the current vigorous antiscience philosophy.[4–6] The laser, by its very nature, can show it belongs to modern science. Passmore[4] calls science "still one of the greatest triumphs of human imagination." The laser scientist, to be formed and those formed already, should see the true potential of lasers in the order of the universe for the glory and good-of-man. Thus oriented and provided with real and possible goals, the laser scientist can move ahead in this brilliant field and make the laser light be a fantastic light for all seasons.

REFERENCES

1. **Morrison, R. S.,** *Ed. Sci.,* December 8, 1972.
2. **Sackler, A. M.,** *Med. Trib.,* January 17, 1973.
3. **Stubbs, P.,** *New Sci.,* 24, August, 1972.
4. **Passmore, J. A.,** Anti-science – a misunderstanding, *Science,* 181, 327, 1973.
5. **Cotgrave, S.,** Antiscience, *New Sci.,* 82, 1973.
6. The Social Responsibility of Scientists, Annals of the New York Academy of Sciences, Philip Siekevitz, Chairman, June 1972, 196.

Chapter 21

THE FUTURE

In his review of the speculations of forecasters about the future, Keith Melville wrote:

"THE FUTURE isn't what is used to be, but then neither is our way of thinking about it. Undoubtedly, one reason why more and more people are enlisting in the science of predictions is the galloping approach of millenial year 2000. A more important impetus: the growing conviction that the accelerating rate of social change demands better techniques for anticipating direction and guiding course. Effective social policy cannot be formulated without a working sketch of alternative futures. Thus, most futurists are trying to heed the injunction of The Whole Earth Catalog: We are as Gods and we'd better get good at it."

The laser industry, although it should be concerned as we indicate also with social variables of the future, is most interested in Herman Kahn, one of the popular prophets for forecasting the future. "Kahn is most convincing when he engages in technological forecasting, which he describes as probably the most sophisticated and highly developed forms of futurology." In this view, Kahn claims that:

"At least for the next ten to twenty years, technological growth and economic progess will be able to make the rest of the world rich; and with...self-restraint, proper policies and designs, and the allocation of sufficient resources, we will be able to cope with problems of pollution. Technology can often solve its problems."

As yet, his specific forecasts about laser technology are not available. The laser industry believes, of course, that such forecasts cannot be anything but highly optimistic. The industry's only concern is whether such progress will be slow or rapid.

Laser Focus, an excellent source for showing the rapid advances in laser technology, has watched the economic outlet for the laser industry for years. In their fifth annual economic review and forecast, they estimate the expansion rate for 1972 was 17% and for 1973, 14.7% to $215.7 million, with slower rise in sales but with a sharp rise in profits and jobs. "Research and development will increase more slowly than production." The military is forecast to continue to dominate the market with "nearly three of every four dollars spent by the military in the laser market will go for R & D." The international market will continue to grow as it should, especially from the U.S.S.R., Britain, France, West Germany, and Japan. The growth will be more rapid than in the U.S. if funding for R & D is reduced in the U.S. Since laser technology is still a new one, R & D rather than production is the greater area of involvement. Without the expertise background of Kahn, Thompson, and the M.I.T. team in the new (?) science of futurology, is it possible to try to guess about specific areas in laser technology? From time to time in this book there have been references to desires, dreams, and needs for the future. Some of these may be repeated for emphasis, for in the many diverse applications of laser technology redundance is inevitable, even incidentally, the redundance of misinformation or inadequate information.

For laser systems, it is necessary to repeat Fred Brech, that in the future power and energy will go up, wavelengths will go up and also go down and similarly, prices will first go up and then go down. As far as high power systems are concerned, the remarkable laser engineering of CGE in France is well known. In regards to Nd glass laser, Solon of Hadron Inc. believes that in 1973, 1,000 J will be available and in the next five years 10,000 to 100,000 J. Chemical lasers, high output reliable, tunable dye lasers, and ultraviolet lasers will be developed. Research and development of laser induced thermonuclear fusion will stimulate the development of high output CO_2, glass, and chemical lasers. The news about an X-ray laser will continue to produce interest and controversy, and hopefully, the future will certainly see a gamma ray laser, the so-called graser. At the other end of laser systems will be increased use and miniaturization of semiconductor laser systems.

Comments have been made previously about the significant economic effects of very rigid, very low, unwarranted permissible exposure levels in regard to laser safety. When there is definite evidence that such hazards have been found, not just suggested, then such restrictions are justified. Again, more funding, not more polemics, is needed to acquire adequate data. Hopefully, a compromise has been made possible through 2-136 of ANSI and the laser expertise for The Technical Electronic Product Radiation Safety Standard Committee of the Bureau of Radiological Health. It is hoped that in the future the laser industry itself will contribute to support for basic research in laser safety.

The two great areas of electrooptics and laser-induced thermonuclear fusion will stimulate, of necessity, continued great advances. That continued and much used postwar phrase, spin-off from these uses, will benefit many applications of the laser in the future.

Research is the chief aspect of the laser in the future. If research is linked with disinterested politics, progress will be even slower than forecast. For the military, in addition to perfection and expansion of current activities including so-called laser H-bombs, laser guidances of missiles and laser anti-missile applications will be investigated. The military all over the world still search for a superlaser.

Since laser-induced thermonuclear fusion is an international affair, more cooperative studies, ideally with less secrecy, are necessary to assure progress in this great field. Energy crises from misuse of natural resources, depletion of these, and population expansion in many countries will require such research and development. Other substitutes for fossil fuels will be in sharp competition.

Moreover, there will be more and more studies, again under the activity of the energy crises, for remote energy transmission by lasers.

With the great expansions of the new discipline of electrooptics, there will be increased activities in the diffuse field called information handling to include especially communications, etc. Laser beams will continue to carry more information than other materials used for transmission such as wires, coaxial cables, and even microwaves. Laser-associated computer technology for information storage and retrieval, video cassettes, and video-audio information discs are some of the features of the new advances to receive more attention in the future. Particularly attractive, at least on paper, is the video disc information storage and retrieval system with the helium-neon laser scanner playback and the T.V. screen readout (MCA video disc). Some applications suggested recently for this include storage for library, catalog, and archival material, teaching, credit card verification, artistic entertainment, etc. It is emphasized that this system is still in the R & D stage. Sonnett[9] is concerned that by 1990 we may not want to go out to the movies. This is true especially if there are no good movies. Will advances in laser technology then help to unite the disintegrated family group? There are some more concerns for the social consequence of the laser for the future. Connor[3] states that new dictated laser systems can be used in the Picture Phone System at Bell Laboratories for transmission displays, processing, etc. The laser switch of Bell Laboratories also will help phone calls and will be an important part of the optical communication system. Advances will continue in holography to make a new world, say "the holographers." How holography storage will compete with the conventional point-by-point techniques is the concern of Huignard.[5] He believes storage density with "conventional holography" is "probably limited to slightly over one billion bits per page. Commercial viability requires 10 times that density in a read-wrote system and 1,000 times as much for a read-only system."

For space technology, the recent three country cooperative lunar project is a model for future laser space research. On January 16, 1973, the Soviet spacecraft Luna 21 put a French-made retroreflector in the moon crater, LeMonnier. On January 25, 1973, a U.S. group from McDonald Observatory on Mt. Locke used the laser on this retroreflector. The reflected beams gave data for lunar radius at the crater and will provide data in the future "for lunar motion, lunar interior, and earth physics." Again, this is an example of how cooperative international laser research can produce progress. Increased communications in space, better instrumentation for lidar and for air speed indicators such as the CO_2 laser air speed indicator of the Global Atmospheric Research Program and laser treatment of air pollution in space laser source mass spectroscopy for lunar ultrastructural research are but some of the space technology

FIGURE 21-1. For the future MCA video-disc technology, video-audio information in 12-in. diameter plastic disc with 40 billion bits per disc, with microgrooves scanned by a low output helium-neon laser whose readout is electronically transmitted to a T.V. screen.

applications for the future. Lasers are truly then a new frontier in avionics not only for the military, but also for space technology. The laser will make the world smaller and perhaps even better.

Laser oceanography, laser earthquake detection, and laser lighthouse systems will all be expanded greatly in the future. Laser surveillance will expand greatly in the future. One example is the portable gallium arsenide laser fence.

Laser chemistry will advance, especially from the insistent demands for analysis, monitoring, and control of environmental pollution. The laser also can be effective in waste disposal. More studies will be done to evaluate environmental exposure to nonionizing radiation.

Laser metalworking and materials processing in almost every medium are expanding rapidly in industry and hopefully will be accompanied by effective laser safety programs. Until the cost is lowered for small industries, the laser establishment will have to take care of such custom work. Such projects should be considered in the beginning as pilot projects, economical to attract more industry participation, reliable for repeated treatment, and accurate. The work in textile processing is now being extended to computer controlled laser pattern cutting in the shoe industry.

For biology, which is most important to advance laser application and, as usual, most likely to receive budget cuts when research monies are scarce, the laser has a glorious future. Laser microirradiation of chromosomes must continue for the future if there is ever to be such a thing as molecular engineering in molecular biology. Cytofluorography, laser microprobe analyses, and laser photobiology are all new important areas of the present and future.

Advances in medicine will provide for increased facilities for diagnosis and treatment of the patient. These will be through advanced laser-associated computer information storage and retrieval for aids in diagnosis, literature surveys, and education with the low light light level T.V. tubes, such as Plumbeam; T.V. diagnosis will be more popular and effective. Cancer diagnosis, in all its phases from the laboratory to the patient, will provide more applications for the laser, again, and monotonously, provided adequate funding is available. Scanned holography of the acoustical image on film will be available for cancer diagnosis. It is said that such detection techniques even now can detect a 2-mm-diameter tumor in a rat and a 4-cm-diameter tumor in the breast. Developments in this phase of holography as well as in the holography of respiration and heart action will again show some more needs of the laser in medicine. More effective laser eye instrumentation, which will include tunable dye lasers, will show that the saturation point has not been reached in laser instrumentation for ophthalmology.

New high output systems will make laser surgery possible in many fields from plastic surgery to cancer surgery. A very important area of medical diagnosis, at present, is fiberoptics. These probe now deep into body tissues, including the heart and far down into the lungs. Laser endoscopy, developing since the single fiber of Nath, now offers laser operative endoscopy with high output laser systems.

Laser industry will also expand if this specialty is given proper sympathy and understanding by granting agencies. Decisions as to value and limitations of laser dentistry should, as in other fields of laser applications, be made by laser expertise. Progress has been slow and tedious but the important fields of dental caries, pulpal cavity treatment, oral surgery, and processing of prosthesis all will be done elsewhere, if there is no interest in the U.S. The famed credo, "If you don't need the laser, don't use it," also extends to dentistry for those who know both the needs and what lasers can do now and may do even more in the future.

The challenge to the young artist and artisan is given by the new visual art of the laser. It is not only, as it has been called, a light of all seasons, but also a light for all media for the artist, musician, and photographer. Hopefully, the industry will cooperate to help, at least to lend safe, reliable instrumentation accompanied by a certificate of understanding and tolerance.

Advances in laser technology will require more peoples to learn more and more about laser technology with all the modern techniques and visual educational aids. This goes all the way from basic science in the lower school to the graduate laser physicist. From all this will develop the laser biomedical engineer.

To develop progress in laser technology, flexible, dynamic safety programs, applications of the laser where needed, and multidisciplined cooperative programs are needed. There are many examples today of the lack of such cooperative programs producing wastes of monies, materials, and time and, consequently, obstructing progress. Especially now, since research is greater than production, an institute of Laser Biology and Medicine should be established, not later after castastrophes, hysteria, and sheer ignorance demand it. In such an institute, a central cooperative multidiscipline facility with adequate funding would contribute rapidly and economically to progress in all phases of laser technology. The staff, with effective lines of communication inward and outward to all concerned, would be made up of physicists of many disciplines, engineers, electronic experts, instrument designers, biologists, physicians, surgeons, and dentists, associated with a medical center; all would be concerned with man. For example, such an institute could rapidly investigate the hazards and the applications in biology and research medicine of the new "phase matched metal vapor cell," which reduces the wavelength laser light to an exit wavelength approximately one third of the original size.

What does all this mean? Again, Keith Melville asks.[7] The future will be:

1. Absolutely terrific
2. An unmitigated disaster
3. Neither of the above

An informed, enlightened, and understanding public can help the laser futurologist, a new (?) specialty, to plan and forecast. Those who have studied matured and controlled laser beams believe that it is possible to make a smaller and, perhaps, even better world with the laser, for the laser is needed often.

We can only quote Rabbi Abraham Joshua. Heschel's heritage of wonder and awe for an "age too haunted by cries of doom, despair and disparagment." He wrote,

> "As civilization advances, the sense of wonder declines. . . . Mankind will not perish for want of information; but only for want of appreciation. . . . Awareness of the divine begins with wonder. It is the result of what man does with his higher incomprehension."

Finally, and briefly, the laser could dispel Galston's[10] "spirit of gloom that pervades American scientific circles proliferating everywhere into a pessimistic outlook for the future."

REFERENCES

1. Air speed indicator, *Opt. Spectra,* February 1973.
2. **Berns, M. W., Cheng, W. K., and Hoover, G.,** Cell division after laser micro-irradiation of mitotic chromosomes, *Nature,* 233, 122, 1972.
3. **Connor, D.,** personal communication, 1972.
4. **Hescher, R. and Joshua, A.,** *Wall St. J.,* February 2, 1973.
5. **Huignard, J. P.,** Holographic storage, *Laser Rep.,* February 1973.
6. **Kahn, H.,** quoted by Melville, K., *Sciences,* December 1972.
7. **Melville, K.,** The future will be, *Sciences,* December 1972.
8. **Solon, L.,** *Laser+Elektro-Optik.,* 2, S-4, 1972.
9. **Sonnett, S.,** In 1990 will we go out to the movies? *New York Times,* February 11, 1973.
10. **Galston, A. W.,** Hard times for American science, *Nat. Hist.,* 82, 24, 1973.

APPENDIX

LASER AND OPTICAL SUPPLIES SOURCES

1. Laser Focus, January 1973, Buyers' Guide, Advanced Technology Publications Inc., 385 Elliot Street, Newton, Mass. 02164.
2. The Optical Industry & Systems Directory, 1972–1973, 19th ed., The Optical Publishing Co. Inc., Lenox Road, Pittsfield, Mass. 01201.
3. Guide to Scientific Instruments, November 1972, Richard G. Sommer, American Association for the Advancement of Science, Guide to Scientific Instruments, 1515 Massachusetts Avenue, N. W., Washington, D. C. 20005.
4. Laser optics directory – 1973, *Electro-Optical Syst. Des.*, February 1973, 26.

LASER LEXICON*

Acousto-optic Diffraction

This is a process whereby the effects of a diffraction grating are produced by an intense sound wave in a solid or liquid. The sound wave sets up a regular pattern of compression and rarefaction which is optically similar to a diffraction grating made by scribing regularly spaced fine lines on a glass plate. A laser beam passed through the liquid has its direction changed by an amount dependent upon the angle between the light wave and the sound wave, on the wavelength of the light wave, and on the wavelength of the sound wave. The sound wavelength can be changed by varying the sound frequency (generally required to be several tens of megacycles to achieve useful deflection). This in turn alters the direction of the diffracted laser beam. The intensity of the diffracted laser beam can be changed by varying the intensity of the sound wave. Acousto-optic diffraction is a useful mechanism for rapid modulation, deflection, or scanning of laser beams.

Active Medium

Laser beams are produced by **stimulated emission** in certain active media. A medium is activated by **pumping**. Typical active media are gases, such as argon, krypton, or a mixture of helium and neon, artificial crystals of ruby or yttrium aluminum garnet (YAG) and certain liquids, such as selenium oxychloride. The crystals and liquids contain small amounts of certain **doping** agents, such as neodymiun, europium, or chromium.

Antireflection (AR) Coatings

Ordinary glass surfaces reflect several percent of the light incident upon them. This reflection loss can be cut to less than one percent through the use of quarterwave thick transparent coatings of a material having a refractive index different from that of the glass. Even smaller reflection losses can be achieved with carefully tailored multiple dielectric films, which form pass-band **optical interference filters.**

Argon Laser

See **ion gas lasers.**

Brewster Angle

Plane-polarized light can pass through a sheet of glass without any of the light being reflected if the light is inclined from the perpendicular to the glass surface by a critical angle, called the "Brewster angle," and if the light is polarized in the plane of incidence (i.e., the plane containing the light ray and a perpendicular to the glass surface).

Brewster Windows

Many gas and liquid lasers use tubes to contain the active material which have **Brewster angle** windows on the ends. These windows permit the laser beam to circulate between the external **laser mirrors** without loss of any light by reflection. In many cases a small reflection loss is enough to prevent **lasing action.** Consequently, the **polariza-**

tion of laser beams tends to be determined by the orientation of the Brewster windows.

Carbon Dioxide Laser

See **Molecular Gas Lasers**.

Coherence Length

The degree of **temporal coherence** of a laser beam is measured by its coherence length, which is the interval of laser beam path length over which mutual coherence is preserved. If a laser beam is divided and subsequently recombined after the two portions have traveled different distances (as in a Michelson interferometer), the beams may combine with interference effects, or they may not, depending on whether the path length difference was less than or greater than the coherence length. The coherence length of a typical gas laser is of the order of a few centimeters to a few tens of centimeters. With **frequency stabilization** the coherence length can be increased several orders of magnitude.

Coherent Radiation

A beam of light is said to be coherent if different portions of the beam can be made to "interfere" when they are superimposed on each other, that is, to tend to reenforce each other where they coincide in the same phase and to tend to cancel each other where they coincide in opposite phase. This does not happen with **incoherent light** from ordinary lamps, except under special circumstances (e.g., in equal-path interferometers).

Dielectric Mirrors

Laser mirrors frequently must be designed to reflect from 99% to 99.95% of the incident laser light. For visible and near-infrared lasers dielectric mirrors are required. These are multilayer **optical interference filters**, consisting of up to 50 low loss transparent dielectric layers of alternating high and low refractive index, each layer being about a quarter of a wavelength thick. These mirrors can be tailored to reflect certain wavelengths and to transmit other wavelengths. In this way they can function as optical filters, permitting laser light of different wavelengths or colors to be either combined in a single beam or separated.

Diffraction-limited Beam Angle

The natural process of diffraction causes narrow beams of light to spread or diverge so that at great distances they form narrow cones of light, even if the light is parallel at the source. The smallest possible cone angle for a given beam diameter at its source is called the "diffraction-limited beam angle," which is proportional to the wavelength to source beam diameter ratio. This small angle is produced by the lowest order **mode of a laser**.

Diffraction-limited Optical System

Projection of narrow laser beams of **diffraction-limited beam angle** or concentration of laser light to a **diffraction-limited spot size** requires that all elements of the optical system (e.g., windows, lenses, polarizers, prisms, modulation crystals) be free of optical distortion. A practical criterion for a diffraction-limited optical system is that the optical wavefront preserve its correct planar or spherical shape to within a quarter of a wavelength (about a tenth of a micron for visible light). Where many components are used in an optical system, the individual components must have even lower distortion (1/10 to 1/40 wavelength being commonly required).

Diffraction-limited Spot Size

When a light beam is brought to a sharp focus by a distortion-free lens (i.e., a **diffraction-limited optical system**) the natural process of diffraction causes the rays to spread apart at the focal point to form a beam of finite size. The diameter of the diffraction-limited spot equals the product of the **diffraction-limited beam angle** and the focal length.

Doping

Most solid liquid lasers use a bulk material, called the "host," with a small amount (or "doping") of active material that produces the **lasing action**. Typical doping agents are chromium, neodymium, and europium.

Doppler Effect

See **Optical Heterodyne Detection**.

Electrooptic Effects

The Pockels and Kerr electooptic effects in crystals or liquids can be used for **optical modulation** or for **optical beam deflection or scanning**. Basically, these effects are minute changes in refractive index produced by applied electric fields. The refractive index changes are different for different directions of **polarization of the laser beam**. This gives rise to changes in the state of polarization of the laser light, or polarization modulation. Refractive index changes can be used in prism devices to control the direction of a laser beam in response to an applied voltage. Typical electrooptic crystals are potassium dihydrogen phosphate (KDP) and lithium niobate. Carbon disulfide is an electrooptic liquid.

Elliptically Polarized Light

Plane-polarized light produced by a laser can be changed to various forms of elliptical polarization by passage through birefringent crystals, such as calcite, or through strained pieces of glass or plastic. Special cases are left- and right-circularly polarized light, each of which has equal amounts of **plane-polarized light** of all orientations.

External-mirror Gas Laser

The more common arrangement in **gas lasers** is for the **laser mirrors** to be external to and separate from the **laser discharge tube**. For such lasers the ends of the discharge tube usually contain **Brewster windows** to allow the laser beam to circulate between the two mirrors. Some lasers have been built in a hybrid construction with one internal mirror and one external mirror, using one Brewster window.

Frequency Stabilization

Many laser applications utilize the principle of **optical heterodyne detection** to achieve discrimination against background light or for precise and sensitive measurement of distance or velocity. If more than one laser is involved, the two lasers must have mutual **temporal coherence**. This is possible only if each of the lasers is extremely stable in frequency. Frequency stabilization is achieved by maintenance of constant ambient conditions and by automatic (i.e., servo-controlled) regulation of the spacing of the **laser mirrors**. This spacing determines the exact laser frequency. A reference optical frequency standard is also needed to determine the amount of frequency deviation. A stable optical cavity is sometimes used as the reference. Some other stabilization schemes depend on identifying the natural center frequency of laser oscillation and measuring the deviation of the actual laser frequency from that value.

Gas Lasers

See **helium-neon laser**, **ion gas lasers**, **molecular gas lasers**.

Giant Pulses

See **Q-switched Lasers**.

Glass Lasers

Pulsed lasers have been made which use a special glass as the **active medium**. The glass contains a **doping** agent, such as neodymium. Like the **liquid laser**, the glass laser does away with the requirement for expensive crystals, such as are used in the **ruby laser** and in the **YAG laser**.

Helium-neon Laser

This was the first laser to produce continuous light and is still the most common and least expensive laser. The most popular forms generate visible red light, but they can also produce infrared light at 1.15 and 3.39 μm wavelength. **Pumping** is achieved by an electric discharge of from 5 to 60 mA in the helium-neon gas mixture. The efficiency is about 0.05%. Small He-Ne lasers produce less than a milliwatt. The largest practical forms produce up to 100 mW.

Hologram

The photographic record produced in the process of **holography** is called a "hologram." It is a photographic plate or film with an intricate pattern of interference fringes. This pattern bears no resemblance to the holographic image, which is viewed by looking through the hologram rather than focusing on the fringes.

Holography

Holography is three-dimensional photography by a process of **wavefront reconstruction**. Although over a decade older than lasers, holography was dependent upon the invention of the laser for it to become a practical form of photography. Holographic images are seen by looking through a special photographic plate or film, called a **hologram**, illuminated by a bright source of light having adequate coherence (not necessarily a laser). Three-dimensional images are seen exactly as they would appear if they were physically present behind the hologram. Looking through the hologram from different vantage points provides different perspective views of the holographed subject. The laser is needed primarily for taking holograms, although it is also much used in viewing holographic images. The image is formed on high-resolution photographic emulsions as a complicated pattern of interference fringes. The fringes are formed by mixing laser light scattered from the subject with a wide "reference beam" of laser light, which is projected directly on the film.

Incoherent Light

Light from sources other than a laser is generally termed "incoherent." However, coherence is a relative quantity, possessed by all light in varying degrees. Laser beams provide many orders of magnitude more **coherent radiation** than beams from ordinary lamps.

Internal-mirror Gas Lasers

Some gas lasers (particularly **helium-neon lasers**) utilize **laser discharge tubes** which contain the **laser mirrors** as an integral part. The **dielectric mirror** coatings are on inside surfaces at the ends of the discharge tube. The first He-Ne lasers were constructed this way, with bellows between the mirrors and the discharge tube to provide for angular adjustment of the mirrors. Some inexpensive gas lasers are currently made with rigidly connected internal mirrors. The proper alignment is achieved when the mirrors are sealed in place. Slight adjustments can be made on some of these lasers by bowing the discharge tube a small amount. Unless internal **Brewster windows** are installed in these tubes, the **polarization of the laser beam** will not be in a prescribed direction. The laser may be accidentally polarized in a fixed direction, but usually the polarization direction changes constantly in a random fashion. This gives rise to polarization noise, which limits the usefulness of such lasers to applications which do not utilize the polarization feature of laser beams.

Ion Gas Lasers

The most important ion lasers are the argon laser, which produces blue and green beams, and the krypton laser, which produces red beams and lesser amounts of blue, green, and yellow. **Pumping** is achieved by a very strong ionic discharge in lasing gas, typically 10 to 30 A. Continuous power of the order of a few watts is produced with an efficiency of about 0.05%.

Krypton Laser

See **Ion Gas Lasers**.

Laser

This name is an acronym for "Light Amplification by Stimulated Emission of Radiation." It has been applied to various devices invented since 1960 which generate narrow beams of **coherent radiation** in the infrared, visible, and ultraviolet portions of the electromagnetic spectrum. All of these portions are loosely referred to as "light." Various lasers use solid, liquid, or gaseous **active media**.

Laser Cavity

Most lasers require an arrangement, called the laser cavity, to trap and recirculate the laser beam in the laser **active medium.** For example, the cavity of a gas laser consists of the gas discharge tube with **Brewster windows** on the ends and a pair of **dielectric mirrors** which reflect most of the laser light back into the tube. One of the mirrors is made slightly transmissive, which permits a small fraction (typically one percent) of the circulating laser light to escape as the output laser beam.

Laser Discharge Tube

Gas lasers contain the gaseous **active medium** in a long thin tube through which is passed an electric discharge. The discharge achieves **pumping** or activation of the gas to make the **lasing action** possible. The discharge can be produced with a DC current between anode and cathode electrodes within or appended to the discharge tube; or it can be produced by a radio-frequency current coupled through the walls by electric or magnetic induction. The far ends of the discharge tube are terminated in **Brewster windows**, in windows with **antireflection (AR) coatings**, or with internal mirrors (in the case of the **internal-mirror gas laser**).

Laser Mirrors

All lasers require cavity mirrors which reflect sufficient light to maintain **lasing action**. The fraction of light lost by the mirrors and windows (if any) must be smaller than the fractional amplification of laser light produced by **stimulated emission** in the **active medium** – this gain being only a few percent in many lasers. Infrared lasers can use front-surface metal mirrors, but most visible lasers require **dielectric mirrors**.

Lasing Action

The process of generating laser beams by

stimulated emission in a **laser cavity** having **laser mirrors** to trap the laser beam is called "lasing action."

Liquid Lasers

Three kinds of liquids have been used to produce laser beams: (1) chelates doped with rare earths; (2) organic dyes; and (3) aprotic (i.e., not containing hydrogen protons) compounds, such as selenium oxychloride and phosphorus oxychloride doped with rare earths. **Pumping** is achieved with intense flash lamps. Commercial lasers are not yet available in these forms.

Modes of a Laser

Laser light is generated in symmetrical beams which vary in amplitude and phase across the beam. High-order modes produce symmetrical patterns of many spots of light. The lowest-order mode is a single spot of light. **Lasing action** in high-order modes provides maximum laser beam power with relatively uncritical adjustment of the **laser mirrors**; however, the lowest-order mode forms a beam which spreads, or diverges, a minimum amount (i.e., it produces a **diffraction-limited beam angle**) and forms the smallest focused spot (**diffraction-limited spot size**).

Molecular Gas Lasers

The most common of these lasers is the **carbon dioxide laser,** which produces continuous light in the far infrared (10.6 μm wavelength). This laser has high efficiency (up to 30%) and high power capability (over a kilowatt). Small CO_2 lasers can produce 10 W from a 100-W pumping discharge.

Nonlinear Optical Materials

Hard steel is saturated by strong magnetic fields, giving rise to nonlinear variation of flux with magnetic intensity. In an analogous fashion, extremely intense laser light (e.g., in the cavity of a high-power laser or in the beam of a **Q-switched laser**) can produce nonlinear refraction in transparent materials. Strongly nonlinear effects are produced in certain electrooptic crystals, such as potassium dihydrogen phosphate (KDP), ammonium dihydrogen phosphate (ADP), barium sodium niobate, and lithium niobate. These effects can give rise to harmonic generation or parametric interactions in which the wavelength of a laser beam is converted from the infrared to the visible or from one color to another.

Optical Beam Deflection or Scanning

The direction of a laser beam can be changed (i.e., the beam can be "steered") by reflection, refraction, or diffraction. All of these processes can be electrically, magnetically, or mechanically controlled by devices variously referred to as "beam deflectors," "beam scanners," or "beam steerers."

Optical Heterodyne Detection

Two laser beams having sufficient **temporal coherence** and **frequency stabilization** can be mixed to produce optical beat frequencies, analogous to beat frequencies produced in superheterodyne radio receivers or to the beats produced by sounds from two musical instruments. Optical heterodyne receivers utilizing this principle can achieve a high degree of selectivity against interfering background light. The beat notes so produced are greatly affected by movement of the laser beams toward or away from the heterodyne receiver. This so-called "Doppler effect" causes the beat frequency to increase with motion of the laser source toward the receiver and to decrease with motion away, just as the pitch of a train whistle increases and decreases as the train approaches and recedes. This effect is useful in optical measurement of relative velocity.

Optical Interference Filters

These filters are formed as thin layers of alternating high and low refractive index. Typically each laser is about a quarter of a wavelength thick. Light is reflected at the top of each laser and the bottom of that layer (which is the top of another layer). The cumulative effect of all N+1 reflections in a N-layer coating can be made to constitute a nearly complete reflection or a nearly zero reflection, depending on the arrangement and thickness of the layers and on the wavelength and direction of the incident light. Optical interference filters are used as **laser mirrors,** for **antireflection coatings,** and as optical spectral filters.

Optical Modulation

The transmitter of an optical communication system uses optical modulation to superimpose a signal on a laser beam. The modulator causes some property of the laser beam (its intensity, frequency, or state of polarization) to vary in accordance with a varying electric signal (e.g., from a microphone or a television camera). The inverse process, optical demodulation, is used in the optical receiver to restore the original electric signal.

Plane-polarized Light

Most lasers produce plane-polarized light. A **Brewster-angle** polarizer (or **Brewster window**) in the **laser cavity** determines the plane of polarization, which is horizontal if the laser beam and a normal to the window are both in the horizontal plane.

Polarization of Laser Beams

Unlike ordinary light, laser beams are generated with a definite polarization. Polarizing elements, such as the lenses of Polaroid® sunglasses, can be made to block the light of the laser. Ninety-degree rotation of the polarizing element allows most of the laser light to pass.

Pulsing

Pumping is achieved in pulsed laser by the formation of an intense short-duration electric discharge either in the laser medium (in the case of a **gas laser**) or in a flash lamp (in the case of a **solid-state laser** or a **liquid laser**). Light from a flash lamp is coupled into the laser medium by means of a **pumping cavity**. Pulsing is achieved electronically through the controlled discharge of an energy storage network (capacitor bank). The gas laser tube or the flash lamp is triggered by a high-voltage, low-energy pulse, generated by a thyratron tube or a silicon-controlled rectifier (SCR). Triggering produces an initial ionization that releases the energy from the capacitor bank in the form of an intense arc discharge of very short duration.

Pump Cavity

Some lasers which are pumped by electric arc or flash lamps require an efficient arrangement for focusing the **pumping** light into the laser **active medium**. A highly reflective cylindrical, spherical, or ellipsoidal mirror, called the pump cavity, surrounds the lamp and the laser material, both carefully arranged so that the pump light is reflected to a focus in the laser rod or cell.

Pumping

This term is used to designate the process of energizing the laser medium so that it can produce **stimulated emission**. **Gas lasers** are generally pumped by an electric discharge in the gas. **Solid-state lasers** and **liquid lasers** are usually pumped by arc lamps or flash bulbs. **Semiconductor lasers** are pumped by an electric current through the semiconductor junction.

Q-switched Lasers

In ordinary lasers **pumping** and **lasing action** occur simultaneously, or nearly so. Lasing requires accurate alignment of the **laser mirrors** so that the laser beam recirculates in the laser medium. Some lasers are equipped with means for "spoiling the cavity Q," or stopping the recirculation, while a great amount of energy is pumped into the laser medium. The cavity is then suddenly "unspoiled" (Q-switched) to release all of the laser energy in a giant pulse lasting much less than a millionth of a second. Billions of watts of laser power can be produced in this way from lasers which otherwise would produce millions of watts with about the same total pulse energy. Q-switching can be accomplished by a spinning **laser mirror**, by polarization modulation in the laser cavity (using **electrooptic effects**), or through the use of saturable dyes (which are bleached by the intense pump light, suddenly becoming transparent when a high intensity is reached).

Ruby Laser

This was the first laser and is still a very important one. It uses a rod of synthetic ruby and a powerful flash lamp as a **pumping** light source. Deep red light is produced with enough energy to burn holes in razor blades. The ruby rod contains chromium as the **doping** agent. Although most commonly used as a pulsed laser, continuous ruby lasers are also available.

Scintillation

Laser light produces a visual effect, called scintillation, which appears as a sparkling of the spot. A spot of laser light projected on a diffusing surface sparkles with fine-grained variations of intensity. The pattern of these variations (not to be confused with the characteristic pattern of the **mode of the laser**) is steady when the observer holds a steady gaze, but it changes when the observer moves his head. Rapid movement of the head, or rapid movement of the diffusing surface, destroys the illusion of scintillation. The effect is produced by the interference of rays from adjacent areas on the laser spot, which travel slightly different distances to the eye. The path length difference for some of the rays is of an amount to cause the interference to be constructive; for some of the rays it is destructive.

Semiconductor Lasers

These lasers are similar in construction to a transistor or a semiconductor diode. Infrared laser light pulses with watts of power can be produced with good efficiency, when the laser is cooled to liquid nitrogen or lower temperatures. **Pumping** is achieved by passage of a current through the junction. That current can be a voice or television signal, thereby producing laser beams modulated with these signals.

Solid-state Lasers

See **ruby laser** and **YAG laser** (the most common crystal laser), **glass lasers**, and **semiconductor lasers.**

Stimulated Emission

Lasers depend on stimulated emission for producing coherent amplification. A photon from the laser beam "stimulates" an excited atom in the laser medium, causing that atom to emit a second photon identical to the first (i.e., "coherent" with it), which becomes part of the laser beam. The growth from billions of stimulated emissions offsets losses from absorption in the laser medium and reflection and light scattering on the **laser mirrors** and the **Brewster windows**, permitting continuous generation of laser light.

Supermode, Mode-locked, and FM Lasers

Ordinary lasers tend to produce a cluster of lines in a narrow wavelength region. For example, a **helium-neon laser** can produce about ten lines spaced about 150 megacycles apart, clustered around an optical frequency of about 500 million megacycles. Each of these lines is incoherent with all of the others. Supermode lasers produce a coherence among these lines through the use of a modulation process which causes the lines to be mutually interdependent. The result can be transfer of all of the laser power to one line, or the production of intense, very short and regular pulses of laser light, or the production of a single line of laser light which scans up and down in frequency over a range of hundreds of megacycles.

Spatial Coherence

Spatial coherence is needed for light to be focused by a distortion-free lens (i.e., a **diffraction-limited optical system**) to a very small spot. Light from lasers and from distant stars share this property to a maximum degree.

Temporal Coherence

A high degree of temporal coherence in a light beam signifies that the light is confined to a narrow wavelength band, or narrow frequency band. Lasers have produced light confined to a bandwidth of a few thousand cycles (about one part in a hundred thousand million of the optical frequency). Discharge lamps can produce narrow-band emission, but their emission is many orders of magnitude broader than the light from a laser. The degree of temporal coherence is measured by the **coherence length** of the laser beam.

Wavefront Reconstruction

Holography produces three-dimensional images by generating a pattern of light from a **hologram** which is identical in its distribution of amplitudes, phases, and ray directions to the light that would be scattered onto the **hologram** surface by an actual three-dimensional object placed behind the hologram. Thus, the process of holography differs from stereoscopic photography in that the hologram can provide a range of perspective views rather than just one. Furthermore, near portions of a holographic image are focused differently by the eye than more remote portions, as is the case in viewing an actual object. Holography does not depend for its illusion on the use of special glasses or other devices to select different images for the left and the right eye.

YAG Laser

This is a well-developed form of pulsed or continuously emitting crystal laser. Commercially available forms can provide continuous power in the near infrared (1.06 μm wavelength) of the order of 250 W with an input power of about 10 kW. **Pumping** is achieved with intense light from arc or metal vapor lamps. The initials "YAG" stand for yttrium aluminum garnet. Sometimes the designation "Nd: YAG" is used, where Nd represents the **doping** agent, neodymium. Intense green light has been obtained by second-harmonic generation in a **YAG laser cavity**. A **nonlinear optical material**, such as a crystal of barium sodium niobate, can produce this result if it is properly aligned inside the laser cavity.

(**Source:* Fowler, V. J., General Telephone & Electronics Laboratories, Bayside, N.Y., General Telephone Co. of Ohio, April 1970. With permission.)

Optical Factors and Conversion Tables

Length – inches to metric

	Millimeters (mm)	Microns (μm)	Angstroms (Å)
1 inch (1.00″)	equals 25.4 (exactly)	25,400.	254,000.000
1 thous. (0.001″)	equals .025 4	25.400	254.000
1 microinch μin (0.000 001″)	equals .000 025 4	.025 4	254

Optical Factors and Conversion Tables

Length – metric to inches

	Inches	Thousandths	Microinches
1 millimeter (mm)	equals .039 37*	39.37	39,370
1 micron (μm)	equals .000 039 37	.039 37	39.37
1 angstrom (Å)	equals .000 000 003 937	.000 003 937	.003 937

Optical Factors and Conversion Tables

Wavelength units, in inches, in kilomegacycles

Wave-length unit	Equivalence in inches	Equivalence in kilomegacycles (KMc)
1 meter	39.37″	0.3
1 decimeter	3.937″	3
1 centimeter	.3937″	30
1 millimeter	.03937″	300
1 micron	.00003937″	300,000
1 millimicron	.00000003937″	300,000,000
1 angstrom unit	.000000003937″	3,000,000,000

Optical Factors and Conversion Tables

Metrology Terms

Prefix	Symbol	Meaning	Shorthand
Micro	μ	One millionth	10^{-6}
Milli	m	One thousandth	10^{-3}
Centi	c	One hundredth	10^{-2}
Deci	d	One tenth	10^{-1}
Deka	dk	Ten	10
Hecto	h	One hundred	10^{2}
Kilo	koek	One thousand	10^{3}
Mega	m	One million	10^{6}
Mega mega	mm	One million million	10^{12}

Optical Factors and Conversion Tables

Conversion – angles to linear measurement

Angles	per in.	per 10 in.	per ft
1 sec	.000 005″	.000 048″	.000 058″
5 sec	.000 024″	.000 242″	.000 291″
10 sec	.000 048″	.000 485″	.000 582″
20 sec	.000 097″	.000 970″	.001 163″
30 sec	.000 145″	.001 454″	.001 745″
1 min	.000 291″	.002 909″	.003 491″

Optical Factors and Conversion Tables

Conversion – linear measurement to angles

Linear measure	per in.	per 10 in.	per ft
.000 001″	0.206 sec**	0.021 sec	0.017 sec
.000 025″	5.157 sec	0.516 sec	0.430 sec
.000 05″	10.3 sec	1.03 sec	0.86 sec
.000 1″	20.6 sec	2.06 sec	1.72 sec
.001″	3 min 26 sec	20.6 sec	17.2 sec
.005″	17 min 11 sec	1 min 43 sec	1 min 26 sec

Optical Factors and Conversion Tables

Wavelengths of monochromatic radiations for interferometry

Color	λ vac. in mm	λ air in mm*	$\frac{\lambda}{2}$ air in in.	Band per in. for λ air*
Kr 86 orange red	.000 605 780 21	.000 605 615 73	.000 011 921 569 58	83,881.5722
Cd red	.000 644 024 91	.000 643 850 37	.000 012 674 219 88	78,900.3197
Hg yellow 1	.000 579 226 84	.000 579 069 34	.000 011 399 002 76	87,726.9724
Hg yellow 2	.000 577 119 83	.000 576 962 88	.000 011 357 537 20	86,047.2572
Hg green	.000 546 227 04	.000 546 078 19	.000 010 749 570 66	93,026.9711
Hg violet	.000 435 956 21	.000 435 836 05	.000 008 579 449 80	115,557.5900

Radiometric and Corresponding Photometric Units*

Radiometric units

Quantity	Symbol	Description	Defining equation	Representative units
Radiant energy	U		$U = \int_0^\tau P dt$	joule (J)
Radiant energy density	u	Radiant energy per unit volume	$u = \frac{\partial U}{\partial V}$	$\frac{\text{joule}}{m^3}$ ($J - m^3$)
Radiant power	P	Radiant energy per unit time	$P = \frac{\partial U}{\partial t}$	watts(W) = joule sec^{-1} ($J\text{-}sec^{-1}$)
Radiant intensity	J	Radiant power per unit solid angle	$J = \frac{\partial P}{\partial \Omega}$	$\frac{\text{watts}}{\text{steradian}}$ ($W - sr^1$)
Radiant emittance	W	Radiant power emitted per unit area of source	$W = \frac{\partial P}{\partial A}$	$\frac{\text{watts}}{\text{meter}^2}$ ($W - m^{-2}$)
Irradiance	H	Radiant power impacting per unit area of receiver	$H = \frac{\partial P}{\partial A}$	$\frac{\text{watts}}{\text{meter}^2}$ ($W - m^{-2}$)
Radiant exposure	R_p	Energy density impacting per unity area of retina, integrated irradiance at retina	$R_p = \int_0^\tau H_{ret} dt$	$\frac{\text{joules}}{cm^2}$ ($J - cm^{-2}$)
Radiance	N	Radiant power per unit solid angle per unit area from a source	$N = \frac{\partial^2 P}{\partial \Omega \partial A}$	$\frac{\text{watts}}{\text{sr - meter}^2}$ ($W - sr^{-1} m^2$)

Photometric units

Quantity	Symbol	Description	Defining equation	Representative units
Luminous energy	Q	Portion of radiant energy in visible region	$Q = \int_o^\tau F dt$	lumen-sec = talbot
Luminous energy density	q	Portion of radiant energy directly in visible region	$q = \frac{\partial Q}{\partial V}$	lumen-sec - m^{-3} = talbot-m^{-3}
Luminous power	F	Luminous energy per unit time	$F = \frac{\partial Q}{\partial t}$	lumen
Luminous intensity	I	Luminous power per unit solid angle, candlepower	$I = \frac{\partial F}{\partial \Omega}$	lumen-sr^{-1} = candela (cd)
Luminous emittance	M	Luminous power emitted per unit area of source	$M = \frac{\partial F}{\partial A}$	$\frac{\text{lumen}}{\text{meter}^2}$ (lumen-m^{-2})
Illuminance	E	Luminous power impacting per unit area of receiver, illumination	$E = \frac{\partial F}{\partial A}$	$\frac{\text{lumen}}{\text{meter}^2}$ (lumen-m^{-2})
Luminance	L	Luminous power per unit solid angle per unit area from a source; brightness	$L = \frac{\partial^2 F}{\partial A \partial \Omega}$	lumen-sr^{-1}-m^{-2} = cd-m^{-2}
Retinal illumination		Product of the luminance and the area of the pupillary opening	$L \cdot A_c$	lumen-mm^2 = troland

LASER LIGHT – A NEW VISUAL ART, CINCINNATI ART MUSEUM AND LASER LABORATORY, NOVEMBER 12, 1970

Philip Adams Richard Boyle Leon Goldman

Exhibit Entries

1. Plexiglass Carving – Martin-Marietta Co.
2. Hologram – Virgin and Child, Spanish XIV Century Alabaster, Richard Boyle, Donovan Coppolk, Barron Krody, and Ralph Wuerker, TRW Holography Lab.
3. Interferogram – Albert Peter
4. Laser beam – Fantasia, Michael Campbell
5. Laser beam – Web, Michael Campbell
6. Laser sculpture, Mack Smith, carbon dioxide laser on Plexiglass
7. Experiment #8 – John Peterson
8. Laser beam synthesizer – R. J. Rockwell, Jr. and Michael Campbell
9. Alpha Chamber – Donovan Coppolk
10. Search – Barron Krody
11. Figure 10 – Barron Krody and Michael Campbell
12. Experiment #2 – John Peterson
13. Experiment #7 – John Peterson

American National Standard for the Safe Use of Lasers

An American National Standard implies a consensus of those substantially concerned with its scope and provisions. An American National Standard is intended as a guide to aid the manufacturer, the consumer, and the general public. The existence of an American National Standard does not in any respect preclude anyone, whether he has approved the standard or not, from manufacturing, marketing, purchasing, or using products, processes, or procedures not conforming to the standard. American National Standards are subject to periodic review and users are cautioned to obtain the latest editions.

CAUTION NOTICE: This American National Standard may be revised or withdrawn at any time. The procedures of the American National Standards Institute require that action be taken to reaffirm, revise, or withdraw this standard no later than five years from the date of publication. Purchasers of American National Standards may receive current information on all standards by calling or writing the American National Standards Institute.

Foreword

(This Foreword is not a part of American National Standard for the Safe Use of Lasers, Z136.1-1973.)

This American National Standard provides guidance for the safe use of lasers and laser systems by defining control measures for each of five laser classifications. Once a laser or laser system is properly classified, there should be no need to carrry out tedious measurements or calculations in order to meet the provisions of the standard. However, technical information on measurements, calculations, and biological effects are also provided within the standard.

American National Standards Committee Z136, under whose jurisdiction this standard was developed, has the following scope:

Protection against hazards which are created by the use of lasers.

It is expected that this standard will be modified as new information and experience in the use of lasers is gained.

Suggestions for improvement of this standard will be welcome. They should be sent to the American National Standards Institute, Inc, 1430 Broadway, New York, N.Y. 10018.

American National Standards Committee on the Safe Use of Lasers, Z136, which processed and approved this standard, had the following members at the time of approval:

George M. Wilkening, Chairman
Sidney S. Charschan, Secretary

Organization Represented	*Name of Representative*
Aerospace Industries Association	Robert Moore
	John Masek (Alt)
American Conference of Governmental Industrial Hygienists	Charles H. Powell
	William Kelley (Alt)
American Federation of Labor & Congress of Industrial Organizations	Edward I. Legan
American Industrial Hygiene Association	Charles McHenry
	Isaac Matelsky (Alt)
American Institute of Chemical Engineers	Mathew M. Braidech
American Insurance Association	J. Philip Berggren
	Samuel A. Kaplan (Alt)
American Iron and Steel Institute	Joseph W. Miller
American Mutual Insurance Alliance	Richard I. Bleier
American Optometric Association	Pawl W. Lappin
American Society for Testing and Materials	Hayes A. Lee
	J. A. Dwyer (Alt)
American Society of Mechanical Engineers	Erwin G. Loewen
Association of American Railroads	John L. McNabb
Association of State and Territorial Health Officers	Thomas Cashman
Automobile Manufacturers Association	Charles C. Wang
	A. C. Doty, Jr (Alt)
The Children's Hospital Research Foundation	R. James Rockwell, Jr
Electronic Industries Association	Robert Mortensen
	Allen M. Wilson (Alt)
	Douglas Worth (Alt)
Health Physics Society	David H. Sliney
Illuminating Engineering Society	John E. Kaufman
	Howard Haynes (Alt)
Industrial Medical Association	Harry E. Tebrock
Industrial Safety Equipment Association, Inc	George Schauweker
	Lawrence A. Medici (Alt)
Institute of Electrical and Electronics Engineers	Samuel Fine
Instrument Society of America	Grover C. Wrenn
Institute for Advancement of Engineering, Inc	Malcolm L. Stitch
	Ralph A. Lamm (Alt)
Laser Institute of America	James F. Smith
	Mason Cox (Alt)
	Sam Fine (Alt)
Medical Center of the University of Cincinnati	Leon Goldman
Medical College of Virginia	William T. Ham, Jr
National Aeronautics and Space Administration	Rudolph M. Marrazzo
	Robert E. Alexander (Alt)

Organization Represented	*Name of Representative*
National Bureau of Standards	Nicholas Winogradoff
National Association of Photographic Manufacturers, Inc	Thomas S. Ely C. E. Hilton (Alt)
National Constructors Association	James R. Milor J. A. Barton, Jr (Alt)
National Electrical Manufacturing Association	H. W. Speicher D. C. Fleckenstein (Alt)
National Safety Council	John A. Janous R. C. Benashski (Alt) Julian Olishifski (Alt) William M. Smith (Alt)
National Society for the Prevention of Blindness, Inc	Charles J. Campbell James E. O'Neil (Alt)
Optical Society of America	Jay M. Enoch Bernard Rosett (Alt)
Polytechnic Institute of Brooklyn	Saul W. Rosenthal Glenn Heimer (Alt)
Roswell Park Memorial Institute	Edmund Klein
Society of Manufacturing Engineers	Robert Willette Glen Kett (Alt)
The Society of the Plastics Industry, Inc	A. E. Sherr J. P. Carrol (Alt)
Underwriters Laboratories, Inc	L. S. Homa A. W. Smott (Alt)
U.S. Army Medical Research Laboratory	Wordie H. Parr
U.S. Department of the Air Force	Donald D. Higgins Lawrence T. Odland (Alt)
U.S. Department of the Air Force, Institute of Technology	Robert W. Ebbers
U.S. Department of the Army	David H. Sliney William A. Palmisano (Alt)
U.S. Department of Health, Education and Welfare	Richard Peterson Robert L. Elder (Alt)
U.S. Department of the Interior, Bureau of Mines	Aurel Goodwin Roy G. Stott (Alt)
U.S. Department of Labor, Bureau of Labor Standards	Floyd A. Van Atta John P. O'Neill (Alt)
U.S. Department of the Navy, Bureau of Medicine and Surgery	D. K. Heffner
U.S. Naval Research Laboratory	Raymond G. Nebelung Henry C. Kennedy, Jr (Alt)
Virginia Commonwealth University	Walter J. Geeraets
WEMA	Herb M. Dwight Eben J. Tisdale (Alt)
Individual Members	John A. Carpenter Leonard Gardnar C. Harry Knowles Grant C. Riggle Howard Rosenthal Hermas Swope Myron L. Wolbarsht H. Christian Zweng Richard C. Honey (Alt) Art Vassiliadis (Alt)

The various subcommittees which participated in developing this standard had the following members:

Biological Effects on the Eye

Myron L. Wolbarsht, Chairman

Herbert E. Bell
R. B. Bedell
George H. Bresnick
Charles J. Campbell
John A. Carpenter
Alexander M. Clarke
Thomas P. Davis
Irving L. Dunsky
Walter J. Geeraets
William R. Godden
William T. Ham, Jr
Jude R. Hayes
Dennis Heffner
Art E. Jones
Edmund Klein
Maurice Landers
Robert Mortensen
Robert Neidlinger
David Randolph
Donald E. Rounds
David H. Sliney
H. G. Sperling
H. E. Tebrock
Arthur Vassiliadis

Biological Effects on The Skin

Wordie H. Parr, Chairman

W. D. Burnett
Glenn E. Conklin
Thomas P. Davis
Samuel Fine
Leon Goldman
William T. Ham, Jr
Edmund Klein
Max Rittenbury
R. James Rockwell, Jr
David H. Sliney

Measurements and Hazard Evaluation

Richard C. Honey, Chairman

Frank S. Barnes
H. S. Boyne
Alex M. Clarke
Mason Cox
Duncan MacVicar
R. James Rockwell, Jr
J. E. Roddy
L. E. Siepietowski
David H. Sliney
Hermas Swope

Control Measures

R. James Rockwell, Jr, Chairman

John A. Carpenter
Mason Cox
Eugene Dynek
Richard C. Honey
Richard A. Jaenicke
Charles J. Koester
Robert W. Neidlinger
Arthur Neilson
John I. O'Donnell
Wordie H. Parr
Charles H. Powell
Grant C. Riggle
Joseph C. Rosenbaum
William Rundle
A. E. Sherr
Richard M. Shotland
David H. Sliney
James F. Smith
Malcom Stitch
Hermas Swope
Charles F. Tedford
Douglas Worth

Elements of a Safety Program

Samuel Fine, Chairman

Herbert E. Bell
Thomas Davis
L. S. Homa
Samuel Levin
Raymond G. Nebelung
Robert Neidlinger
Gerald Parker
George T. Petchel
David Randolph
Henry J. L. Rechen
David H. Sliney
Victor E. Whitehouse

Medical Surveillance

H. Christian Zweng, Chairman

Walter J. Geeraets
Leon Goldman
Edmund Klein
Robert Neidlinger
Heinrich Rose

Special Considerations

David H. Sliney, Chairman

Harry Ackerman
Paul Dietz
Samuel Fine
John A. Janous
Walter Johnson
L. M. Kline
Rudolph Marrazzo
Charles McHenry
Henry J. L. Rechen

Terminology

Nicholas Winogradoff, Chairman

James C. Edwards, Jr
J. Longfellow
Art. H. Neill
J. C. Richmond
C. Wang

Contents

American National Standard for the Safe Use of Lasers

1. General

1.1 Scope. This standard provides reasonable and adequate guides for the safe use of lasers and laser systems with output wavelengths between 0.2 μm and 1 mm.

1.2 Application. The objective of this standard is to provide reasonable and adequate guidance for the safe use of lasers and laser systems. A practical means for accomplishing this is to first classify lasers and laser systems according to their relative hazards and then to specify appropriate controls for each classification.

The basis of the hazard classification scheme in Section 3 of this standard is the ability of the primary laser beam or reflected primary beam to cause biological damage to the eye or skin. For example, a Class I laser is one that is considered to be incapable of producing damaging radiation levels and is, therefore, exempt from any control measures or other forms of surveillance. A Class II laser (or low-power system) may be viewed directly under carefully controlled exposure conditions, but must have a cautionary label affixed to the external surfaces of the device. A Class III laser (or medium-power system) requires control measures to prevent viewing of the direct beam. A Class IV laser (or high-power system) requires the use of controls which prevent exposure of the eye and skin to the direct and diffusely reflected beam. Class V laser systems are comprised of Class II, Class III, or Class IV lasers contained in a protective housing and operated in such manner as to be incapable of producing damaging radiation. Stringent controls must be installed and maintained in order for any laser system to qualify as a Class V.

It must be recognized that the classification scheme given in this standard relates specifically to the laser device and its potential hazard, based on operating characteristics. However, the conditions under which the laser is used, the level of safety training of persons using the laser, and other environmental and personnel factors are important considerations in determining the full extent of safety control measures. Since such situations require informed judgments by responsible persons, major responsibility for such judgments has been assigned to a qualified person, namely, the Laser Safety Officer.

The recommended stepwise procedure for using this standard is as follows:

(1) Determine the appropriate class of laser or laser system using the manufacturer's specification and the hazard evaluation and classification descriptions given in Section 3.

(2) Comply with the measures specified for that class of laser or laser system using the following table as a guide. This procedure will in most cases eliminate the need for the measurement of laser radiation, or the quantitative analysis of hazard potential, or the use of the Maximum Permissible Exposure (MPE) values given in Section 8 of the standard.

Class	*Control Measures*	*Medical Surveillance*
I	Not applicable	Not applicable
II	Applicable	Not applicable
III	Applicable	Applicable
IV	Applicable	Applicable
V	Not applicable*	Not applicable*

*In normal operation only. Alignment and maintenance procedures shall require programs appropriate to the unenclosed laser.

For quantitative evaluation of the hazard associated with a given laser or laser system, Sections 8 and 9 should be consulted. To use the MPE values in Section 8, first determine whether there is a possibility of intrabeam viewing (in which case the intrabeam MPE criteria in 8.2 are to be used) or whether a hazardous diffuse reflection is being viewed (in which case the MPE values for an extended source as given in 8.3 are to be used). (See Figs. B1, B2, and B3 in Appendix B for illustrated viewing conditions.)

For the purpose of this standard, except for focused Class III lasers (see 3.4.1.3), only Class IV lasers are capable of producing hazardous diffuse reflections; hence calculations for viewing diffuse reflections from Class I, II, III, and V systems are normally unnecessary.

2. Definitions

The definitions of the terms listed below are based on a pragmatic rather than a basic approach. The terms defined are therefore limited to those actually used in this standard and are in no way intended to constitute a dictionary of terms used in the laser field as a whole.

absorption. Transformation of radiant energy to a different form of energy by interaction with matter.

accommodation. The ability of the eye to change its power and thus focus for different object distances.

aerosol. A suspension of small solid or liquid particles in a gaseous medium. Typically, the particle sizes may range from 100 μm to 0.01 μm or less.

α_{min}. The limiting angular subtense. (See 8.1 for criteria.)

apparent visual angle. The angular subtense of the source as calculated from the source size and distance from the eye. It is *not* the beam divergence of two sources. (See 8.1 for criteria.)

aperture. An opening through which radiation can pass.

attenuation. The decrease in the radiant flux as it passes through an absorbing or scattering medium.

back focal length. The distance from the last optical surface of a lens to the focal point.

beam. A collection of rays which may be parallel, divergent, or convergent.

beam diameter. The distance between diametrically opposed points in that cross section of a beam where the power per unit area is 1/e times that of the peak power per unit area.

beam divergence, ϕ. The full angle of the beam spread between diametrically opposed 1/e-irradiance points; usually measured in milliradians (one milliradian $\approx$ 3.4 minutes of arc).

beam expander. A combination of optical elements which will increase the diameter of a laser beam.

beam splitter. An optical device which uses controlled reflection to produce two beams from a single incident beam.

bolometer. A radiation detector of the thermal type in which absorbed radiation produces a measurable change in the physical property of the sensing element. The change in state is usually that of electrical resistance.

calorimeter. A device for measuring the total amount of energy absorbed from a source of electromagnetic radiation.

carcinogenic. Capable of causing cancer in a biological system.

coherent. A light beam is said to be coherent when the electric vector at any point in it is related to that at any other point by a definite, continuous sinusoidal function.

collimated beam. Effectively, a "parallel" beam of light with very low divergence or convergence. (See 8.1 for criteria.)

collimator. An optical device for converting a diverging or converging beam of light into a collimated or "parallel" one.

conjunctival discharge (of the eye). Increased secretion of mucus from the surface of the eyeball.

continuous wave (cw). The output of a laser which is operated in a continuous rather than pulsed mode. In this standard, a laser operating with a continuous output for a period greater than 0.25 second is regarded as a cw laser.

controlled area. An area where the occupancy and activity of those within is subject to control and supervision for the purpose of protection from radiation hazards.

cornea. The transparent outer coat of the human eye which covers the iris and the crystalline lens. It is the main refracting element of the eye.

cryogenics. The branch of physics dealing with very low temperatures.

cycloplegic. An agent which causes paralysis of the ciliary muscle of the eye, thereby preventing its accommodation.

depigmentation. The removal of the pigment of melanin granules from human tissues.

diffraction. Deviation of part of a beam, determined by the wave nature of radiation, and occurring when the radiation passes the edge of an opaque obstacle.

diffuse reflection. Change of the spatial distribution of a beam of radiation when it is reflected in many directions by a surface or by a medium.

diffusion. Change of the spatial distribution of a beam of radiation when it is deviated in many directions by a surface or by a medium.

edema. The swelling of tissues in the human body due to the presence of abnormal amounts of fluid within the extracellular spaces.

electromagnetic radiation. The flow of energy consisting of orthogonally vibrating electric and magnetic fields lying transverse to the direction of propagation. X-rays, ultraviolet, visible, infrared, and radio waves occupy various portions of the electromagnetic spectrum and differ only in frequency and wavelength.

energy, *Q*. The capacity for doing work. Energy content is commonly used to characterize the output from pulsed lasers, and is generally expressed in joules.

epithelium (of the cornea). The layer of cells forming the surface or skin of the cornea.

erythema. A redness of the skin due to congestion of the capillaries.

exposure. The product of an irradiance and its duration.

extended source. An extended source of radiation can be resolved by the eye into a geometrical image, in contrast to a point source of radiation, which cannot be resolved into a geometrical image. (See 8.1 for criteria.)

focal length. The distance from the secondary nodal point of a lens to the primary focal point. In a thin lens, the focal length is the distance between the lens and the focal point.

focal point. The point toward which radiation converges or from which radiation diverges or appears to diverge.

hertz (Hz). The unit which expresses the frequency of a periodic oscillation in cycles per second.

infrared radiation. Electromagnetic radiation with wavelengths which lie within the range 0.7 μm to 1 mm.

integrated radiance, *L*. The integral of the radiance over the exposure duration. Also known as pulsed radiance ($W \cdot s \cdot cm^{-2} \cdot sr^{-1}$).

intrabeam viewing. The viewing condition whereby the eye is exposed to all or part of a laser beam. (See Figs. B1 and B2 in Appendix B.)

iris. The circular pigmented membrane which lies behind the cornea of the human eye. The iris is perforated by the pupil.

irradiance (at a point of a surface), *E*. Quotient of the radiant flux incident on an element of the surface containing the point, by the area of that element. Unit: watt per square centimeter ($W \cdot cm^{-2}$).

joule (J). A unit of energy; one (1) joule = 1 watt · second.

lambertian surface. An ideal surface whose emitted or reflected radiance is independent of the viewing angle.

laser. A device which produces an intense, coherent, directional beam of light by stimulating electronic or molecular transitions to lower energy levels. Also, an acronym for *Light Amplification by Stimulated Emission of Radiation.*

laser safety officer. One who is knowledgeable in the evaluation and control of laser hazards and has authority for supervision of the control of laser hazards.

laser system. An assembly of electrical, mechanical, and optical components which includes a laser.

lasing medium. A material emitting coherent radiation by virtue of stimulated electronic or molecular transitions to lower energy levels.

limiting angular subtense (α_{min}). The apparent visual angle which divides intrabeam viewing from extended-source viewing. (See 8.1 for criteria.)

limiting aperture. The maximum circular area over which radiance and radiant exposure can be averaged.

lossy medium. A medium which absorbs or scatters radiation passing through it.

MPE. Maximum Permissible Exposure.

maser (optical). Another, older name for a laser, with *microwave* (of optical frequency) replacing *light* in laser.

meter (m). A unit of length in the international systems of units; currently defined as a fixed number of wavelengths, in vacuum, of the orange-red line of the spectrum of krypton 86. Typically, the meter is subdivided into the following units:

Centimeter = 10^{-2}m (cm)
Millimeter = 10^{-3}m (mm)
Micrometer = 10^{-6}m (μm)
Nanometer = 10^{-9}m (nm)

micrometer (μm). A unit of length equal to 10^{-6} meter.

mydriatic. An agent which is used to dilate the pupil of the human eye.

optical density, D_λ. Logarithm to the base ten of the reciprocal of the transmittance: $D_\lambda = -\log_{10} \tau_\lambda$, where τ is transmittance.

optically pumped laser. A laser in which the electrons are excited into an upper energy state by the absorption of light from an auxiliary light source.

P_{exempt}. That output power (Q_{exempt} that output energy per pulse) of a laser such that no applicable MPE for exposure of the eye (see 8.2 or 8.3) may be exceeded, with or without optical instruments. (See 3.2.3.)

photophobia. Aversion to light, usually caused by physical discomfort upon exposure to light.

photosensitizers. Substances which increase the sensitivity of a material to irradiation by electromagnetic radiation.

pigment epithelium (of the retina). The layer of cells which contain brown or black pigment granules next to and behind the rods and cones.

point source. A source of radiation whose dimensions are small enough compared with the distance between source and receptor for them to be neglected in calculations. (See 8.1 for criteria.)

power, Φ. The time rate at which energy is emitted, transferred, or received; usually expressed in watts (or in joules per second).

prf. Abbreviation for pulse-repetition frequency. High prf = more than 1 Hz.

pulse duration. The time duration of a laser pulse; usually measured as the time interval between the half-power points on the leading and trailing edges of the pulse.

pulsed laser. A laser which delivers its energy in the form of a single pulse or a train of pulses. In this standard, the duration of a pulse ≤ 0.25 s.

pupil. The variable aperture in the iris through which light travels toward the interior regions of the eye.

Q-switch. A device for producing very short (≈ 30 ns), intense laser pulses by enhancing the storage and dumping of electronic energy in and out of the lasing medium, respectively.

Q-switched laser. A laser which emits short (≈ 30 ns), high-power pulses by utilizing a Q-switch.

radian. A unit of angular measure equal to the angle subtended at the center of a circle by an arc whose length is equal to the radius of the circle. One (1) radian ≈ 57.3 degrees; 2π radians = 360 degrees.

radiance, L. Radiant flux or power output per unit solid angle per unit area ($W \cdot sr^{-1} \cdot cm^{-2}$).

radiant energy, Q. Energy emitted, transferred, or received in the form of radiation. Unit: joule (J).

radiant exposure, H. Surface density of the radiant energy received. Unit: joules per centimeter squared ($J \cdot cm^{-2}$).

radiant flux (Φ), radiant power. Power emitted, transferred, or received in the form of radiation. Unit: watt (W).

radiant intensity, I (of a source in a given direction). Quotient of the radiant flux leaving the source, propagated in an element of solid angle containing the given direction, by the element of solid angle. Unit: watt per steradian ($W \cdot sr^{-1}$).

Rayleigh scattering. Scattering of radiation in the course of its passage through a medium containing particles, the sizes of which are small compared with the wavelength of the radiation.

reflectance, reflectivity, ρ. The ratio of total reflected radiant power to total incident power.

reflection. Deviation of radiation following incidence on a surface.

repetitively pulsed laser. A laser with multiple pulses of radiant energy occurring in a sequence.

retina. That sensory membrane which receives the incident image formed by the cornea and lens of the human eye. The retina lines the inside portion of the eye.

scintillation. This term is used to describe the rapid changes in irradiance levels in a cross section of a laser beam.

shall. The word "shall" is understood to mean mandatory.

should. The word "should" is understood to mean that which is advisable.

solid angle, ω. The ratio of the area on the surface of a sphere to the square of the radius of that sphere. It is expressed in steradians.

source. The term source is taken to mean either laser or laser-illuminated reflecting surface.

specular reflection. A mirrorlike reflection.

steradian (sr). The unit of measure for a solid angle. There are 4π steradians in a sphere.

stromal haze (of the cornea). Cloudiness in the connective tissue or main body of the cornea.

surface exfoliation (of the cornea). A stripping or peeling off of the surface layer of cells from the cornea.

transmission. Passage of radiation through a medium.

transmittance, τ. The ratio of total transmitted radiant power to total incident radiant power.

ultraviolet radiation. Electromagnetic radiation with wavelengths smaller than those for visible radiation. For the purposes of this standard, 0.2–0.4 μm.

visible radiation (light). Electromagnetic radiation which can be detected by the human eye. It is commonly used to describe wavelengths which lie in the range between 0.4 μm and 0.7 μm.

att (W). The unit of power, or radiant flux.

avelength, λ. The distance between two points in a riodic wave which have the same phase is called a avelength.

. Hazard Evaluation and Classification

1 General. Three aspects of a laser application influ-ce the total hazard evaluation and thereby influence e application of control measures, namely:

(1) The laser or laser system's capability of injuring rsonnel

(2) The environment in which the laser is used

(3) The personnel who may use or be exposed to ser radiation

The laser classification scheme is based on aspect). Laser and laser systems classified in accordance ith this standard shall be labeled with the appropriate azard classification (see 3.3). This hazard classifica-on should be used unless the laser is modified to gnificantly change its output power or energy, or is closed.

Aspects (2) and (3) vary with each laser applica-on and cannot be readily standardized. The total azard evaluation procedure must consider all three pects, although in most cases only aspect (1) influ-ces the control measures which are applicable.

2 Laser Considerations. The laser hazard classifica-on is based upon laser-output parameters unless the ser is "enclosed." The individual responsible for laser assification must ensure that laser-output data are alid and in accord with Section 9, Measurements. Clas-fication shall be based on the maximum output which ay be expected during operation.

3.2.1 Multiwavelength Lasers. The classification of sers or laser systems which can potentially emit umerous wavelengths shall be based on the most azardous possible operation. (See Appendix B3, Ex-nples 12 and 13.)

3.2.1.1 A multiwavelength laser which by design n operate only as a single-wavelength laser shall be assified as if it were a single-wavelength laser.

3.2.1.2 A multiwavelength laser which by design n operate over two or more wavelength regions (as efined in Section 8) shall require classification in each gion of operation. The appropriate control measures r each region shall be required.

3.2.2 Repetitively Pulsed Lasers. The evaluation f lasers with pulse-repetition frequencies (prf) greater an 1 Hz requires the determination of both the con-nuous-wave (cw) and pulsed Maximum Permissible Exposure (MPE) levels as given in Section 8. (See Appendix B3, Examples 3 to 7.)

3.2.3 Radiometric Parameters of the Laser Required for Determining Laser Hazard Classification

3.2.3.1 Classification of essentially all lasers requires the following parameters:

(1) Wavelength(s) or wavelength range

(2) For cw or repetitively pulsed lasers: average power output and limiting exposure duration, τ_{max}, inherent in the design of the laser or laser system

(3) For pulsed lasers: total energy per pulse (or peak power), pulse duration, prf, and emergent beam radiant exposure

3.2.3.2 Classification of extended-source lasers or laser systems (such as laser arrays, injection laser diodes, and those lasers having a permanent diffuser within the output optics) requires, in addition to the parameters listed in 3.2.3.1, the laser source radiance or integrated radiance and the maximum viewing angular subtense.

3.2.3.3 Exempt laser output power (P_{exempt}) or exempt laser output energy (Q_{exempt}). To determine the laser potential for producing injury, it is necessary to consider not only the laser output irradiance or radiant exposure, but also whether a hazard would exist if the total laser output were concentrated within the limiting aperture for applicable MPE. (See Table 9.)

These terms are defined in two different ways, depending upon whether the laser is considered a point source or an "extended source" (an unusual case).

3.2.3.3.1 Most lasers can be considered point sources. For such, P_{exempt} and Q_{exempt} are each the product of two factors, namely:

(1) The intrabeam MPE for the eye (see Figs. 4 and 5) for the limiting exposure time

(2) The area of the limiting aperture for the MPE (see Table 9) in cm^2

3.2.3.3.2 For extended-source lasers or laser systems which emit in the spectral range 0.4 to 1.4 μm, P_{exempt} and Q_{exempt} are determined by that power or energy output such that:

(1) The source radiance does not exceed the MPE (see Table 6) if the source were viewed at the minimum viewing distance

(2) A theoretically perfect optical viewing system (80 mm limiting entrance aperture, 7 mm exit aperture) collected the entire laser beam output

If this definition is difficult to apply, the definition in 3.2.3.3.1 may be applied and will result in a conservative P_{exempt} or Q_{exempt}.

3.3 Laser and Laser System Hazard Classification Definitions. Tables 1 and 2 offer a summary of levels for laser and laser systems classification; Table 1 for con-

Table 1
Summary of Levels of Power Emissions for Continuous-Wave* Laser and Laser System Classification

Wavelength Range (μm)	MPE Limiting Aperture (mm)	Emission Duration (s)	Class I†	Class II	Class III	Class IV
Ultraviolet 0.2–0.4	1	3×10^4	$\leq 0.8 \times 10^{-9}$ W to $\leq 8 \times 10^{-6}$ W depending on wavelength (see Table 5)	–	> Class I but $\leq$ 0.5 W depending on wavelength (see Table 5)	> 0.5 W
Visible 0.4–0.7	7	3×10^4	$\leq 0.4 \times 10^{-6}$ W	> Class I but $\leq 1 \times 10^{-3}$ W	> Class II but $\leq$ 0.5 W††	> 0.5 W
Near-Infrared 0.7–1.06	7	3×10^4	$\leq 0.4 \times 10^{-6}$ W to $\leq 200 \times 10^{-6}$ W depending on wavelength (see Fig. 7)	–	> Class I but $\leq$ 0.5 W depending on wavelength (see Fig. 7)	> 0.5 W
1.06–1.4	7	3×10^4	$\leq 200 \times 10^{-6}$ W	–	> Class I but $\leq$ 0.5 W	> 0.5 W
Far-Infrared 1.4–10^2	1	> 10	$\leq 0.8 \times 10^{-3}$ W	–	> Class I but $\leq$ 0.5 W	> 0.5 W
Submillimeter 10^2–10^3	11	> 10	$\leq$ 0.1 W	–	> Class I but $\leq$ 0.5 W	> 0.5 W

*Emission duration $\geq$ 0.25 second.

†When the design of the laser or laser system assures personnel exposures of less than 10^4 seconds in any 24-hour period, the limiting exposure duration may establish a higher exempt power level (P_{exempt}) as discussed in 3.3.3.

††For 1–5 mW cw laser systems see 4.2.3.10 and 4.6.2.

tinuous-wave lasers with an emission duration > 0.25 second, and Table 2 for pulsed lasers with an emission duration $\leq$ 0.25 second. (See Appendix A, Examples of Classification of Lasers or Laser Systems.)

3.3.1 Class I – Exempt Lasers and Laser Systems. Any laser, or laser system containing such a laser, that cannot emit laser radiation levels in excess of P_{exempt} or Q_{exempt} for the maximum possible duration inherent in the design of the laser or laser system. The exemption strictly applies to emitted laser radiation hazards and not to other potential hazards (see Section 7, Special Considerations).

3.3.2 Class II – Low-Power Visible Lasers and Laser Systems

3.3.2.1 Visible (0.4–0.7 μm) cw lasers or laser systems which can emit a power exceeding P_{exempt} for the maximum possible duration inherent in the design of the laser or laser system (0.4 μW for emission duration greater than 3×10^4 s), but not exceeding 1 mW.

3.3.2.2 Visible (0.4–0.7 μm) repetitively pulsed lasers or laser systems which can emit a power exceeding the appropriate P_{exempt} for the maximum possible duration inherent in the design of the laser or laser system but not exceeding P_{exempt} for a 0.25-second exposure.

3.3.3 Class III – Medium-Power Lasers and Laser Systems

3.3.3.1 Infrared (1.4 μm–1 mm) and ultraviolet (0.2–0.4 μm) lasers and laser systems which can emit a radiant power in excess of P_{exempt} for the maximum possible duration inherent in the design of the laser or laser system, but cannot emit: 1) an average radiant power in excess of 0.5 W for τ_{max} > 0.25 second; or 2) a radiant exposure of 10 J · cm^{-2} within an exposure time $\leq$ 0.25 second.

3.3.3.2 Visible (0.4–0.7 μm) cw or repetitively pulsed lasers or laser systems producing a radiant power in excess of P_{exempt} for a 0.25-second exposure (1 mW for a cw laser), but cannot emit an average radiant power greater than 0.5 W. For cw visible (0.4–0.7 μm) lasers having a power output of from 1 to 5 mW, see 4.2.3.10 and 4.6.2.

3.3.3.3 Visible and near-infrared (0.4–1.4 μm) single-pulsed lasers which can emit a radiant energy in excess of Q_{exempt}, but which cannot emit a radiant exposure that exceeds 10 J · cm^{-2} or that required to produce a hazardous diffuse reflection as given in Table 3.

3.3.3.4 Near-infrared (0.7–1.4 μm) cw lasers or single repetitively pulsed lasers which can emit power in excess of P_{exempt} for the maximum duration in-

Table 2
Summary of Levels (Energy and Radiant Exposure Emissions) for Single-Pulsed Laser and Laser System Classification*

Wavelength Range (μm)	MPE Limiting Aperture (mm)	Emission Duration (s)	Class I	Class III	Class IV
Ultraviolet† 0.2–0.4	1	$> 10^{-2}$	$\leq 24 \times 10^{-6}$ J to 7.9×10^{-3} J	> Class I but ≤ 10 J · cm^{-2}	> 10 J · cm^{-2}
Visible 0.4–0.7	7	10^{-9} to 0.25	$\leq 0.2 \times 10^{-6}$ J	> Class I but $\leq 31 \times 10^{-3}$ J · cm^{-2}	$> 31 \times 10^{-3}$ J · cm^{-2}
			$\leq 0.25 \times 10^{-3}$ J	> Class I but ≤ 10 J · cm^{-2}	> 10 J · cm^{-2}
Near-Infrared†† 0.7–1.06	7	10^{-9} to 0.25	$\leq 0.2 \times 10^{-6}$ to 2×10^{-6} J	> Class I but $\leq 31 \times 10^{-3}$ J · cm^{-2}	$> 31 \times 10^{-3}$ J · cm^{-2}
			$\leq 0.25 \times 10^{-3}$ to 1.25×10^{-3} J	> Class I but ≤ 10 J · cm^{-2}	> 10 J · cm^{-2}
1.06–1.4	7	10^{-9} to 0.25	$\leq 2 \times 10^{-6}$ J	> Class I but $\leq 31 \times 10^{-3}$ J · cm^{-2}	$> 31 \times 10^{-3}$ J · cm^{-2}
			$\leq 1.25 \times 10^{-3}$ J	> Class I but ≤ 10 J · cm^{-2}	> 10 J · cm^{-2}
Far-Infrared 1.4–10^2	1	10^{-9} to 0.25	$\leq 80 \times 10^{-6}$ J	> Class I but ≤ 10 J · cm^{-2}	> 10 J · cm^{-2}
			$\leq 3.2 \times 10^{-3}$ J	> Class I but ≤ 10 J · cm^{-2}	> 10 J · cm^{-2}
Submillimeter 10^2–10^3	11	10^{-9} to 0.25	$\leq 10 \times 10^{-3}$ J	> Class I but ≤ 10 J · cm^{-2}	> 10 J · cm^{-2}
			≤ 0.4 J	> Class I but ≤ 10 J · cm^{-2}	> 10 J · cm^{-2}

*There are no Class II single-pulsed lasers.

†Wavelength dependent (see Table 5).

††Diffuse reflection criteria (Table 3) apply from 10^{-9} to 33×10^{-3}s for Class III. For $> 33 \times 10^{-3}$s exposure, the maximum radiant exposure is 10 J · cm^{-2}. Class I and III values are wavelength dependent (see Fig. 7).

Table 3
Maximum Radiant Exposure Incident upon a Diffuse Surface Which Will Not Produce Hazardous Reflections

Exposure Duration (s)	Maximum Radiant Exposure ($J \cdot cm^{-2}$) Visible (400–700 nm)	Near-Infrared† (700–1060 nm)	Near-Infrared (1060–1400 nm)
10^{-9}	3.1×10^{-2}	$CF_{8.5.2}(3.1 \times 10^{-2})$	1.5×10^{-1}
10^{-8}	6.8×10^{-2}	$CF_{8.5.2}(6.8 \times 10^{-2})$	3.1×10^{-1}
10^{-7}	1.5×10^{-1}	$CF_{8.5.2}(1.5 \times 10^{-1})$	8.0×10^{-1}
10^{-6}	3.1×10^{-1}	$CF_{8.5.2}(3.1 \times 10^{-1})$	1.5
10^{-5}	6.8×10^{-1}	$CF_{8.5.2}(6.8 \times 10^{-1})$	3.1
10^{-4}	1.5	$CF_{8.5.2}(1.5)$	8.0
10^{-3}	3.1	$CF_{8.5.2}(3.1)$*	15*
10^{-2}	6.8	$CF_{8.5.2}(6.8)$*	31*
10^{-1}	15*	$CF_{8.5.2}(15)$*	80*
0.25	20*	$CF_{8.5.2}(20)$*	100*
General expression for duration t	$10\pi t^{1/3}$	$10\pi CF_{8.5.2}t^{1/3}$	$50\pi CF_{8.5.2}t^{1/3}$

*Values for classification are limited to 10 $J \cdot cm^{-2}$ (see 3.3.3.3 and 3.3.4.2).

†Values for $CF_{8.5.2}$ will be found in 8.5.2.

herent in the design of the laser or laser system, but cannot emit an average power of 0.5 W or greater for periods in excess of 0.25 second.

3.3.4 Class IV – High-Power Lasers and Laser Systems

3.3.4.1 Ultraviolet (0.2–0.4 μm) and infrared (1.4 μm–1 mm) lasers and laser systems which emit: 1) an average power in excess of 0.5 W for periods greater than 0.25 second; or 2) a radiant exposure of 10 $J \cdot cm^{-2}$ within an exposure duration of 0.25 second or less.

3.3.4.2 Visible (0.4–0.7) μm) and near-infrared (0.7–1.4 μm) lasers and laser systems which emit: 1) an average power of 0.5 W or greater for periods greater than 0.25 second; or 2) a radiant exposure in excess of 10 $J \cdot cm^{-2}$, or that required to produce a hazardous diffuse reflection as given in Table 3, for periods less than 0.25 second.

3.3.5 Class V – Enclosed Lasers and Laser Systems. Any Class II, III, or IV laser or laser system which, by virtue of appropriate design or engineering controls, cannot directly irradiate the eye with levels which are in excess of P_{exempt} or Q_{exempt}.

3.4 Environment in Which the Laser is Used. Following laser or laser system classification, environmental factors require consideration. Their importance in the total hazard evaluation depends upon the laser classification. The decision to employ additional hazard controls not specifically required in Section 4 of this standard is influenced by environmental considerations principally for Class III and Class IV lasers or laser systems.

The probability of personnel exposure to hazardous laser radiation must be considered and is influenced by whether the laser is used indoors (for example, in a classroom, in a machine shop, in a closed research laboratory, or on a factory production line) or outdoors (for example, in a mining tunnel, a highway construction site, a military laser range, in the atmosphere above occupied areas, in a pipe-line construction trench, or in outer space). Other environmental hazard (see Section 7) must be considered. If exposure of unprotected personnel to the primary or specularly reflected beam is possible, calculations or measurements of either irradiance or radiant exposure of the primary or specularly reflected beam at that specific location, or radiance of an extended source, are required. These are discussed in Appendix B.

Viewing the main beam or a specular laser target th an optical instrument is potentially hazardous due the viewing instrument's light gathering capability. e 4.2.3.8 and Appendix B4, Examples 21 and 22.)

3.4.1 Indoor Laser Operations. In general, only the er is considered in evaluating an indoor laser opera- n if the beam is enclosed or is operated in a con- lled area. The step-by-step procedure described in .1.1 through 3.4.1.4 is recommended for evaluating ss III lasers indoors when this is necessary (due to tential exposure of unprotected personnel).

3.4.1.1 Step 1. Evaluate permanence of laser am path(s). If laser is not mounted in a fixed posi- n, determine the hazardous beam path(s).

3.4.1.2 Step 2. Determine extent of hazardous cular reflection (as from lens' surfaces).

3.4.1.3 Step 3. Determine the extent of hazard- s diffuse reflections if the emergent laser beam is cused. Hazardous diffuse reflections are possible m a focused beam of a Class III laser. However, the gular subtense of the source is normally sufficiently all at all practical viewing distances so that intra- am MPEs apply.

3.4.1.4 Step 4. Determine if other (nonlaser) zards exist (see Section 7, Special Considerations).

3.4.2 Outdoor Laser Operations Over Extended stances. The total hazard evaluation of a particular er system depends on defining the extent of several tentially hazardous conditions. This may be done in tep-by-step manner as given in 3.4.2.1 through .2.5.

3.4.2.1 Step 1. Estimate the nominal hazardous ge of the laser. Calculations of radiant exposure or am irradiance as a function of range can be made th the range equation for a circular beam. (See pendix B.)

These calculated ranges are only estimates beyond ew hundred meters since uncertainties arise from nospheric effects (for example, scintillation due to bulence). (See 7.5.3.)

3.4.2.2 Step 2. Evaluate potential hazards from cular reflections. Specular surfaces ordinarily en- untered (for example, windows and mirrors in vehi- s and windows in buildings) are oriented vertically d will usually reflect a horizontal beam in a hori- ntal plane.

As much as 8% of the beam's original irradiance radiant exposure can be reflected toward the laser m a clear glass window which is oriented perpen- cular to the beam. If the beam strikes a flat, specu- surface at an angle, a much greater percentage the beam can be reflected beyond or to the side of target area. If the beam strikes a still pond or other nilar surface at a low angle of incidence, effective reflectivity may also be high. Specular reflective surfaces, such as raindrops, wet leaves, and most other shiny natural objects, seldom reflect hazardous intensities beyond a meter from these reflectors.

3.4.2.3 Step 3. Determine whether hazardous diffuse reflections exist. (See Table 3 and Appendix B3, Examples 8 and 9.)

3.4.2.4 Step 4. Evaluate the stability of the laser platform to determine the extent of lateral range control and the lateral constraints that should be placed upon the beam traverse.

3.4.2.5 Step 5. Consider the likelihood of people being present in the area of the laser beam.

3.5 Personnel Who May Be Exposed. The personnel who may be in the vicinity of a laser and its emitted beam(s) and the operator can influence the total hazard evaluation and hence influence the decision to adopt additional control measures not specifically required for the class of laser being employed. This depends upon the classification of the laser or laser system.

3.5.1 If children or others unable to read or understand warning labels may be exposed to potentially hazardous laser radiation, the evaluation of the hazard is affected and control measures may require appropriate modification.

3.5.2 The type of personnel influences the total hazard evaluation. It must be kept in mind that for certain lasers or laser systems (for example, military laser rangefinders and some Class III lasers used in the construction industry), the principal hazard control rests with the operator; that is, his responsibility not to aim the laser at personnel or flat mirrorlike surfaces.

The following are considerations regarding operating personnel and those who may be exposed:

(1) Maturity of judgment of the laser user(s)

(2) General level of training and experience of laser user(s), (that is, whether high school students, soldiers, scientists, etc)

(3) Awareness of onlookers that potentially hazardous laser radiation may be present, and relevant safety precautions

(4) Degree of training in laser safety of all individuals involved in the laser operation

(5) Reliability of individuals to follow control procedures

(6) Number and location of individuals relative to the primary beam or reflections, and potential for accidental exposure

(7) Other hazards not due to laser radiation which may cause the individuals to react unexpectedly or which influence the choice of personnel protective equipment

NOTE: Examples of typical lasers classified in accordance with this standard are given in Appendix A. Examples of calculations which may be useful in applying this standard are given in Appendix B.

4. Control Measures

4.1 General Considerations. Control measures shall be devised to reduce the possibility of exposure of the eye and skin to hazardous laser radiation, and to other hazards associated with operation of laser devices (see Section 7, Special Considerations).

For all uses of lasers and laser systems it is recommended that the minimum radiation level be used for the required application.

The control measures outlined herein shall not be considered to restrict or limit in any way the use of laser radiation of any type which may be intentionally administered to an individual for diagnostic, therapeutic, or medical research purposes, by or under the direction of those qualified professionals engaged in the healing arts. However, those administering the laser radiation shall be protected by the control measures as outlined herein.

The Laser Safety Officer has the responsibility for the establishment and surveillance of appropriate safety measures.

4.2 Control Measures for Laser Classifications

4.2.1 Class I – Exempt Lasers and Laser Systems. No control measures or warning label shall be required, although any needless direct exposure of the eye should be avoided as a matter of good practice.

4.2.2 Class II – Low-Power Visible Lasers and Laser Systems. The control measures specified in 4.2.2.1 shall apply.

4.2.2.1 Warning Label. Lasers shall have an appropriate warning label (see 4.6) affixed to a conspicuous place on the laser housing or control panel, or on both the laser housing and control panel.

4.2.3 Class III – Medium-Power Lasers and Laser Systems. The control measures specified in 4.2.3.1 through 4.2.3.11 shall apply.

4.2.3.1 Education and Training. All persons using a laser or laser system shall be duly informed about the potential hazards of laser operations.

4.2.3.2 Engineering Controls. Priority shall be given to the incorporation of appropriate safety mechanisms as an integral part of the laser or laser system. Examples include beam stops, beam enlarging systems, enclosures, shutters, interlocks, etc.

4.2.3.3 Laser Controlled Area. Consideration should be given to operation of laser devices in a controlled area. Special emphasis shall be placed on control of the path of the laser beam.

4.2.3.4 Laser Operators. Only authorized personnel shall operate laser systems.

4.2.3.5 Spectators. Spectators shall not be permitted into the laser controlled area (see 4.2.3.3) unless appropriate supervisory approval has been obtained and protective measures taken.

4.2.3.6 Beam Enclosures. The emitted laser radiation should be contained within enclosures whenever practicable. Laser beams emitted by nonenclosed systems should be terminated at the end of the useful beam path when the exposure level is greater than the MPE for direct irradiation of the eye (see 8.2) and a possibility of human exposure exists.

4.2.3.7 Alignment Procedures. Alignment of laser optical systems (mirrors, lenses, beam deflectors, etc) shall be performed in such a manner that the primary beam, or a specular reflection of the primary beam, does not expose the eye to a level above the MPE for direct irradiation of the eye (see Section 8).

4.2.3.8 Optical Viewing Aids. Optical systems such as lenses, telescopes, microscopes, etc, may increase the hazard to the eye when viewing a laser beam; therefore special care shall be taken in their use. (See Appendix B4, Examples 21 and 22.)

Microscopes and telescopes may be used as optical instruments for viewing, but shall be provided with an interlock or filter, if necessary, to prevent ocular exposures above the appropriate MPE for irradiation of the eye.

4.2.3.9 Eye Protection. Eye-protection devices which are specifically designed for protection against radiation from the laser system should be used when engineering and procedural controls are inadequate to eliminate potential exposure in excess of the applicable MPE (see 4.5).

For cases in which long-term intentional exposure to the eye by visible lasers (only) is not intended, the applicable MPE may be based upon a 0.25-second exposure (see 8.2).

4.2.3.10 Equipment Labeling. Lasers shall have warning labels (see 4.6 and 4.7) with the appropriate cautionary statement affixed to a conspicuous place on the laser housing or control panel, or on both the housing and control panel.

For the case of visible (0.4–0.7 μm) cw lasers whose output power is in the range 1 to 5 mW and whose beam irradiance is $<$ 2.5 mW per square centimeter, all other provisions of 4.2.3 may be considered advisory rather than mandatory, with the exception that such cw laser systems shall be labeled in the manner given in Fig. 1, but shall contain the following additional precautionary language on the label: "DO NOT VIEW

LASER BEAM WITH OPTICAL INSTRUMENTS." (See 4.6.2.)

4.2.3.11 Diffusely Reflecting Materials. In addition to beam stops, shields, and enclosures, materials which will diffusely reflect any stray or incidental laser beams should be used in laser areas whenever possible.

4.2.4 Class IV – High-Power Lasers and Laser Systems. The probability of injury and the extent of injury increases with increasing laser output power. High-power lasers require more rigid control measures, not only because there is a greater likelihood that specular reflections will have sufficient power to cause injury, but because of the greater risk of injury from hazardous diffuse reflections. The entire beam path capable of producing hazardous diffuse reflections shall be controlled. Controls shall rely primarily on more positive methods, such as enclosures and interlocks, and secondarily upon procedural safeguards.

In addition to the control measures outlined in 4.2.3.1, 4.2.3.2, 4.2.3.4, 4.2.3.5, 4.2.3.7, 4.2.3.8, 4.2.3.10, and 4.2.3.11, the control measures specified in 4.2.4.1 through 4.2.4.6 shall be applied.

4.2.4.1 Laser Controlled Area. Laser devices shall be isolated in an area solely designed for laser operations. Access to such an area shall require appropriate authorization.

Under conditions where the entire beam path is not enclosed, as described in 4.2.4.2, safety latches or interlocks shall be used to prevent unexpected entry into laser controlled areas. Such measures shall be designed to allow both rapid egress by the laser personnel at all times and admittance to the laser controlled area in an emergency condition. For such emergency conditions, a "panic button" (control-disconnect switch or equivalent device) shall be available for deactivating the laser.

During tests requiring continuous operation, the person in charge of the controlled area shall be permitted to momentarily override the safety interlocks to allow access of other authorized personnel if it is clearly evident that there is no radiation hazard at the point of entry, and if the necessary protective devices are worn by the entering personnel.

Should removal of the protective covers or the overriding interlocks become necessary for special training, service adjustments, or maintenance procedures, a temporary laser controlled area shall be devised following specific procedures approved by the Laser Safety Officer which will outline all safety requirements during the service and maintenance procedures. Such a temporary laser controlled area, which by its nature will not have the built-in protective features, as defined above, for a laser controlled area, shall nevertheless provide for all safety requirements for all personnel, both within and without the temporary laser controlled area during the service or maintenance procedure.

Under conditions where the entire beam path is not completely enclosed, access to the laser controlled area shall be limited to persons wearing proper laser protective eyewear when the laser is capable of emission (see 4.2.4.2 and 4.5). In this case all other optical paths (for example, windows) from the facility shall be covered or restricted in such a material way as to reduce the transmitted intensity of the laser radiation to levels at or below the MPE for direct irradiation of the eye (see Section 8). Specularly reflecting surfaces which are not required when using the laser shall be removed from the beam path.

When the laser beam must exit the controlled area (as in the case of exterior atmospheric beam paths, for example, Lidar), the operator shall be responsible for assuring that the beam path is limited to controlled air space (see 7.5.1) or ground space when the beam irradiance is above the appropriate MPE.

4.2.4.2 Enclosed Beam Path. Whenever possible, the entire beam path, including the interaction area, that is, the area in which irradiation of materials by the primary or secondary beams occurs, should be enclosed. Enclosures should be equipped with interlocks so that the laser system will not operate unless such enclosures are properly installed. For pulsed systems, interlocks shall be designed so as to prevent firing of the laser, by dumping the stored energy into a dummy load. For cw lasers, the interlocks shall turn off the power supply or interrupt the beam by means of shutters. Interlocks shall not allow automatic reenergizing of the power supply, but shall be designed so that after tripping the interlock, the power supply or shutter must be reset manually.

4.2.4.3 Eye Protection. Eye-protection devices (see 4.5) which are designed for protection against radiation from a specific laser system shall be used when engineering and procedural controls are inadequate to eliminate potential exposure in excess of the applicable MPE.

For cases in which long-term intentional exposures to the eye by visible lasers (only) are not intended, the applicable MPE may be based upon a 0.25-second exposure (see 8.2).

4.2.4.4 Remote Firing and Monitoring. Whenever possible, the laser system should be fired and monitored from remote positions.

4.2.4.5 Warning Systems. An alarm system, for example, an audible sound, or a warning light visible through protective eyewear, or a verbal "countdown" command, should be used prior to laser activation.

The audible system may consist of a bell or chime which commences when a pulsed-laser power supply is charged for operation, for example, during the charging of capacitor banks. Systems should be used in which a warning will sound intermittently during the charging procedure (pulsed systems) and continuously when fully charged.

4.2.4.6 Key-Switch Master Interlock. Any laser or laser system designated as Class IV shall be provided with an operative keyed master interlock or switching device. The key shall be removable and the device shall not be operable when the key is removed.

4.2.5 Class V–Enclosed Lasers and Laser Systems. In order to assure the fail-safe nature of an enclosed laser system, the criteria specified in 4.2.5.1 through 4.2.5.4 shall be required. Compliance with these criteria shall result in an exemption from "Laser Safety Program" and "Medical Surveillance" requirements.

4.2.5.1 Protective Housing. The laser or laser system shall have a protective housing that prevents emission of radiation at levels above the MPE for either direct exposure of the eye (see 8.2) or extended-source viewing for exposure times of 3×10^4 seconds (see 8.3). The protective housing is considered to prevent human access during normal operation. No special control measures shall be required for personnel outside of the protective housing.

Personnel who require access to the protective housing for the purpose of maintenance shall comply with the control measures specified for the class laser or laser system contained therein.

4.2.5.2 Safety Interlocks. Each laser or laser system shall be provided with safety interlocks for any portion of the protective housing which, when removed or displaced, allows human access to radiation in excess of the applicable MPE limits.

A minimum of two operative safety interlocks, one of which must be concealed, shall be provided for any portion of the protective housing which, by design, can be removed or displaced during normal operation and thereby allow access to radiation in excess of the applicable MPE limits. Interruption of these safety interlocks shall insure that the beam cannot leave the laser above the MPE.

Failure of any single mechanical or electrical component in the redundant interlock system shall not prevent the total interlock system from functioning.

Service adjustments or maintenance procedures on the laser contained within the enclosure shall not cause safety interlocks to become inoperative or the radiation levels outside of the enclosure to exceed the MPE limits unless performed in a temporary controlled area as defined in 4.2.4.1.

4.2.5.3 Fail-Safe System. Adjustments or failure of any part of the laser, or the enclosure containing a laser or laser system, shall not foreseeably cause the laser or laser system to be in noncompliance with the requirements of an enclosed-laser operation.

4.2.5.4 Viewing Windows. In accordance with the requirements of 4.2.5.1, all viewing portals incorporated into an enclosed laser or laser system shall incorporate a suitable filter material which attenuates the laser radiation transmitted through the portal to levels below the specified MPE levels under any conditions of operation of the laser or laser system.

4.3 Infrared Laser Systems – Special Control Measures. Since infrared radiation is invisible, particular care must be taken when using infrared laser systems. Thus, in addition to the control measures which apply to the laser hazard classifications, the requirements given in 4.3.1 through 4.3.3 shall also apply.

4.3.1 Termination of the Beam. The beam from a Class III laser or laser system should be terminated by a highly absorbent backstop wherever practicable.

NOTE: Many surfaces which appear "dull" visually can act as reflectors of infrared radiation.

The beam from a Class IV laser or laser system should be terminated in a fire-resistant material wherever practicable. Periodic inspection of the absorbent material is required since many materials degrade with use.

4.3.2 Control of Beam Path. Areas which are exposed to reflections from Class III lasers or laser systems shall be protected by appropriately screening the beam or target area with infrared-absorbent material.

In the case of beams from Class IV lasers or laser systems the screening material shall be fire-resistant.

4.3.3 Personal Protection. Personal protective equipment shall be used whenever the radiation exposure exceeds the applicable MPE levels.

4.4 Ultraviolet Laser Systems – Special Control Measures. Since ultraviolet radiation is invisible, particular care must be taken when using ultraviolet systems. Thus, in addition to the control measures which apply to the laser hazard classifications, the requirements given in 4.4.1 and 4.4.2 shall also apply.

4.4.1 Beam Shields. Exposure to ultraviolet radiation shall be minimized by using shield material which attenuates the radiation to levels below the MPE for the specific ultraviolet wavelength.

4.4.2 Hazardous By-Products. Special attention shal be given to the possibility of producing undesirable reactions in the presence of ultraviolet radiation, for example, formation of skin sensitizing agents, ozone, etc (see 7.3).

4.5 Laser Protective Eyewear. Protective eyewear shall be worn whenever hazardous conditions may result

Table 4
Calculation of Minimum Optical Densities Required for Protective Eyewear

H_o/MPE	D_λ
$1 = 10^0$	0
$10 = 10^1$	1
$100 = 10^2$	2
$1\,000 = 10^3$	3
$10\,000 = 10^4$	4
$100\,000 = 10^5$	5
$1\,000\,000 = 10^6$	6

from laser radiation or laser related operations (flash tubes, chemicals, etc). American National Standard Practice for Occupational and Educational Eye and Face Protection, Z87.1-1968, should be consulted for guidance.

The following factors shall be considered in determining the appropriate protective eyewear:

(1) Wavelength of laser output
(2) Radiant exposure or irradiance
(3) Maximum permissible exposure (MPE) (see Section 8, Criteria for Exposure of the Eye and Skin)
(4) Optical density of eyewear at laser output wavelength
(5) Visible light transmission requirement
(6) Radiant exposure or irradiance at which laser safety eyewear damage occurs
(7) Need for prescription glasses
(8) Comfort
(9) Degradation of absorbing media
(10) Strength of materials (resistance to shock)
(11) Need for peripheral vision

4.5.1 Specification of Optical Density, D_λ. The attenuation, D_λ, of laser protective eyewear at a specific wavelength shall be specified. Many lasers radiate at more than one wavelength; thus eyewear designed to have an adequate D_λ for a particular wavelength could have a completely inadequate D_λ at another wavelength radiated by the same laser. This problem may become particularly serious with lasers that are tunable over broad frequency bands.

If the actual eye exposure is given by H_o, then the optical density, D_λ, required of protective eyewear to reduce this exposure to the MPE is given by

$$D_\lambda = \log_{10} \frac{H_o}{\text{MPE}}$$

where the units of H_o are the same as those of the appropriate MPE. (See Section 8, Criteria for Exposure of the Eye and Skin.)

Table 4 provides values for minimum optical density for given values of H_o/MPE. It should be noted that optical densities greater than 3 or 4 (depending on exposure time) could reduce eye exposures below the MPEs in 8.2 and 8.3 but leave the unprotected skin surrounding the eyewear exposed to values in excess of the MPEs in 8.4.

Attenuation through the protective material shall be determined in all anticipated viewing angles and wavelengths.

4.5.2 Visible Transmission. Adequate optical density, D_λ, at the laser wavelengths of interest shall be weighed with the need for adequate visible transmission.

4.5.3 Identification of Eyewear. All laser protective eyewear shall be clearly labeled with optical density values and wavelengths for which protection is afforded. (See 4.5.1.)

4.5.4 Comfort and Fit. Protective eyewear shall provide a comfortable and snug fit so that laser radiation is satisfactorily attenuated before reaching the viewer's eyes.

4.5.5 Inspection. Periodic inspections shall be made of protective eyewear to ensure the maintenance of satisfactory conditions. This shall include: 1) inspection of the attenuator material for pitting, crazing, cracking, etc; and 2) inspection of the goggle frame for mechanical integrity and light leaks.

4.5.6 Responsibility of Manufacturer of Laser Safety Protective Eyewear. Manufacturers of laser safety protective eyewear shall provide the following information with each item:

(1) Wavelength(s) and corresponding optical density where protection is afforded

(2) Pertinent data (obtained by the manufacturer during his evaluation of the protective eyewear) for laser safety purposes

4.6 Warning Signs and Labels

4.6.1 Symbol. The laser hazard symbol shall be represented by a sunburst pattern, consisting of two sets of radial spokes of different length and one longer spoke radiating from a common center.

The color, dimensions, and location of the symbol within the sign shall be as specified in American National Standard Specifications for Accident Prevention Signs, Z35.1-1972.

4.6.2 Signal Words. The signal word "CAUTION" shall be used with all signs and labels associated with Class II—low-powered lasers and Class III—cw visible (0.4–0.7 μm) lasers having an output power between 1 and 5 mW and a maximum emergent beam irradiance

Fig. 1
Sample Warning Sign for Class II Laser

of 2.5 mW per square centimeter as measured over a 7-mm aperture. The signal word "DANGER" shall be used with all signs and labels associated with Class III lasers other than those defined above, and Class IV–high-powered lasers.

4.6.3 Inclusion of Pertinent Information. Signs and labels shall conform to the specifications of 4.6.3.1 and 4.6.3.2.

4.6.3.1 The appropriate signal word shall be written in the upper panel.

4.6.3.2 Appropriate space will be left on all signs and labels to allow the inclusion of pertinent information. Such information may be included during the printing of the sign or label or may be handwritten in a legible manner to include the following information:

(1) Above the tail on the sunburst: type of laser (Pulsed Ruby, CW Helium-Neon, etc)

(2) Below the tail on the sunburst: special precautionary instructions or protective actions required by the reader

4.6.4 Display of Signs and Labels. All signs and labels shall be conspicuously displayed in locations which serve to warn onlookers.

4.6.5 Design of Signs. The dimensions of the sign, letter sizes, colors, etc, shall be in accordance with American National Standard Specifications for Accident Prevention Signs, Z35.1-1972. Figs. 1 and 2 show sample signs for Class II and Class IV lasers or laser systems.

4.6.6 Class V – Warning Signs and Labels for Enclosed Lasers and Laser Systems. Warning labels shall not be required on the exterior protected enclosure of a Class V enclosed laser or laser system. The appropriate warning signs or label which designate the class

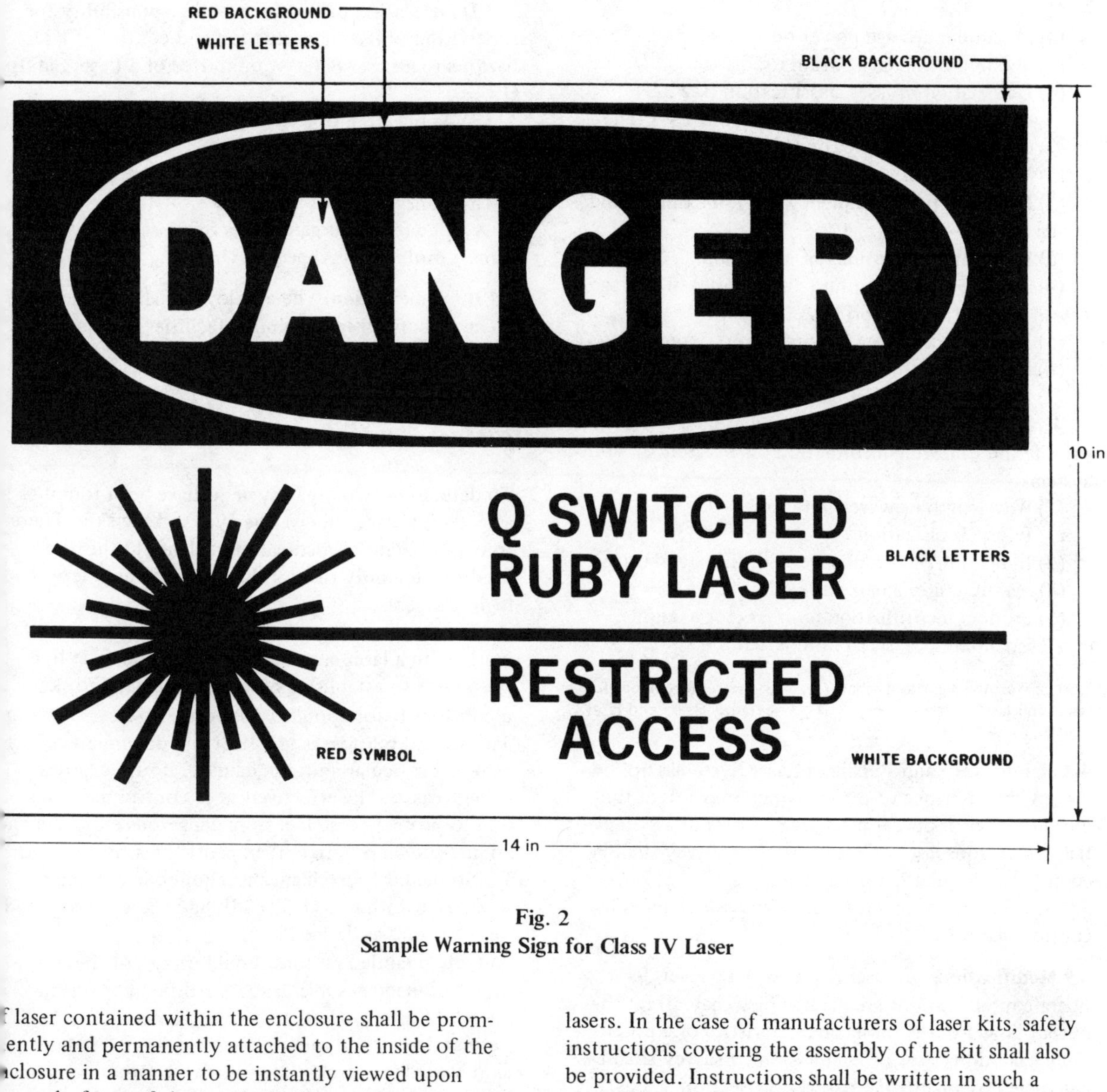

Fig. 2
Sample Warning Sign for Class IV Laser

f laser contained within the enclosure shall be prom-ently and permanently attached to the inside of the closure in a manner to be instantly viewed upon moval of any of the interlocked access panels or re-ovable covers which allow exposure to the laser beam.

7 Responsibility of Laser Equipment Manufacturers

4.7.1 Equipment Design. The manufacturers of laser vices shall provide suitable safeguards on the laser uipment which comply with the principles and actice of industrial safety and the protection of the er of the equipment. The design shall keep exposure laser radiation as low as practicable.

4.7.2 Safety Instructions. Manufacturers of laser uipment shall prepare adequate safety instructions vering the use and operator maintenance of their uipment. This includes Class I, II, III, IV, and V lasers. In the case of manufacturers of laser kits, safety instructions covering the assembly of the kit shall also be provided. Instructions shall be written in such a manner that they can be readily understood by persons minimally informed in laser technology.

The safety instructions shall accompany the sale or shipment, or both the sale and shipment, of all laser equipment.

4.7.3 Equipment Characteristics. Manufacturers of laser equipment shall provide information on the equipment's characteristics. The minimum information to be included with the sale or shipment of laser equip-ment shall be as specified in 4.7.3.1 or 4.7.3.2.

4.7.3.1 Continuous-Wave (CW) Laser. The following minimum information shall be provided:

(1) Wavelength or wavelength range

(2) Emergent beam size

(3) Beam divergence
(4) Maximum average power output
(5) Maximum emergent beam irradiance
(6) Designation of laser classification (see 3.3)

4.7.3.2 Pulsed Lasers (All Lasers not CW). The following minimum information shall be provided:
(1) Wavelength or wavelength range
(2) Pulse-repetition frequency (prf), maximum and minimum
(3) Shape and dimensions of exit beam
(4) Average energy output (single pulse) or average power output (for high-prf systems)
(5) Maximum emergent beam radiant exposure
(6) Pulse duration, maximum and minimum
(7) Designation of laser classification (see 3.3)

4.7.4 Equipment Labeling. The manufacturer shall provide the following information on all lasers or laser systems:
(1) Wavelength or wavelength range
(2) Laser classification
(3) Indication of compliance with this standard
(4) Manufacturer's name and address
(5) Product identification number (for example, date of manufacture, serial number, etc)

NOTE: Warning signs and labels (see 4.6) are required for all lasers and laser systems except Class I–exempt lasers and laser systems.

4.8 Maintenance and Repair of Laser Systems. Following any maintenance or repairs which may affect the output power or operating characteristics of a potentially hazardous laser or laser system, a survey shall be conducted by qualified persons (see Section 5, Laser Safety Program) to ascertain any required changes in control measures.

4.9 Modifications of Laser Systems. Whenever deliberate modifications are made which may affect the output power or operating characteristics of a potentially hazardous laser or laser system, a survey shall be conducted by qualified persons (see Section 5, Laser Safety Program) to ascertain any required changes in control measures.

5. Laser Safety Program

5.1 Organization. The management (employer) shall establish and maintain an adequate program for the control of laser hazards. The program shall include provisions for the following:
(1) Education of authorized personnel (operators and others) in the assessment and control of laser hazards
(2) Delegation of authority and responsibility for supervision of hazard evaluation and control of laser hazards to a Laser Safety Committee or a Laser Safety Officer
(3) Application of adequate protective measures for the control of laser hazards
(4) Medical surveillance of laser users
(5) Management of accidents

A guide for the organization of a laser safety program is outlined in Appendix D.

5.2 Implementation. The employer shall provide adequate supervision, training, facilities, equipment, and supplies to control laser hazards.

6. Medical Surveillance

To date, only two organ systems have been found to b affected by laser energy: the eyes and the skin. Therefore, a program of medical surveillance at this time need include only the ophthalmological and dermatological aspects.

The purpose of medical surveillance of personnel working in a laser environment is twofold. The first purpose is to establish a baseline of ocular and skin conditions before potential exposure to laser radiation The second purpose is to detect and document as early as possible ocular and skin damage. Both purposes serve to assess the effectiveness of control measures and to protect personnel from dangerous exposure and to institute appropriate therapeutic measures promptly. No medical surveillance is required in personnel associated with Class I, Class II, and Class V lasers and laser systems as defined in 3.3.

Only qualified persons should carry out these programs of medical surveillance. Qualified persons include ophthalmologists to do eye examinations and dermatologists and other physicians trained in laser skin hazards to do skin examinations. Other personnel may carry out portions of the examinations under medical supervision.

6.1 Risk Classification. The risk classification shall be determined by the Laser Safety Officer in charge of th installation involved. He shall be responsible for havin the appropriate examinations carried out and records kept. Those persons who should be under laser medica surveillance for possible eye or skin damage are define in 6.1.1 through 6.1.3.

6.1.1 Minimal-Risk Personnel. Those whose work makes it possible but unlikely that they are exposed to laser energy sufficient to damage their eyes or skin.

6.1.2 Moderate-Risk Personnel. Those who work routinely in laser environments but who are ordinarily ılly protected by safety features built into machines nd procedures.

6.1.3 **High-Risk Personnel**. Those who run a high sk of exposure to laser energy sufficient to do eye or kin damage.

.2 General Procedures

6.2.1 Minimal-risk personnel shall have an eye xamination for visual acuity. (See Appendix E1 for ecommended protocol.)

6.2.2 Moderate-risk personnel shall be subject to a odified protocol (see Appendix E1 for recommended rotocol) covering the following:

(1) Ocular history

(2) Visual acuity

(3) Examination of the ocular fundus with an phthalmoscope

If best corrected visual acuity in either eye is less ıan 20/20, or if any pathologic phenomenon is seen, ıe potential worker is advised to have a complete eye xamination done. If any eye pathology is found, the ye examination protocol specified for personnel in .1.3 is recommended.

6.2.3 For high-risk personnel, both a complete eye xamination (see Appendix E1 for recommended pro- ocol) and a complete skin examination (see Appendix 2 for recommended protocol) are required.

.3 Frequency of Medical Examinations

6.3.1 For minimal-risk personnel, required examina- ons shall be done previous to participation in laser ork.

6.3.2 For moderate-risk personnel, required exami- ations shall be done as follows:

(1) Previous to participation in laser work

(2) On discharge from laser environment

(3) Immediately after suspected laser eye or skin amage

6.3.3 For high-risk personnel, required examinations all be done as follows:

(1) Previous to participation in laser work

(2) On discharge from laser environment

(3) Immediately after suspected laser eye or skin amage

(4) Every 3 years while working with lasers

. Special Considerations

1 Industrial Hygiene Considerations. In some laser plications, particularly in the research laboratory, dustrial hygiene aspects may require consideration. ssociated hazards shall be evaluated and appropriate ntrol measures shall be taken. Examples of industrial hygiene aspects are hazards associated with compressed gas, cryogenic material, toxic material, noise, and ionizing radiation. Appendix F provides reference material to aid in the control of such hazards.

7.2 Explosion Hazards. High-pressure arc lamps and filament lamps in laser equipment shall be enclosed in housings which can withstand the maximum explosive pressures resulting from lamp disintegration. The laser target and elements of the optical train which may shatter during laser operation shall also be enclosed or equivalently protected to prevent injury to operators and observers.

7.3 Optical Radiation Hazards Excluding the Laser Beam. Ultraviolet radiation emitted from laser discharge tubes and pumping lamps (that is, not part of the primary laser beam) shall be suitably shielded.

7.4 Electrical Hazards

7.4.1 **Installation**. The intended application of the laser equipment determines the method of electrical installation and connection to the power supply circuit (for example, conduit vs flexible cord). All equipment shall be installed as outlined in American National Standard National Electrical Code, C1-1971 (NFPA 70-1971), Articles 300 and 400. Such installed equipment is acceptable to Occupational Safety and Health Administration if accepted, certified, listed, labeled, or otherwise determined safe by a nationally organized testing laboratory, such as, but not limited to, Underwriters' Laboratories, Inc, and Factory Mutual Corporation.

7.4.2 **Shock Hazard**. Live parts of circuits and components with peak open-circuit potentials over 42.5 volts are considered hazardous, unless limited to less than 0.5 mA. Such circuits require positive protection against contact. For equipment intended for general use, interlock switches (and capacitor bleeder resistors if applicable) or their equivalent shall be installed to remove the voltage from accessible live parts to permit servicing operation. Bleeder resistors shall be of such size and rating as to carry the capacitor discharge current without burnout or mechanical injury. Circuits and components with peak open-circuit potentials of 2500 volts or more shall be adequately covered or enclosed if an appreciable capacitance is associated with the circuits.

If servicing of equipment requires entrance into an interlocked enclosure within 24 hours of the presence of high voltage within the unit, a solid metal grounding rod shall be utilized to assure discharge of high-voltage capacitors. The grounding rod shall be firmly attached to ground prior to contact with the potentially live point. A resistor grounding rod (for example, a large-wattage ceramic resistor) may be used prior to application of the aforementioned solid conductor grounding rod to protect circuit components from overly rapid discharge, but not as a replacement.

7.4.3 Grounding. The frames, enclosures, and other accessible metal non-current-carrying metallic parts of laser equipment shall be grounded. Grounding shall be accomplished by providing a reliable, continuous, metallic connection between the part or parts to be grounded and the grounding conductor of the power wiring system.

7.4.4 Electrical Fire Hazards. Components in electrical circuits shall be evaluated with respect to fire hazards. Circuit components of combustible material, such as transformers, that do not pass a short-circuit test without ignition (see American National Standard Safety Standard for Radio and Television Receiving Appliances, C33.55-1969 (UL 492 – June 1969)) shall be provided with individual noncombustible enclosures. Power supply circuit wiring shall be completely enclosed in noncombustible material.

7.4.5 Electrical Hazards from Explosion. Gas laser tubes and flash lamps shall be supported to ensure that their terminals cannot make any contact which will result in a shock or fire hazard in the event of a tube or lamp failure. Components such as electrolytic capacitors may explode if subjected to voltages higher than their ratings, with the result that ejected metal may bridge live electrical parts. Such capacitors shall be tested to make certain that they can withstand the highest probable potentials should other circuit components fail, unless the capacitors are adequately contained so as not to create a hazard.

7.4.6 Marking. The user shall ensure that each laser device is permanently marked with its primary electrical rating in volts, frequency, and watts or amperes. If the laser device is intended for use by the public or by personnel untrained in laser safety, and is provided with electrical safety interlocks, warning notices instructing the user not to defeat the interlock should be applied to the device immediately adjacent thereto.

7.5 Laser Operation in the Outdoor Environment

7.5.1 Use of Lasers in the Airspace. The Federal Aviation Administration (FAA) is responsible for the use of navigable airspace to insure the safety of aircraft, protection of persons and property on the ground, and its efficient utilization. Laser experiments or programs contemplating use of the airspace should be coordinated with the FAA, Washington, D.C. 20590, or with any FAA regional office, in the planning stages to assure proper control of any attendant hazard to airborne personnel or equipment.

7.5.2 Laser Beam Attenuation. The effects of attenuation in a laser beam traveling through a homogeneous but lossy medium can be handled at visible wavelengths by an exponential attenuation coefficient. For propagation at longer wavelengths, individual lines must be separately determined, particularly at wavelengths where the effects of water vapor become important.

7.5.3 Scintillation. Scintillation arises from local variations in the refractive index of the medium through which a laser beam is propagated. It causes random pointing, spreading, blurring, and energy redistribution of a beam. A "worst-case" analysis shall be used in evaluating the potential ocular hazard from such scintillation effects.

NOTE: References useful for the special considerations covered in 7.1 through 7.5 may be found in Appendix F.

8. Criteria for Exposure of the Eye and Skin

Maximum permissible exposure (MPE) values are below known hazardous levels. The MPE values given may be uncomfortable to view. Thus, it is good practice to maintain exposure levels as far below the MPE values as is practicable.

When a laser emits radiation at several widely different wavelengths, or where pulses are superimposed upon a cw background, computation of the MPE is complex. Exposures from several wavelengths in the same time domain are additive on a proportional basis of spectral effectiveness with due allowance for all correction factors. The simultaneous exposure to pulses and cw radiation is not strictly additive (there may be synergism) and caution should be used in these situations until more data is available.

An appropriate aperture must be used for measurements and calculations with all MPE values. This is the limiting aperture and is the maximum circular area over which irradiance and radiant exposure can be averaged. See Table 9 for proper application of this aperture.

In the tabulations for MPE, values for irradiance can be obtained by dividing the radiant exposure by t in seconds, and values for radiant exposure can be obtained by multiplying the irradiance by t in seconds. Only the simplest values will be listed.

Appendix G provides reference material on this subject.

8.1 Intrabeam or Extended-Source Ocular Exposures. For the purposes of this standard, sources such as laser arrays, diodes, and diffuse reflecting surfaces are considered extended sources if their angular subtense (apparent visual angle) is equal to or greater than α_{min} in Fig. 3. All other lasers, such as those with collimated

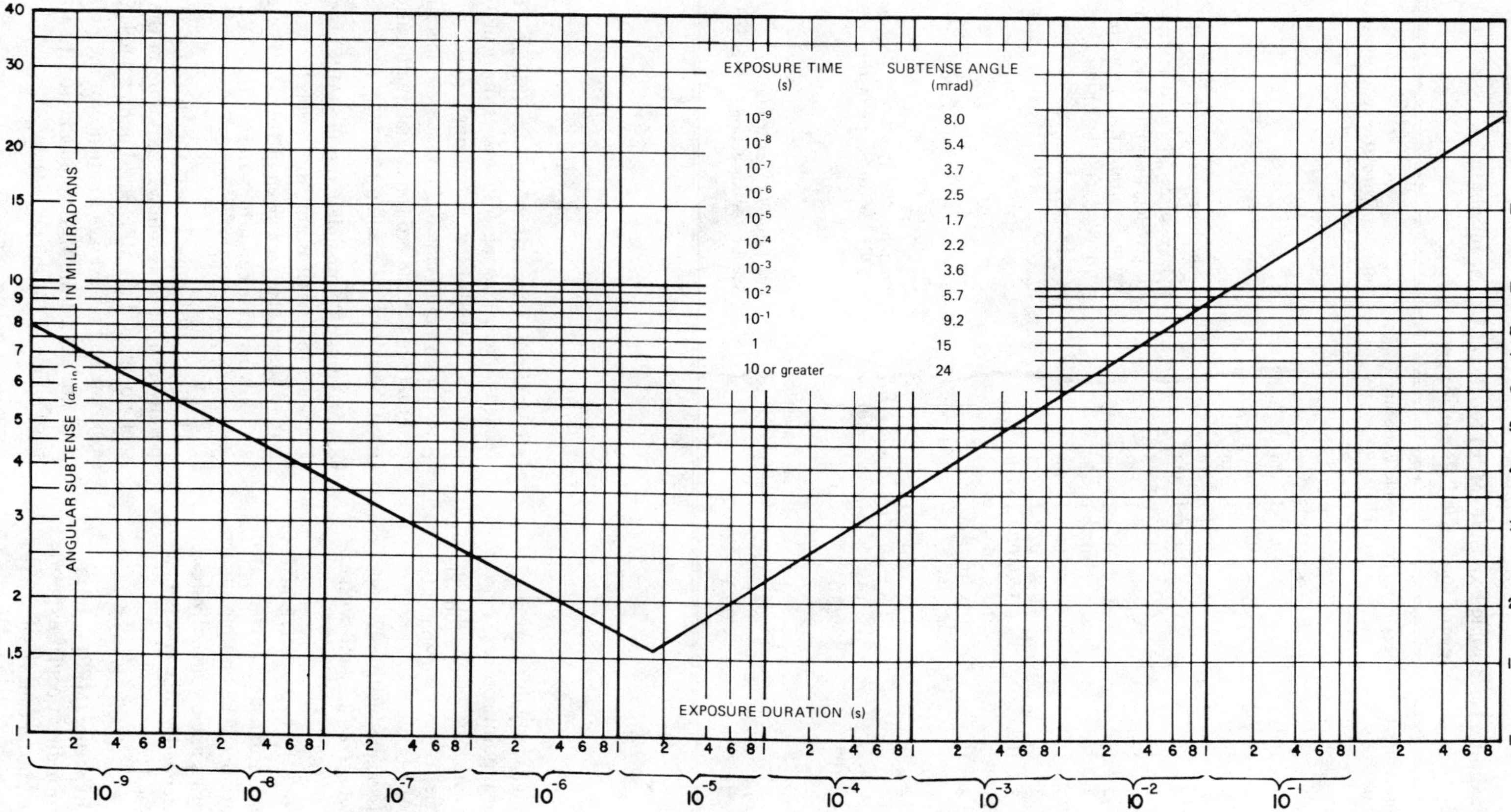

NOTE: Extended sources have an angular subtense or apparent visual angle $\geqq \alpha_{min}$. Angular subtenses (apparent visual angles) $<\alpha_{min}$ are considered intrabeam viewing.

Fig. 3
Limiting Angular Subtense (Apparent Visual Angle), α_{min}

Table 5
Maximum Permissible Exposure (MPE) for Direct Ocular Intrabeam Viewing for Single Pulses or Exposures

Wavelength, λ (μm)	Exposure Time, t (s)	Maximum Permissible Exposure (MPE)	Notes for Calculation and Measurement
Ultraviolet			
0.200–0.302	10^{-2}–3×10^4	3×10^{-3} J · cm^{-2}	1-mm limiting aperture. In no case shall the total irradiance, over al the wavelengths within the UV spectral region, be greater than 1 watt per square centimeter upon the cornea
0.303	10^{-2}–3×10^4	4×10^{-3} J · cm^{-2}	
0.304	10^{-2}–3×10^4	6×10^{-3} J · cm^{-2}	
0.305	10^{-2}–3×10^4	1.0×10^{-2} J · cm^{-2}	
0.306	10^{-2}–3×10^4	1.6×10^{-2} J · cm^{-2}	
0.307	10^{-2}–3×10^4	2.5×10^{-2} J · cm^{-2}	
0.308	10^{-2}–3×10^4	4.0×10^{-2} J · cm^{-2}	
0.309	10^{-2}–3×10^4	6.3×10^{-2} J · cm^{-2}	
0.310	10^{-2}–3×10^4	1.0×10^{-1} J · cm^{-2}	
0.311	10^{-2}–3×10^4	1.6×10^{-1} J · cm^{-2}	
0.312	10^{-2}–3×10^4	2.5×10^{-1} J · cm^{-2}	
0.313	10^{-2}–3×10^4	4.0×10^{-1} J · cm^{-2}	
0.314	10^{-2}–3×10^4	6.3×10^{-1} J · cm^{-2}	
0.315–0.400	10^{-2} to 10^3	1 J · cm^{-2}	
0.315–0.400	10^3–3×10^4	1×10^{-3} W · cm^{-2}	
Visible and Near-Infrared*			
0.4–1.4	10^{-9}–2×10^{-5}	5×10^{-7} J · cm^{-2}	7-mm limiting aperture. See 8.5 for correction factors
	2×10^{-5}–10	$1.8 \times 10^{-3} t^{3/4}$ J · cm^{-2}	
	10–10^4	10^{-2} J · cm^{-2}	
	10^4–3×10^4	10^{-6} W · cm^{-2}	
Far-Infrared			
1.4–10^3	10^{-9}–10^{-7}	10^{-2} J · cm^{-2}	See Table 9 for apertures. See 8.6 for correction factors and Fig. 5
	10^{-7}–10	$0.56 t^{1/4}$ J · cm^{-2}	
	> 10	0.1 W · cm−2	

*See Fig. 4 for graphic presentation.

beams which produce a small (nearly diffraction-limited) retinal image and also point sources, are intrabeam viewing cases. In this category the angular subtense (apparent viewing angle) is less than α_{min} in Fig. 3.

Appropriate MPEs are associated with each type, that is, with intrabeam viewing (see 8.2) and with extended-source viewing (see 8.3).

The angular subtense is not the beam divergence of the source. It is the apparent visual angle as calculated from the source size and distance. The limiting angular subtense is that apparent visual angle which divides intrabeam viewing from extended-source viewing. See Appendix B3.2 for examples.

8.2 Maximum Permissible Exposure (MPE) – Intrabeam Viewing. MPE values for direct ocular exposure to single pulses or exposures, intrabeam viewing, are given in Table 5. (See 8.5 and 8.6 for special qualifications and use.) (See also Figs. 4 and 5.)

8.3 Maximum Permissible Exposure (MPE) – Extended-Source Viewing. MPE values for ocular exposure to extended sources for single pulses or exposures are given in Table 6. All values are specified at th cornea. (See 8.5 and 8.6 for special qualifications and use.) (See also Figs. 5 and 6.)

8.4 Maximum Permissible Exposure (MPE) for Skin Exposure to a Laser Beam. MPE levels for skin exposure to a laser beam are given in Table 7. These levels are for worst-case conditions and are based on the bes available information.

For repetitive-pulsed lasers the MPEs for skin exposure are applied as follows: Exposure of the skin shall not exceed the MPE based upon a single-pulse exposure and the average irradiance of the pulse train shall not exceed the MPE applicable for the total exposure time (P) of the pulse train.

8.5 Special Qualifications – Visible and Near-Infrared

8.5.1 When limiting aperture (pupil) is not 7 mm.

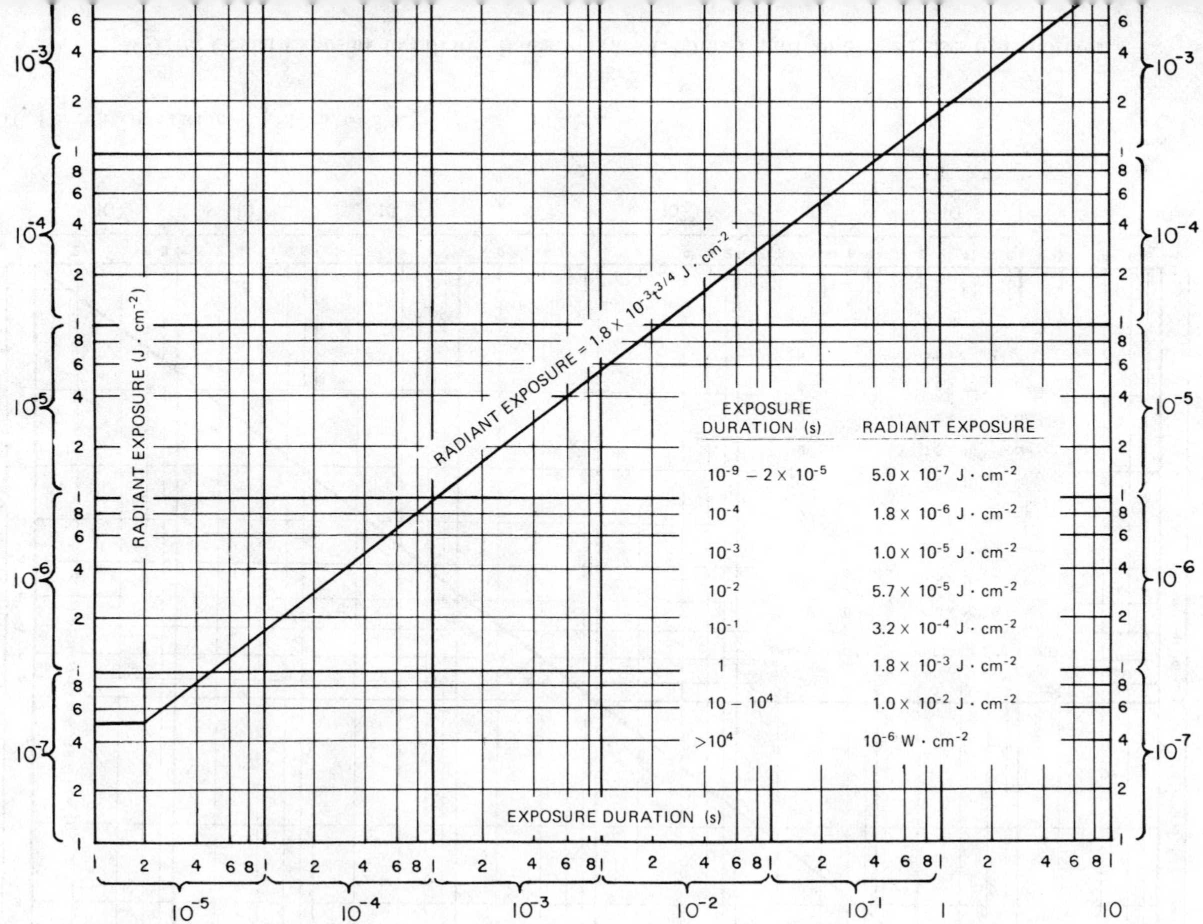

NOTE: For correction factors at wavelengths between 0.7 and 1.4 μm, see 8.5.

Fig. 4
MPE for Direct Ocular Exposure to Visible Radiation (λ = 0.4–1.4μm), Intrabeam Viewing,*
for Single Pulses or Exposures

*Angular subtense less than α_{min} in Fig. 3.

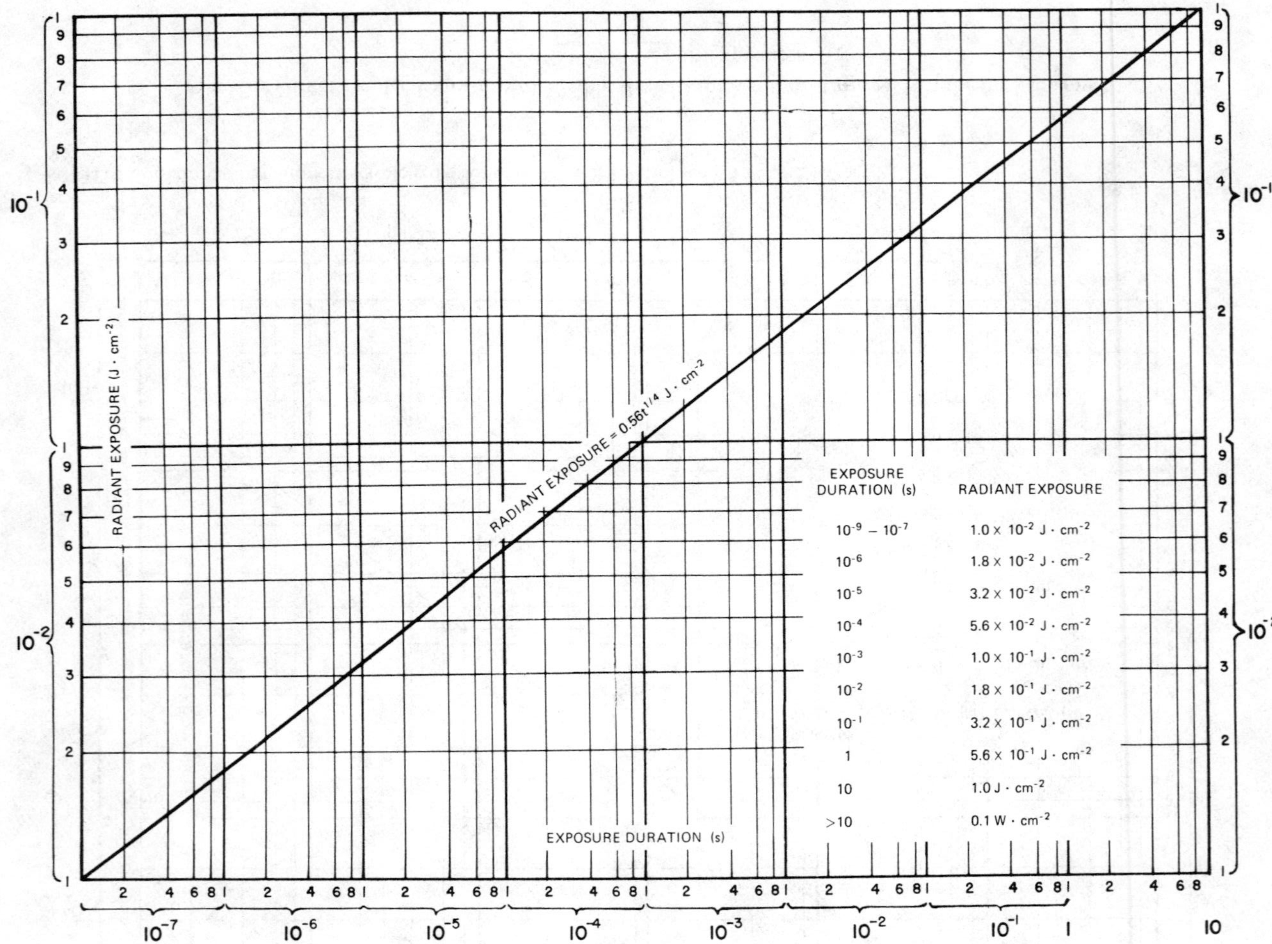

NOTE: For correction factors at 1.54 μm, see 8.6.

Fig. 5
MPE for Ocular Exposure to Infrared Radiation (λ = 1.4 μm–1 mm) for Single Pulses or Exposures

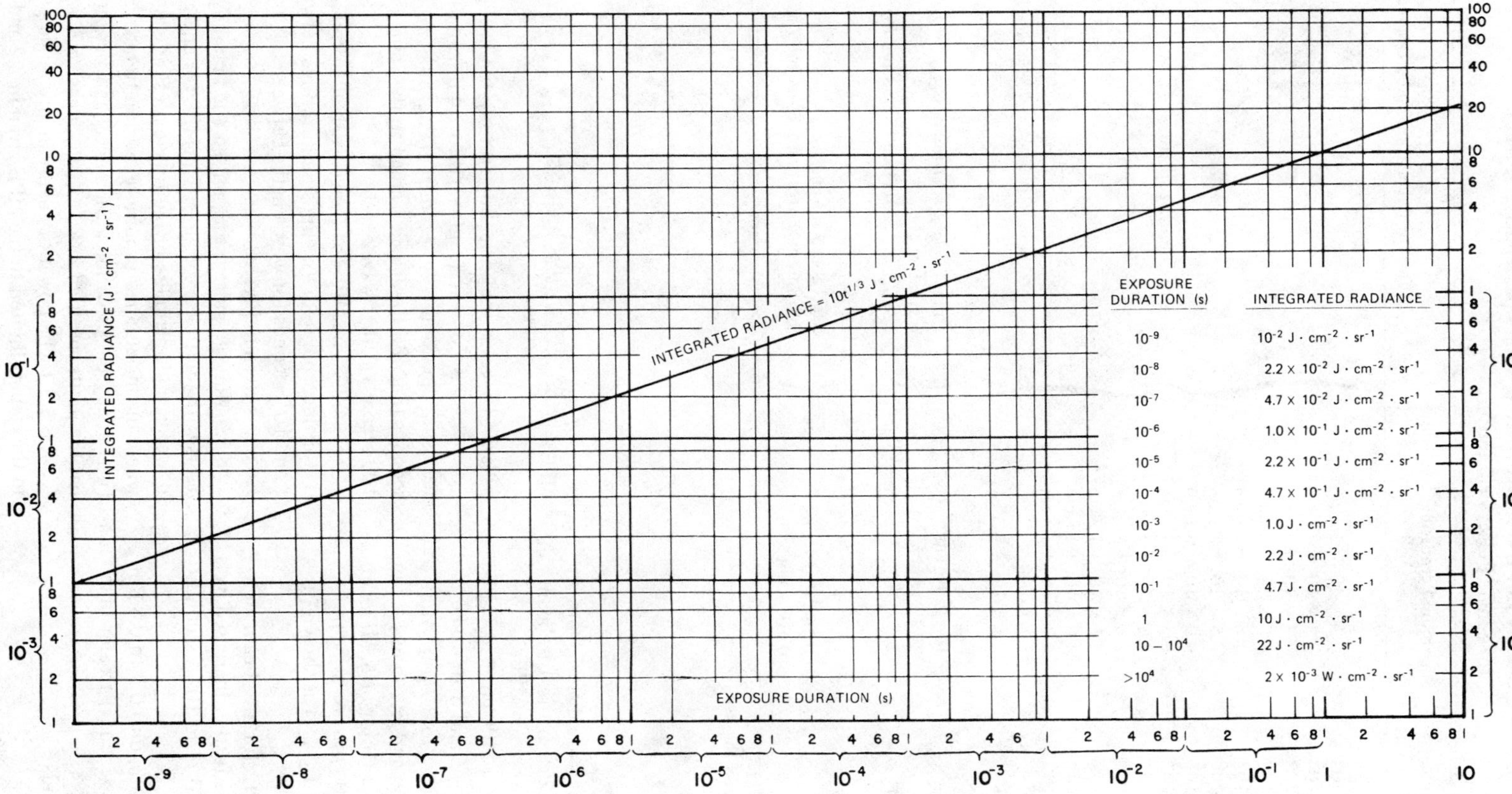

EXPOSURE DURATION (s)	INTEGRATED RADIANCE
10^{-9}	10^{-2} J · cm^{-2} · sr^{-1}
10^{-8}	2.2×10^{-2} J · cm^{-2} · sr^{-1}
10^{-7}	4.7×10^{-2} J · cm^{-2} · sr^{-1}
10^{-6}	1.0×10^{-1} J · cm^{-2} · sr^{-1}
10^{-5}	2.2×10^{-1} J · cm^{-2} · sr^{-1}
10^{-4}	4.7×10^{-1} J · cm^{-2} · sr^{-1}
10^{-3}	1.0 J · cm^{-2} · sr^{-1}
10^{-2}	2.2 J · cm^{-2} · sr^{-1}
10^{-1}	4.7 J · cm^{-2} · sr^{-1}
1	10 J · cm^{-2} · sr^{-1}
$10 - 10^{4}$	22 J · cm^{-2} · sr^{-1}
$>10^{4}$	2×10^{-3} W · cm^{-2} · sr^{-1}

NOTE: For correction factors at wavelengths between 0.7 and 1.4 μm, see 8.5.

Fig. 6
MPE for Direct Ocular Exposure to Visible Radiation ($\lambda = 0.4$–$0.7\ \mu$m) from Extended Sources*
for Single Pulses or Exposures

*Angular subtense $\geqq \alpha_{min}$ in Fig. 3.

Table 6
Maximum Permissible Exposure (MPE) for Ocular Exposure to Extended Sources for Single Pulses or Exposures

Wavelength, λ (μm)	Exposure Time, t (s)	Maximum Permissible Exposure (MPE)	Notes for Calculation and Measurement
Ultraviolet			
0.200–0.302	10^{-2}–3×10^4	3×10^{-3} J · cm^{-2}	
0.303	10^{-2}–3×10^4	4×10^{-3} J · cm^{-2}	
0.304	10^{-2}–3×10^4	6×10^{-3} J · cm^{-2}	
0.305	10^{-2}–3×10^4	1.0×10^{-2} J · cm^{-2}	
0.306	10^{-2}–3×10^4	1.6×10^{-2} J · cm^{-2}	
0.307	10^{-2}–3×10^4	2.5×10^{-2} J · cm^{-2}	
0.308	10^{-2}–3×10^4	4.0×10^{-2} J · cm^{-2}	
0.309	10^{-2}–3×10^4	6.3×10^{-2} J · cm^{-2}	1 = mm limiting aperture
0.310	10^{-2}–3×10^4	1.0×10^{-1} J · cm^{-2}	
0.311	10^{-2}–3×10^4	1.6×10^{-1} J · cm^{-2}	
0.312	10^{-2}–3×10^4	2.5×10^{-1} J · cm^{-2}	
0.313	10^{-2}–3×10^4	4.0×10^{-1} J · cm^{-2}	
0.314	10^{-2}–3×10^4	6.3×10^{-1} J · cm^{-2}	
0.315–0.400	10^{-2} to 10^3	1 J · cm^{-2}	
0.315–0.400	10^3–3×10^4	1×10^{-3} W · cm^{-2}	
Visible and Near-Infrared*			
0.4–1.4	10^{-9}–10	$10t^{1/3}$ J · cm^{-2} · sr^{-1}	1-mm limiting aperture or α_{min}, whichever is greater
	10–10^4	20 J · cm^{-2} · sr^{-1}	
	10^4–3×10^4	2.0×10^{-3} W · cm^{-2} · sr^{-1}	See 8.5 for correction factors and Fig. 6
Far-Infrared			
1.4–10^3	10^{-9}–10^{-7}	10^{-2} J · cm^{-2}	See Table 9 for apertures
	10^{-7}–10	$0.56t^{1/4}$ J · cm^{-2}	See 8.6 for correction factors and Fig. 5
	> 10	0.1 W · cm^{-2}	

*See Fig. 6 for graphic presentation and Fig. B3.

When the limiting aperture (pupil) size differs from 7 mm, no corrections are to be made for the intrabeam-viewing MPE or for the extended-source MPE.

8.5.2 Wavelength correction for MPE given in Tables 5 and 6. All MPEs given in Tables 5 and 6 are to be used as given for wavelengths of 0.4 to 0.7 μm. At all wavelengths greater than 1.06 μm and less than 1.4 μm the MPEs are to be increased by a factor of 5. MPEs at wavelengths between 0.7 and 1.06 μm are to be increased by a correction factor ($CF_{8.5.2}$) as shown in Fig. 7.

8.5.2.1 Corrections to MPEs for intrabeam viewing (see 8.2), using 8.5.2, may be further increased by a factor of 2 for exposure times 10^{-9} to 10^{-4} seconds at wavelengths of 1.0 to 1.4 μm.

8.5.2.2 For wavelengths (λ in μm) 0.7 to 0.8 μm and for exposure durations (t in seconds) greater than

$$t = \left(\frac{10\ \mu\text{m}}{(\lambda - 0.699\ \mu\text{m})}\right) \text{ seconds}$$

the MPE irradiance (see 8.2) is given by

$$\text{MPE} = \left(\frac{\lambda - 0.699\ \mu\text{m}}{\mu\text{m}}\right) 10^{-3}\ CF_{8.5.2}\ \text{W} \cdot \text{cm}^{-2}$$

where $CF_{8.5.2}$ is the correction factor given in Fig. 7.

8.5.2.3 For wavelengths 0.8 to 1.4 μm and for exposure durations (t in seconds) greater than 100 seconds the MPE irradiance is equal to the value for 100 seconds at that wavelength.

8.5.2.4 The MPE irradiance, with all correction factors incorporated, for exposure durations between 1 and 10^4 seconds for wavelengths between 0.4 and 1.4 μm is given in Fig. 8.

8.5.3 Multiple-pulse trains, pulsed and scanning lasers with multiple exposures. The MPE for energy or power in multiple-pulse or multiple-exposure trains where the instantaneous pulse-repetition frequency (prf) of any pulses within a train exceeds 1 per second has the limits specified in 8.5.3.1 through 8.5.3.5.

8.5.3.1 The power or energy in any single pulse in the train is limited to the MPE given in 8.2 and 8.4 for a comparable pulse. The irradiance or radiant ex-

Table 7
Maximum Permissible Exposure (MPE) for Skin Exposure to a Laser Beam

avelength, λ (μm)	Exposure Time, t (s)	Maximum Permissible Exposure (MPE)	Notes for Calculation and Measurement
traviolet			
200 –0.302	10^{-2}–3×10^{4}	3×10^{-3} J · cm^{-2}	1-mm limiting aperture In no case shall the total irradiance, over all the wavelengths within the UV spectral region, be greater than 1 watt per square centimeter upon the cornea
303	10^{-2}–3×10^{4}	4×10^{-3} J · cm^{-2}	
304	10^{-2}–3×10^{4}	6×10^{-3} J · cm^{-2}	
305	10^{-2}–3×10^{4}	1.0×10^{-2} J · cm^{-2}	
306	10^{-2}–3×10^{4}	1.6×10^{-2} J · cm^{-2}	
307	10^{-2}–3×10^{4}	2.5×10^{-2} J · cm^{-2}	
308	10^{-2}–3×10^{4}	4.0×10^{-2} J · cm^{-2}	
309	10^{-2}–3×10^{4}	6.3×10^{-2} J · cm^{-2}	
310	10^{-2}–3×10^{4}	1.0×10^{-1} J · cm^{-2}	
311	10^{-2}–3×10^{4}	1.6×10^{-1} J · cm^{-2}	
312	10^{-2}–3×10^{4}	2.5×10^{-1} J · cm^{-2}	
313	10^{-2}–3×10^{4}	4.0×10^{-1} J · cm^{-2}	
314	10^{-2}–3×10^{4}	6.3×10^{-1} J · cm^{-2}	
315–0.400	10^{-2}–10^{3}	1 J · cm^{-2}	
315–0.400	10^{3}–3×10^{4}	1×10^{-3} W · cm^{-2}	
sible and Near-Infrared			
4–1.4	10^{-9}–10^{-7}	2×10^{-2} J · cm^{-2}	1-mm limiting aperture
	10^{-7}–10	$1.1t^{1/4}$ J · cm^{-2}	
	10–3×10^{4}	0.2 W · cm^{-2}	
r-Infrared			
4–10^{3}	10^{-9}–10^{-7}	10^{-2} J · cm^{-2}	1-mm limiting aperture for 1.4 to 100 μm 11-mm limiting aperture for 0.1 to 1 mm
	10^{-7}–10	$0.56t^{1/4}$ J · cm^{-2}	
	> 10	0.1 W · cm^{-2}	

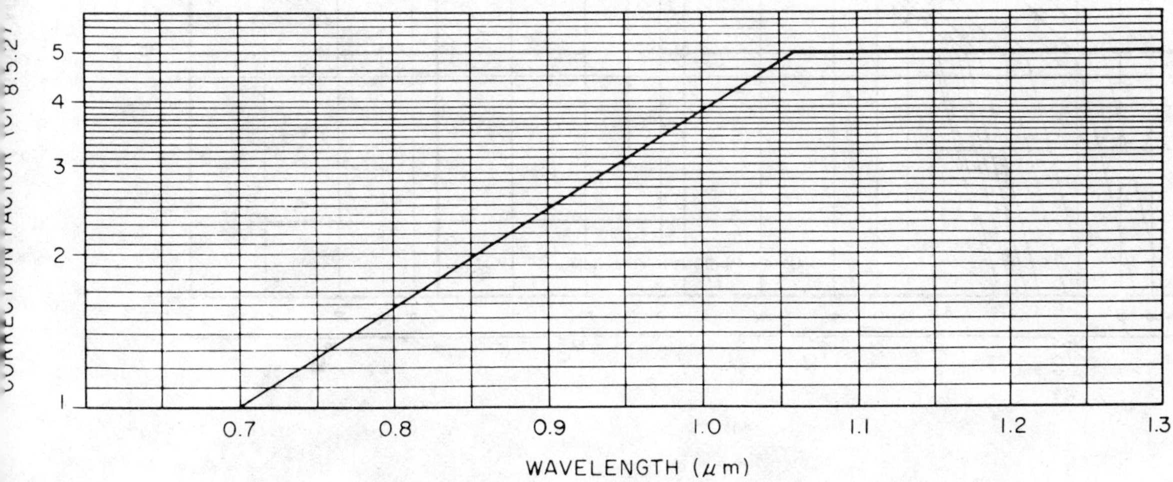

Fig. 7
Correction Factors for Wavelengths 0.7–1.4 μm ($CF_{8.5.2}$)

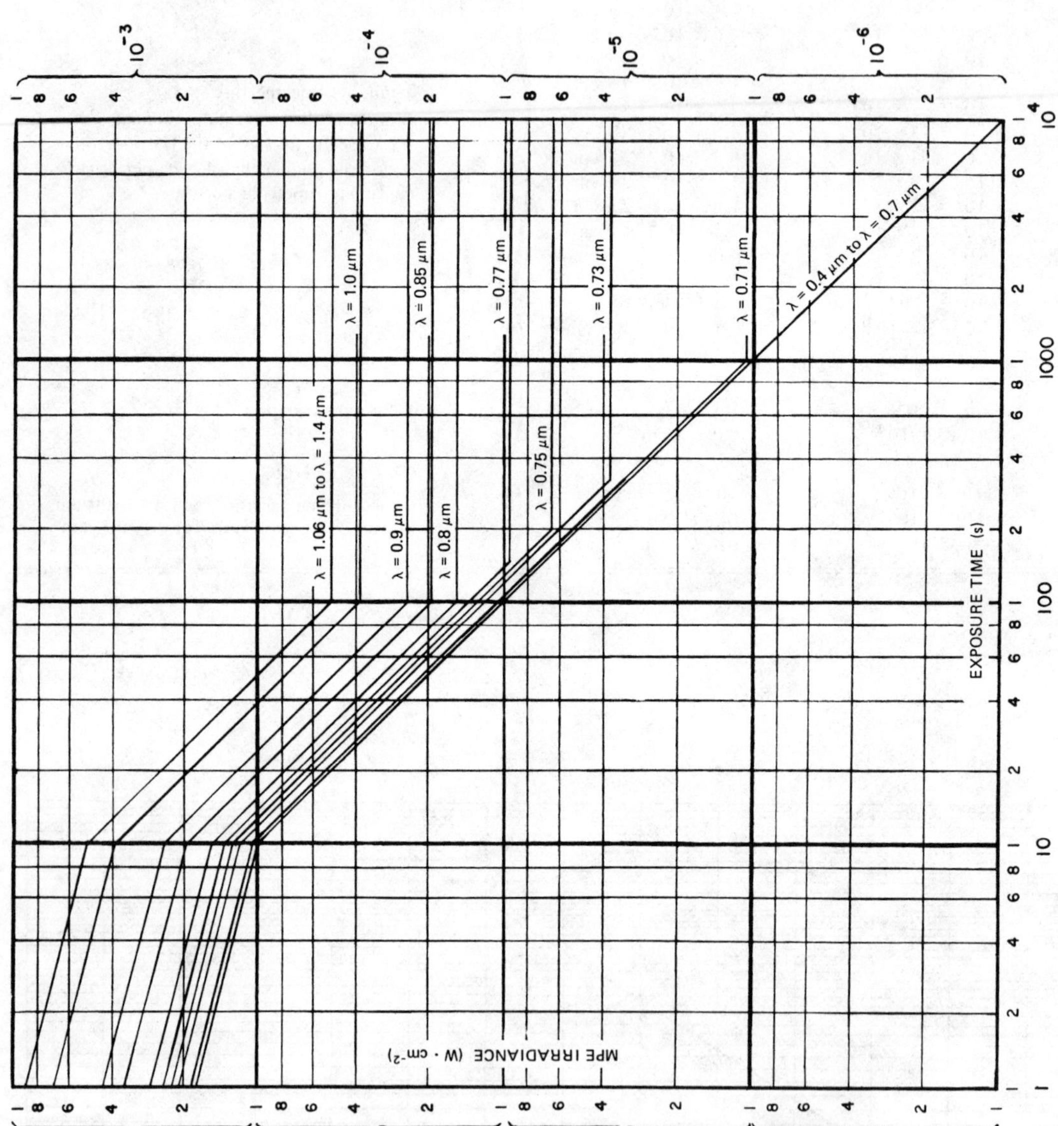

Fig. 8
Ocular MPE as a Function of Exposure Time and Wavelength

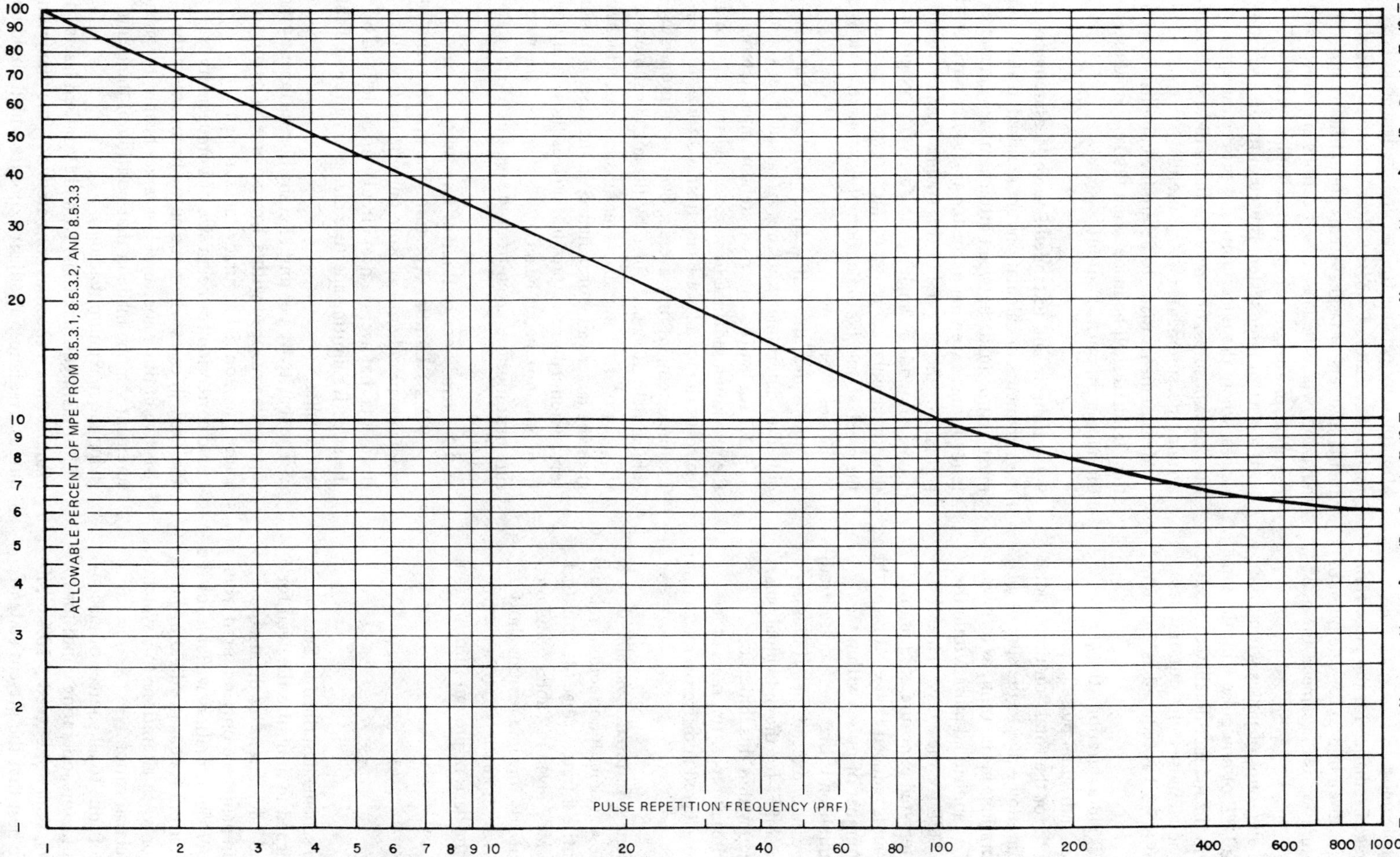

NOTE: For repetition rates greater than 1000 pps, the percent MPE = 6.

Fig. 9
Reduction of MPE for Repetitively Pulsed or Multiple Exposures from Scanning Lasers, Individual Pulse or Exposure Less than 10 μs

posure in any individual pulse in a train is limited to the MPE for a comparable pulse.

8.5.3.2 The average irradiance for the pulse train is limited to the MPE (as given in 8.2) for one pulse of this irradiance whose duration is the same as the pulse train.

8.5.3.3 Any and all groups of pulses within the train are limited to the MPE of one pulse with the same duration as the group in the manner detailed in 8.5.3.2.

8.5.3.4 For individual pulses with duration less than 10 μs, the MPE of 8.5.3.1 is reduced as shown in Fig. 9.

8.5.3.5 Individual pulse duration 10 μs or greater. The MPE for an individual pulse in a train is calculated from the MPE for the total "on time" pulse (TOTP), which has a duration equal to the sum of all the individual pulse durations in the train, as follows: The MPE irradiance of an individual pulse within the train is reduced to the MPE for the TOTP. The MPE radiant exposure or integrated exposure of an individual pulse within the train is reduced to the MPE for the TOTP divided by the number of pulses within the train. An additional limitation is that the average irradiance in the pulse train cannot exceed the MPE as defined in 8.5.3.2 and the MPEs for the individual pulses must be reduced to keep within this limitation.

8.5.4 Repetitive pulses at repetition rates of less than 1 Hz should be considered additive over a 24-hour period.

8.6 Special Qualifications – Infrared. Available data is not sufficient to define wavelength corrections relative to 10.6 μm over the entire infrared range (1.4 μm to 1 mm). At 1.54 μm, the MPE given in Tables 5 and 6 is increased by a factor of 10^2 for time periods shorter than 1 μs. However, no extrapolation to other wavelengths is justified on the basis of present information.

9. Measurements

The laser classification scheme described in Section 3 of this standard is designed to minimize the need for laser measurements and calculations by the user. Generally, such measurements are only required when there is no manufacturer's information available, or the laser or laser system's output is near the dividing line between laser classes, or when alterations to a system may have changed its classification.

The cumulative errors due to all sources of inaccuracy, including human factors, operating conditions, and instrumental errors, shall not exceed ± 20% or, if this is not possible, the best that the present state of the art will reasonably permit.

9.1 General Principles. Measurements should only be attempted by persons trained or experienced in laser technology and radiometry. Routine survey measurements of lasers or laser systems are neither required nor advisable when the laser classifications are known and the appropriate control measures implemented.

If a laser or laser system is used outdoors over long ranges, where the uncertainties of propagation influence exposures, or where the beam divergence is uncertain, measurements may be useful.

Measurements made to determine compliance with this standard shall be made when the laser is adjusted for maximum normal output.

9.2 Intrabeam and Extended-Source Measurements. If measurements or calculations are required, distinction must first be made between intrabeam viewing and extended-source viewing in the 0.4-to-1.4-μm-wavelength region. For the purpose of this standard, an extended source subtends an angle at the observer's eye greater than the angular subtense, α_{min}, shown in Fig. 3 across the greatest angular dimension of the source as viewed by the observer.

Table 8 gives the radiometric parameter that is required to determine exposures to various laser sources.

9.2.1 Radiance. This measurement must determine the maximum radiance of an extended source, such as the scattering of a laser beam from a diffuse surface. The measurement may average over the appropriate conical field of view defined by the angular subtense, α_{min}, or over a 1-mm-diameter circular area, whichever gives the larger value of radiance. In the case of nonuniform extended-source profiles, such as those resulting from inhomogeneous beams or "hot spots," the measurement must be taken from the regions of greatest radiance.

9.2.2 Irradiance or Radiant Exposure

9.2.2.1 Limiting Aperture. The measurement of irradiance or radiant exposure must be made with instruments that average over circular areas defined by the effective limiting aperture diameters given in Table 9 or smaller.

The sensitivity per unit area shall be sufficiently uniform, when mapped with a 1-mm-diameter beam, to assure the required accuracy measurement.

No correction for beam size or homogeneity is necessary in cases where the entire beam enters the effective limiting aperture. For larger beams, the measurement shall be made in that position of the beam giving the maximum reading.

For distinguishing Class III and IV pulsed lasers (if manufacturer's specifications are used), the maximum output irradiance or radiant exposure which could be measured through a 1-mm circular aperture shall be used.

Table 8
Required Radiometric Parameters

Laser Source	Eye: Visible (0.4–1.4 μm)	Eye: Ultraviolet (0.2–0.4 μm)	Eye: Infrared (1.4–10^3 μm)	Skin: All Wavelengths (0.2–10^3 μm)
Extended Sources (Angular subtense > α_{min})	Radiance	Irradiance	Irradiance	Irradiance
Intrabeam (Angular subtense < α_{min})	Irradiance	Irradiance	Irradiance	Irradiance

Table 9
Maximum Aperture Diameters (Limiting Aperture) for Measurement Averaging

Measurement	Exposure Time, t (s)	Wavelength Range: Ultraviolet (0.2–0.4 μm)	Visible and Near-Infrared (0.4–1.4 μm)	Medium and Far-Infrared (1.4–10^2 μm)	Submillimeter (0.1–1 mm)
Eye MPE	10^{-9}–10^3	1 mm	7 mm	1 mm	11 mm
Skin MPE	10^{-9}–10^3	1 mm	1 mm	1 mm	11 mm
Laser Classification*	10^{-9}–10^3	80 mm	80 mm	80 mm	80 mm

*The apertures are used for the measurement of total output power or output energy for laser classification purposes, that is, to distinguish between all classes of cw lasers or between Class I and Class III pulsed lasers.

NOTE: For the specific case of optical viewing (beam collecting) instruments, the apertures listed for eye MPE and skin MPE apply to the exit beam of such devices.

9.2.2.2 Field of View (0.2–0.4-μm and 1.4–10^3-μm Wavelengths). In measuring the irradiance or radiant exposure from ultraviolet and far-infrared extended sources, care must be taken to insure that the field of view of the instrument is sufficiently large to assure the required accuracy of measurement.

9.3 Instruments. Many optical power, energy, pulse shape, and pulse-repetition frequency measuring devices available commercially can be used to determine classification and compliance with this standard. Instruments shall be calibrated sufficiently well to permit overall measurement accuracies of ± 20% wherever possible.

Measurements with instruments having smaller effective limiting apertures than those in Table 9 are permitted, provided the appropriate correction factors are applied to assure the required accuracy of measurement.

A variety of such instruments is described in the references of Appendix H4.

Appendixes

(These Appendixes are not a part of American National Standard for the Safe Use of Lasers, Z136.1-1973, but are included for information purposes only.)

Appendix A

Examples of Classification of Lasers or Laser Systems

Since the laser classification was designed to include all types of lasers operating at essentially any wavelength or pulse duration, the rules of classification (see 3.2) may appear complicated. To assist in the classification of commonly available lasers, Tables A1 and A2 have been prepared to aid the user in rapidly determining the required radiometric parameters needed to classify his laser and the applicable class once he has obtained the required output parameters. Table A1 applies to cw lasers (potential exposure time greater than 0.25 s) and Table A2 applies to pulsed lasers. To classify a repetitively pulsed laser the values in Tables A1 and A2 are not generally applicable, but may be used as a first step in estimating approximately what class the laser will fall into; the user should then apply the rules given in Section 3 of the standard.

Calculations required for some classifications are presented in Appendix B.

Table A1
Typical Laser Classification – Continuous-Wave (CW) Lasers

Wavelength Range	Laser	Wavelengths	Exempt – Class 1*	Low-Power – Class II	Medium-Power – Class III	High-Power – Class IV
Ultraviolet (100–280 nm)	CW Neodymium: YAG (Quadrupled)	266 nm only	≦ 0.8 × 10^{-9} W for 8 hours	–	> Class I but ≦ 0.5 W	> 0.5 W
Ultraviolet (315–400 nm)	Helium-Cadmium Argon Krypton	325 nm only 351.1, 363.8 nm only 350.7, 356.4 nm only	≦ 8 × 10^{-6} W	–	> Class I but ≦ 0.5 W	> 0.5 W
Visible (400–700 nm)	Helium-Cadmium Argon (Visible) Helium-Selenium CW Neodymium: YAG (Doubled) Helium-Neon Krypton	441.6 nm only 457.9, 476.5, 488, 514.5 nm, etc 460.4–700 nm only (30 lines) 532 nm 632.8 nm 647.1, 530.9, 676.4 nm, etc	≦ 0.4 × 10^{-6} W	> Class I but ≦ 1 × 10^{-3} W	> Class II but ≦ 0.5 W	> 0.5 W
Near-Infrared (0.7–1.4 μm)	CW Galium-Aluminum Arsenide	850 nm (20°C)	≦ 80 × 10^{-6} W	–	> Class I but ≦ 0.5 W	> 0.5 W
	CW Galium-Arsenide	905 nm (20°C)	≦ 0.1 × 10^{-3} W	–	> Class I but ≦ 0.5 W	> 0.5 W
	CW Neodymium: YAG	1064 nm	≦ 0.2 × 10^{-3} W	–	> Class I but ≦ 0.5 W	> 0.5 W
	Helium-Neon	1.08, 1.152 μm only	≦ 0.2 × 10^{-3} W	–	> Class I but ≦ 0.5 W	> 0.5 W
Far-Infrared (1.4–100 μm)	Hydrogen-Fluoride Carbon-Monoxide Carbon-Dioxide Helium-Neon	4–6 μm 5.0–5.5 μm 10.6 μm 3.39 μm only	≦ 0.8 × 10^{-3} W	–	> Class I but ≦ 0.5 W	> 0.5 W
Far-Infrared (0.1–1 mm)	Water Vapor Hydrogen-Cyanide	118 μm 337 μm	≦ 0.1 W	–	> Class I but ≦ 0.5 W	> 0.5 W

*Assumes no mechanical or electrical design incorporated into laser system to prevent exposures from lasting to τ_{max} = 8 hours (one workday); otherwise P_{exempt} could be larger than tabulated.

Table A2
Typical Laser Classification – Single-Pulse Lasers

Wavelength Range	Laser	Wavelengths	Pulse Duration	Exempt – Class I	Medium-Power – Class III	High-Power – Class IV
Ultraviolet	Neodymium: YAG Q-sw (Quadrupled)	266 nm	$10 - 30 \times 10^{-9}$s (Q-sw)	No criteria	≤ 10 J · cm^{-2}	> 10 J · cm^{-2}
	Ruby (Doubled)	347.1 nm	$10 - 30 \times 10^{-9}$s (Q-sw)	No criteria	≤ 10 J · cm^{-2}	> 10 J · cm^{-2}
Visible (400-700 nm)	Neodymium: YAG (Doubled)	532 nm }	$\sim 20 \times 10^{-9}$s (Q-sw)	$\leq 0.2 \times 10^{-6}$ J	$>$ Class I but $\leq 74 \times 10^{-3}$ J · cm^{-2}	$> 75 \times 10^{-3}$ J · cm^{-2}
	Ruby	694.3 nm }				
	Ruby (Long Pulse)	694.3 nm	$\sim 1 \times 10^{-3}$s	$\leq 4 \times 10^{-6}$ J	$>$ Class I but ≤ 3.1 J · cm^{-2}	> 3.1 J · cm^{-2}
	Rhodamine 6G (Dye Laser)	450-650 nm	$\sim 1 \times 10^{-6}$s	$\leq 0.2 \times 10^{-6}$ J	$>$ Class I but ≤ 0.31 J · cm^{-2}	> 0.31 J · cm^{-2}
Infrared	Neodymium: YAG	1064 nm	$\sim 20 \times 10^{-9}$s (Q-sw)	$\leq 2 \times 10^{-6}$ J	$>$ Class I but ≤ 0.16 J · cm^{-2}	> 0.16 J · cm^{-2}
	Erbium: Glass	1.54 μm	$\sim 10\text{-}100 \times 10^{-9}$s (Q-sw)	$\leq 8 \times 10^{-3}$ J	$>$ Class I but ≤ 10 J · cm^{-2}	> 10 J · cm^{-2}
	Carbon Dioxide	10.6 μm	$1\text{-}100 \times 10^{-9}$s (Q-sw)	$\leq 80 \times 10^{-6}$ J	$>$ Class I but ≤ 10 J · cm^{-2}	> 10 J · cm^{-2}

Appendix B

Calculations for Hazard Evaluation and Classification

B1. General

Calculations are not necessary for hazard evaluation and classification in many applications; however, in outdoor applications and other specialized uses where eye exposure is contemplated, several types of calculations permit the important quantitative study of potential hazards.

Mathematical symbols used here are defined in Appendix B2. Hazard classification and MPE determination may require the use of formulas in Appendix B3. Formulas useful in estimating exposure at significant distances from the laser and optically aided viewing are presented in Appendix B4.

Figs. B1, B2, and B3 illustrate the three conditions of ocular exposure to laser radiation.

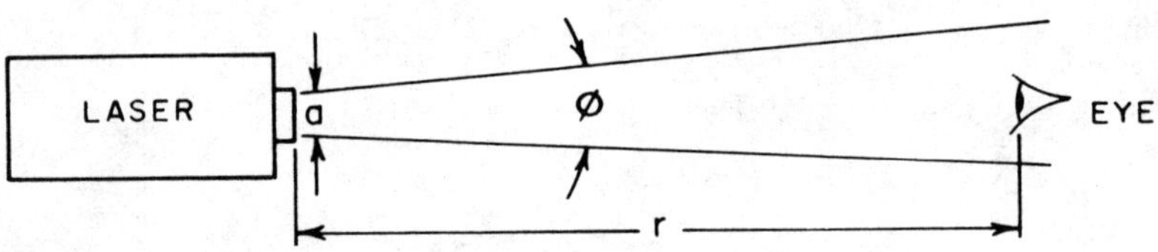

Fig. B1
Intrabeam Viewing – Direct (Primary) Beam

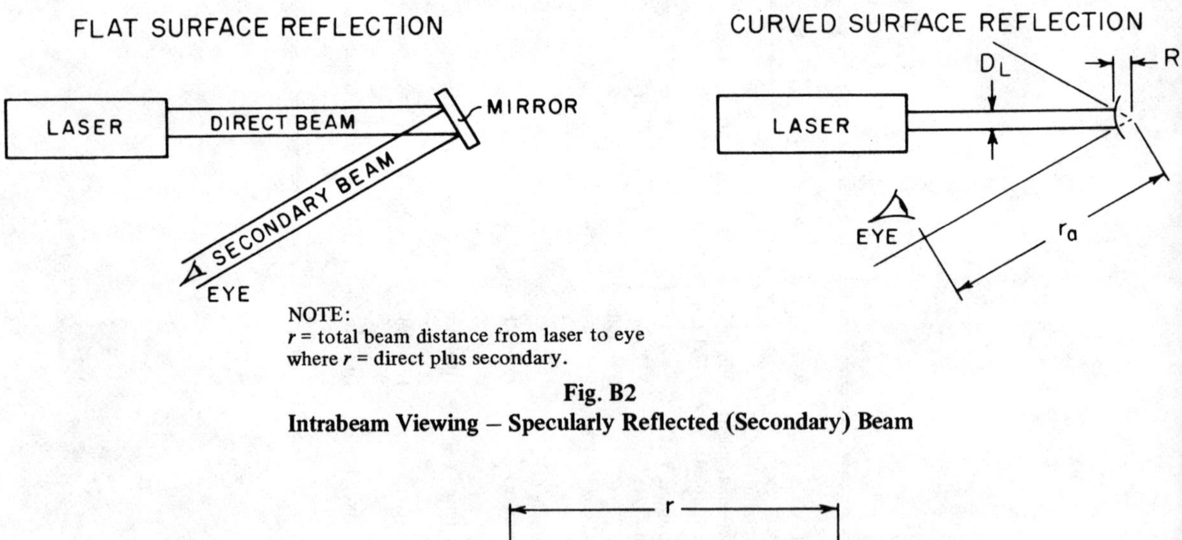

NOTE:
r = total beam distance from laser to eye
where r = direct plus secondary.

Fig. B2
Intrabeam Viewing – Specularly Reflected (Secondary) Beam

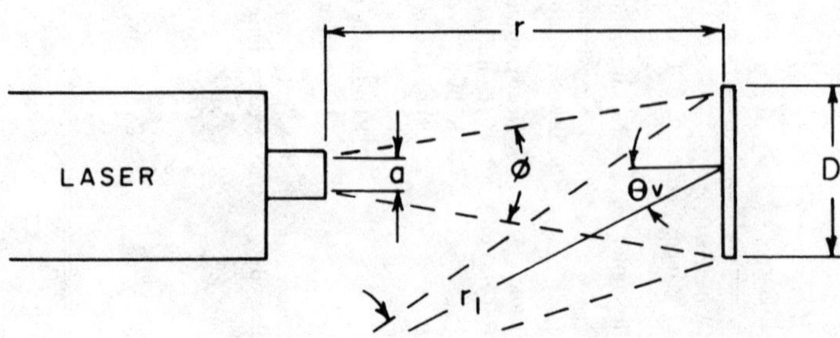

Fig. B3
Extended-Source Viewing – Normally Diffuse Reflection

B2. Symbols

The following symbols are used in the formulas of this Appendix.

a = Diameter of emergent laser beam (cm)
d_e = Diameter of the pupil of the eye (varies from approximately 0.2 to 0.7 cm)
d_{min} = Limiting object size of extended object (cm)
D_e = Diameter of the exit pupil of an optical system (cm)
D_L = Diameter of laser beam at range r (cm)
D_o = Diameter of objective of an optical system (cm)
e = Base of natural logarithms
E,H = Radiant exposure (H) or irradiance (E) at range r, measured in $J \cdot cm^{-2}$ for pulsed lasers and $W \cdot cm^{-2}$ for cw lasers
E_o, H_o = Emergent beam radiant exposure (H_o) or irradiance (E_o) at zero range (units as for E, H)
f = Effective focal length of eye (1.7 cm)
F = Pulse-repetition frequency, prf (s^{-1})
G = Ratio of retinal irradiance or radiant exposure received by optically aided eye to that received by unaided eyes
L = Radiance of an extended source ($W \cdot cm^{-2} \cdot sr^{-1}$)
L_p = Integrated radiance of an extended source ($J \cdot cm^{-2} \cdot sr^{-1}$)
P = Magnifying power of an optical system
Q = Total radiant energy output of a pulsed laser, measured in joules
r = Range from the laser to the viewer or to a diffuse target (cm)
r_1 = Range from the laser target to the viewer (cm)
$r_{1\,max}$ = Maximum range from the laser target to the viewer where extended-source MPE applies (cm)
R = Radius of curvature of a specular surface (cm)
S = Scan rate of a scanning laser (number of scans across eye per second)
T = Total exposure duration (in seconds) of a train of pulses
T_i = Integrated "on time" of a train of pulses (TOTP)
t = Duration of single pulse(s)
α = Viewing angle subtended by an extended source
α_{min} = Minimum angle subtended by a source for which extended-source MPE applies (rad)
μ = Atmospheric attenuation coefficient (cm^{-1}) at a particular wavelength
ϕ = Emergent beam divergence measured in radians
Φ = Total radiant power output of a cw laser, or average radiant power of a repetitively pulsed laser, measured in watts
ρ_λ = Spectral reflectance of a diffuse object at wavelength λ
θ_s = Maximum angular sweep of a scanning beam (rad)
θ_v = Viewing angle (see Fig. B3)

B3. Examples of MPE Determination and Laser Classification

B3.1 Determining the MPE for Intrabeam Viewing for Particular Exposures

B3.1.1 Single-Pulse Laser MPEs. MPEs for single-pulse lasers are determined by reading the values directly from the graphs of Figs. 4, 5, or 6, or they may be calculated as in Examples 1 and 2.

Example 1: Single-Pulse Visible Laser. Determine the MPE for a direct intrabeam exposure to a 694.3-nm ruby laser pulse having a duration of 8×10^{-4}s (0.8 ms).

The MPE for such a laser exposure, as given in Table 5, is

$$\text{MPE: } H = 1.8 \times 10^{-3}\, t^{3/4}\ \text{J} \cdot \text{cm}^{-2}$$

Substituting the values in the example gives

$$\text{MPE: } H = \frac{1.8 \times 10^{-3} t}{\sqrt[4]{t}} = \frac{(1.8 \times 10^{-3})(8 \times 10^{-4})}{\sqrt[4]{8 \times 10^{-4}}}$$

$$= \frac{1.44 \times 10^{-6}}{1.68 \times 10^{-1}} = 8.6 \times 10^{-6}\ \text{J} \cdot \text{cm}^{-2}$$

Since $E \cdot t = H$, we may also express the MPE as:

$$\text{MPE: } E = \frac{H}{t} = \frac{8.6 \times 10^{-6}\ \text{J} \cdot \text{cm}^{-2}}{8 \times 10^{-4}\text{s}}$$

$$= 1.1 \times 10^{-2}\ \text{W} \cdot \text{cm}^{-2}$$

Example 2: Single-Pulse Near-Infrared Laser. Determine the intrabeam direct-viewing MPE for a 1.064-μm (Nd:YAG) laser having a pulse duration of 8×10^{-4} s. As explained in 8.5, the direct-beam MPE for this laser is five times that for the visible laser having the same exposure duration (see Example 1). Therefore, the MPE for this exposure is

$$\text{MPE: } H = 5 \times (8.6 \times 10^{-6}\ \text{J} \cdot \text{cm}^{-2})$$
$$= 4.3 \times 10^{-5}\ \text{J} \cdot \text{cm}^{-2}$$

or

$$\text{MPE: } E = 5 \times (1.1 \times 10^{-2}\ \text{W} \cdot \text{cm}^{-2})$$
$$= 5.5 \times 10^{-2}\ \text{W} \cdot \text{cm}^{-2}$$

B3.1.2 Repetitively Pulsed Laser MPE. To determine the MPE applicable for an exposure to a repetitively pulsed laser one must know the wavelength, prf, duration of a single pulse, and duration of a complete exposure. In every instance this process requires two analyses or approaches and a conclusion, Steps 1 and 2 and Step 3, respectively, in the following examples.

Step 1. Individual Pulse Limitation. This requires the calculation of the MPE based upon the limitation that a single-pulse exposure shall not exceed the single-pulse MPE (see 8.5.3.1 and 8.5.3.5) for pulses greater than 10^{-5} second and may not exceed this single-pulse MPE multiplied by the correction factor in Fig. 9 for pulses less than or equal to 10^{-5} second (see 8.5.3.4).

Step 2. Average Irradiance or Average Radiant Exposure Limitation. The average-power limitation requires the calculation of the average irradiance or total radiant exposure for the entire pulse train for comparison with the MPE applicable for the duration of the entire exposure (see 8.5.3.5).

Step 3. Conclusion. Compare the results of Step 1 and Step 2; the limitation which provides the lowest total exposure is applied.

Example 3: Repetitively Pulsed Visible Laser with Very High PRF. Determine the direct intrabeam MPE of a 514.5-nm (Argon) laser for a 0.25-s total exposure T, operating at $F = 10$ MHz and $t = 10$ ns (10^{-8}s). Following 8.5 we must consider at least two analyses:

Step 1. Individual Pulse Limitation (see 8.5.3.1). First determine the reduced MPE for an individual pulse. For a prf greater than 1 kHz the reduction factor from Fig. 9 is 0.06 for a 10-ns pulse.

$$\text{MPE/Pulse: } H_i \leq 0.06 \times (5 \times 10^{-7}\ \text{J} \cdot \text{cm}^{-2})$$
$$= 3 \times 10^{-8}\ \text{J} \cdot \text{cm}^{-2}$$

On this basis the MPE for the total exposure is:

$$\text{MPE/Train: } H \leq H_i \cdot F \cdot T$$
$$= (3 \times 10^{-8}\ \text{J} \cdot \text{cm}^{-2})\ (10^7\ \text{Hz})\ (0.25\ \text{s})$$
$$= 7.5 \times 10^{-2}\ \text{J} \cdot \text{cm}^{-2}$$

Step 2. Average Radiant Exposure Limitation (see 8.5.3.2). For a single exposure of 0.25 s, the MPE (from Fig. 4) is

$$\text{MPE (avg): } H \leq 6.3 \times 10^{-4}\ \text{J} \cdot \text{cm}^{-2}$$

Step 3. Conclusion. Since Step 2 gives the limiting case (more restrictive), the correct MPE is:

$$\text{MPE: } H = 6.3 \times 10^{-4}\ \text{J} \cdot \text{cm}^{-2} \text{ in } 0.25\ \text{s}$$

or

$$E = \frac{H}{T} = \frac{6.3 \times 10^{-4}\ \text{J} \cdot \text{cm}^{-2}}{0.25\ \text{s}}$$
$$= 2.5 \times 10^{-3}\ \text{W} \cdot \text{cm}^{-2}$$

Example 4: Repetitively Pulsed Near-Infrared Laser with Moderate PRF. Determine the intrabeam direct-viewing MPE for a 905-nm (GaAs) laser which has $t =$ 100 ns (10^{-7}s) and $F = 1$ kHz. Since the 905-nm wavelength will not provide a natural aversion response, as a visible-wavelength laser would, assume a 10-s exposure duration T for this particular laser application. From Fig. 9 we find the reduction factor for $F = 1$ kHz is 0.06 and from Fig. 7 the wavelength correction factor is 2.5 at 905 nm.

Step 1. Individual Pulse Limitation. The MPE for an individual pulse is

$$\text{MPE/Pulse: } H \leq (2.5)\ (0.06)\ (5 \times 10^{-7}\ \text{J} \cdot \text{cm}^{-2})$$
$$= 7.5 \times 10^{-8}\ \text{J} \cdot \text{cm}^{-2}$$

and on this basis the MPE for the entire train would be

$$\text{MPE/Train: } H \leq (10\ \text{s})\ (10^3 \text{Hz})\ (7.5 \times 10^{-8}\ \text{J} \cdot \text{cm}^{-2})$$
$$= 7.5 \times 10^{-4}\ \text{J} \cdot \text{cm}^{-2}$$

Step 2. Average Power Limitation. From 8.2 and 8.5 the MPE is

$$\text{MPE (avg): } E \leq (2.5)\ (10^{-3}\ \text{W} \cdot \text{cm}^{-2})$$
$$= 2.5 \times 10^{-3}\ \text{W} \cdot \text{cm}^{-2}$$

or

$$H \leq 2.5 \times 10^{-2}\ \text{J} \cdot \text{cm}^{-2}$$

Step 3. Conclusion. Clearly the limitation in Step 1 determines the MPE.

$$\text{MPE/Train: } H = 7.5 \times 10^{-4}\ \text{J} \cdot \text{cm}^{-2}$$

or

$$E = \frac{H}{T} = \frac{7.5 \times 10^{-4}\ \text{J} \cdot \text{cm}^{-2}}{10\ \text{s}}$$
$$= 7.5 \times 10^{-5}\ \text{W} \cdot \text{cm}^{-2}$$

Example 5: Low-PRF, Long-Pulse, Repetitively Pulsed Visible Laser. Determine the MPE for a 632.8-nm (He-Ne) laser where $T = 0.25$ s, $t = 10^{-3}$s, and $F = 100$ Hz.

Step 1. Individual Pulse Limitation. Following 8.5,

we must find the total "on time" T_i for the 0.25-s exposure.

$$T_i = t \cdot F \cdot T \qquad \text{(Eq B1)}$$

$$= (10^{-3}\text{s})(100\text{ Hz})(0.25\text{ s}) = 2.5 \times 10^{-2}\text{s}$$

From Fig. 4 the MPE for the exposure time T_i is:

MPE (T_i): $H \leq 10^{-4}$ J · cm^{-2}

MPE/Train: $H \leq 10^{-4}$ J · cm^{-2}

or

MPE/Pulse: $H_i \leq (10^{-4}$ J · cm$^{-2})/(25$ pulses)

$$= 4 \times 10^{-6} \text{ J} \cdot \text{cm}^{-2}$$

As a comparison, the MPE for a single 1-ms pulse not in a train is 10^{-5} J · cm^{-2}. Thus, on a single-pulse basis, the MPE has been reduced by a factor of 2.5.

Step 2. Average Power Limitation. For a 0.25-s exposure the MPE from Fig. 4 is:

MPE: $H \leq 6.3 \times 10^{-4}$ J · cm^{-2}

Step 3. Conclusion. We see that Step 1 defines the MPE in this case.

MPE (T_i): $H = 10^{-4}$ J · cm^{-2}

$$\text{MPE (average power): } E_{avg} = H_i \cdot F \qquad \text{(Eq B2)}$$

$$= (4 \times 10^{-6} \text{ J} \cdot \text{cm}^{-2})(100\text{ Hz})$$

$$= 4 \times 10^{-4} \text{ W} \cdot \text{cm}^{-2}$$

Example 6: One-Pulse Group, Short-Pulse Laser. Find the MPE of a Q-switched ruby laser (694.3 nm) which has an output of three 20-ns pulses, each separated by 100 ns.

Since this is not a repetitively pulsed laser in the usual sense (that is, one having a continuous train of pulses lasting of the order of 0.25 s or more with the pulses being reasonably equally spaced), the following rule should be applied. Since the intrabeam visible MPE is a total radiant exposure of 5×10^{-7} J · cm^{-2} between 10^{-9} and 2×10^{-5} s, regardless of how fast the radiant energy is delivered, this group of pulses should be considered as one pulse. The MPE per pulse is

$$\frac{5 \times 10^{-7}}{3} = 1.7 \times 10^{-7} \text{ J} \cdot \text{cm}^{-2}$$

Example 7: Repetitively Pulsed Pulse Groups. Find the MPE for an Argon laser (488 nm) used in a pulse-code-modulated (pcm) communications link. The laser presents 10^4 "words" per second (that is, 10^4 pulse groups per second) and each word consists of five 20-ns pulses spaced at coded intervals such that each pulse group lasts no longer than 1 μs.

Using the approach of Example 6, each pulse group is considered a single pulse and the MPE for each group is obtained using the prf correction factor of 0.06 for a prf of 10 kHz.

Step 1. Individual Pulse (Group) Limitation. The MPE is

MPE (pulse group): $H \leq (5 \times 10^{-7}$ J · cm$^{-2})(0.06)$

$$= 3 \times 10^{-8} \text{ J} \cdot \text{cm}^{-2}$$

and, from Eq B2,

MPE (average power): $E_{avg} \leq 3 \times 10^{-4}$ W · cm^{-2}

Step 2. Average Power Limitation. For a 0.25-s exposure, the MPE from Section 8 for a nonmodulated laser is

MPE: $E = 2.5 \times 10^{-3}$ W · cm^{-2}

Step 3. Conclusion. Since Step 1 gives the more restrictive case, it is applicable.

MPE (average power) = 3×10^{-4} W · cm^{-2}

B3.2 Determining When to Use the Extended-Source MPEs. The intrabeam MPEs are used in all situations of intrabeam viewing of the direct beam or specularly reflected beam, except for close viewing of laser diodes or diode arrays. The intrabeam MPEs are also used when viewing an extended source at a distance greater than r_{max}.

B3.2.1 Extended-Source MPEs Application. The extended-source MPEs are applied only in the special region of 0.4–1.4 μm where the source size is significantly larger than a "point" and where the corresponding retinal image in the viewer's eye is definitely not a "minimal spot." Diffuse reflections are extended sources at close viewing distances; therefore, depending upon environmental considerations and the laser, the extended-source MPEs may be necessary to consider. Class I and II lasers are not capable of producing hazardous diffuse reflections and only the direct ocular intrabeam-viewing MPEs are applied (except in the case of intrabeam viewing of semiconductor diode lasers and laser arrays). Class IV lasers are always capable of producing hazardous diffuse reflections at close viewing distances. Class III lasers will not produce hazardous diffuse reflections unless focused, and direct-beam MPEs would normally be used (except for laser diode arrays).

B3.2.2 The Brightness Units. The radiometric quantities of "radiance" (W · cm^{-2} · sr^{-1}) and "integrated radiance" (J · cm^{-2} · sr^{-1}) are used for extended sources since these quantities, which describe the source directly, determine the irradiance distribution

Table B1
Typical Diffuse Reflection Cases for Visible Radiation ($\lambda = 0.4–0.7\ \mu m$)*

Duration of Exposure (s)	MPE at Cornea† ($J \cdot sr^{-1} \cdot cm^{-2}$)	Limiting Angle‡, α_{min}	Permissible Laser Beam Radiant Exposure Incident on the Diffusely Reflecting Surfaces, MPE × π/ρ ($J \cdot cm^{-2}$) Reflectance (ρ) = 100%	Reflectance (ρ) = 50%	Reflectance (ρ) = 10%
1	2	3	4	5	6
10^{-9}	1.0×10^{-2}	8.0	3.1×10^{-2}	6.3×10^{-2}	3.1×10^{-1}
10^{-8}	2.2×10^{-2}	5.4	6.8×10^{-2}	1.4×10^{-1}	6.8×10^{-1}
10^{-7}	4.6×10^{-2}	3.7	1.5×10^{-1}	2.9×10^{-1}	1.5
10^{-6}	1.0×10^{-1}	2.5	3.1×10^{-1}	6.3×10^{-1}	3.1
10^{-5}	2.2×10^{-1}	1.7	6.8×10^{-1}	1.4	6.8
10^{-4}	4.6×10^{-1}	2.2	1.5	2.9	15
10^{-3}	1.0	3.6	3.1	6.3	31
10^{-2}	2.2	5.7	6.8	14	68
10^{-1}	4.6	9.2	15	29	150
1	10	15	31	63	310
$10–10^{4}$	22	24	68	140	680

*For wavelengths between 0.7 and 1.4 μm, use the appropriate correction factors given in 8.5.

†See Fig. 6 for a graphic presentation of these values.

‡For extended sources, angular subtense $\geq \alpha_{min}$ (see 8.1).

of the retinal image due to an extended source. The radiance of a diffuse reflection is directly related to the incident beam irradiance and the latter quantity may be more easily applied in practical hazard evaluation. Table B1 presents the incident beam radiant exposures corresponding to integrated radiance MPEs.

B3.2.3 Applying the Limiting Angle (α_{min}). This angle is used to determine if viewing distances (r_1) in a given situation may be sufficiently close to apply extended-source criteria. Fig. B3 shows the relation of r_1, D_L, and α_{min}. From the general case

$$\alpha = \frac{D_L \cdot \cos \Theta_v}{r_1} \qquad \text{(Eq B3)}$$

we get

$$\alpha_{min} = \frac{D_L \cdot \cos \Theta_v}{r_{1max}} \quad \text{for } \Theta_v \lesssim 0.37 \text{ rad } (21°)$$

Example 8: Finding the Maximum Distance Where the Extended-Source MPE Applies. Find the maximum distance r_1 for a visible laser having an emergent beam diameter $a = 1$ cm, a beam divergence $\phi = 10^{-4}$ rad, and a pulse duration of 20 μs. A diffuse matte target is placed 100 cm from the beam exit of the laser, that is, the target distance $r = 100$ cm.

The relation of D_L to the emergent beam divergence and diameter is

$$D_L = a + r \cdot \phi \qquad \text{(Eq B4)}$$

At short target distances, D_L is clearly the same as a, or 1 cm, and using Eq B3 and finding α_{min} from Fig. 3, we have

$$r_{1max} = \frac{a}{\alpha_{min}} = \frac{1 \text{ cm}}{(1.6 \times 10^{-3} \text{ rad})} = 625 \text{ cm}$$

Therefore, for distances less than 625 cm, the applicable MPE from Fig. 6 is $2.8 \times 10^{-1}\ J \cdot cm^{-2} \cdot sr^{-1}$ and for greater distances the applicable MPE from Fig. 4 is $5 \times 10^{-7}\ J \cdot cm^{-2}$.

This example illustrates that for most pulsed lasers, the extended-source MPEs are applicable in such indoor areas as laboratories. Exceptions would be focused-beam diffuse reflections, such as those occurring in microdrilling. For visible cw lasers where the exposure time could be 0.25 s or greater and where a is often only 0.1 to 0.2 cm, α_{min} is sufficiently great (10–24 mrad) that the maximum r_1 to apply these MPEs would be only a few centimeters and the intrabeam MPEs would be more relevant. For $a = 0.1$ cm, and $\alpha_{min} = 24$ mrad, $r_{1max} = 4.2$ cm.

Example 9: Extended-Source MPEs for Diffuse Reflections Expressed as Incident-Beam Irradiance or Radiant Exposure. In most cases it is simplest to consider the incident-beam irradiance or radiant exposure as being capable or not capable of producing a hazardous diffuse reflection rather than have to deal with less familiar radiometric quantities such as radiance.

Consider the laser defined in Example 8. Since the extended-source MPE is expressed as an integrated radiance, we need to know the beam radiant exposure at the target which produces this integrated radiance. The relation is

$$L_p = \frac{H \cdot \rho_\lambda}{\pi} \qquad \text{(Eq B5)}$$

or

$$H = \frac{\pi \cdot L_p}{\rho_\lambda}$$

Hence, the MPE expressed for a 100%-reflectance white diffuse target is

$$H = \frac{(3.14)\,(2.8 \times 10^{-1}\ \text{J} \cdot \text{cm}^{-2} \cdot \text{sr}^{-1})}{1.0}$$

$$= 0.88\ \text{J} \cdot \text{cm}^{-2}$$

NOTE: Eq B5 is strictly true only for a theoretically perfect Lambertian surface; however, unless a surface has a highly glossy sheen, it may be considered sufficiently "diffuse" to apply this formula and the diffuse-surface MPEs. The above result could have been obtained by interpolating the values in Column 4 of Table B1 between 10^{-5} s and 10^{-4} s.

Example 10: Spectral Corrections for Near-Infrared Laser MPEs. A gallium-arsenide laser operating at room temperature has a peak wavelength at 904 nm. What is the MPE for a single pulse of 200-ns duration?

Using Fig. 7, we note that the spectral correction factor is 2.5. Using Fig. 4, the MPE for direct viewing of the source is $(2.5)\,(5 \times 10^{-7}\ \text{J} \cdot \text{cm}^{-2}) = 1.3 \times 10^{-6}\ \text{J} \cdot \text{cm}^{-2}$ for a single pulse. Similarly, the extended-source MPE for sources subtending an angle greater than 3.3 mrad (from Fig. 6 for 200 ns) is $(2.5)\,(5.8 \times 10^{-2}\ \text{J} \cdot \text{cm}^{-2} \cdot \text{sr}^{-1}) = 1.5 \times 10^{-1}\ \text{J} \cdot \text{cm}^{-2} \cdot \text{sr}^{-1}$.

NOTE: Unlike gas or solid-state lasers, this semiconductor laser could be an extended source at close viewing range within the beam if the line source were magnified by a projection lens.

B3.3 Central-Beam Irradiance or Radiant Exposure. Often the beam irradiance or radiant exposure is not provided in a laser's specification. However, if the laser is single-mode and has a Gaussian beam profile the central-beam values may be obtained from the beam diameter specified at 1/e points and the beam radiant power or energy. The relations are

$$E_o = \frac{4\Phi}{\pi a^2} = \frac{1.27\ \Phi}{a^2} \qquad \text{(Eq B6)}$$

or

$$H_o = \frac{4Q}{\pi a^2} = \frac{1.27\ Q}{a^2}$$

Often the beam divergence or diameter is specified at at $1/e^2$ points rather than 1/e points so that the above relations would calculate average values rather than central-beam values. In this case, multiply these specifications by $\sqrt{2}$ to obtain the 1/e values. Do not forget, however, that the MPEs may be specified as averaged over a 1-mm or 7-mm aperture, in which case the calculated values of H_o or E_o may not be relevant.

B3.4 Laser Classification. Examples 11 through 16 show methods of calculating parameters necessary for classifying lasers in accordance with Section 3 of the standard.

Example 11: Classify a Q-switched ruby laser having an output peak power specified by the manufacturer as 20 MW, a pulse duration of 25 ns, and a laser rod diameter of 5/8 inch. The output energy is

$$Q = P \cdot t = (2 \times 10^7\,\text{W})\,(2.5 \times 10^{-8}\,\text{s}) = 0.5\ \text{J}$$

Since the emergent beam diameter is not specified, it is only possible to estimate H_o. The value of a is clearly less than (5/8) (2.54) = 1.6 cm. Hence, H_o can be no less than

$$H_o \geqq \frac{(1.27)\,(0.5\ \text{J})}{(1.6\ \text{cm})^2} = 0.25\ \text{J} \cdot \text{cm}^{-2}$$

Since H_o is well above the permissible radiant exposure incident upon a diffuse surface for $\rho = 100\%$ (which must be a value between 0.063 and 0.16 $\text{J} \cdot \text{cm}^{-2}$ from Column 4, Table B1), the device is classified as a high-power laser (see 3.2).

Example 12: Classify a Rhodamine 6G Dye Laser that has a peak output at a wavelength of 590 nm. The energy output is 10 mJ in a 5-mm-diameter beam for a duration of 1 μs.

Since the emergent beam radiant exposure is 0.051 $\text{J} \cdot \text{cm}^{-2}$, which is less than 0.31 $\text{J} \cdot \text{cm}^{-2}$ (in Column 4, Table B1), a diffuse reflection hazard does not exist; and since 0.051 $\text{J} \cdot \text{cm}^{-2}$ is greater than $5 \times 10^{-7}\ \text{J} \cdot \text{cm}^{-2}$, the laser is a Class III–medium-power laser.

Example 13: Classify a tunable laser which could emit between 0.7 and 2 μm at an output of 10 mJ in a 5-mm-diameter beam for a duration of 1 μs; the device has been altered to lase *only* at a wavelength of 750 nm.

From Fig. 7, the wavelength correction factor is 1.5 and the applicable MPEs used in Example 12 are increased by this factor, which still defines the laser as a Class III—medium-power device.

Example 14: A 1-watt Argon laser operating at 514.5 nm is to be used in a communications link. Determine under what conditions the emergent beam would not be considered a skin hazard. Then determine what hazard classification would apply.

Since the total output power is greater than 0.5 W, the beam would have to be sufficiently large to reduce the irradiance to 0.2 W • cm^{-2} or less (see 8.4).

$$H_o = \frac{4\Phi}{\pi a^2} = 0.2 \text{ W} = \frac{(4)(0.5)}{\pi a^2}$$

and

$$a = \sqrt{\frac{(4)(0.5)}{(\pi)(0.2)}} = 1.8 \text{ cm}$$

Therefore, the beam diameter must be greater than 1.8 cm to preclude a skin hazard problem. The laser could fall into one of three classifications: Class V, if the entire laser beam path were enclosed, as in a sealed optical pipe; Class IV, if more than 0.5 W exited from the laser system as an unenclosed beam; or Class III, if after passing through beam-forming optics the total optical power in the beam were less than 0.5 W (see 3.2).

Example 15: Find the beam diameter to be used in calculations in this standard. A laser beam diameter is specified as being 3 mm in diameter as measured at $1/e^2$-of-peak-irradiance points. The beam is further specified to be single-mode and Gaussian.

Since the beam is Gaussian, we may use the relation that the beam diameter measured at $1/e^2$ points is greater by a factor of $\sqrt{2} = 1.41$; hence

$$a = \frac{0.3 \text{ cm}}{\sqrt{2}} = 0.21 \text{ cm}$$

This exercise however was purely academic, since the calculation of diameters below 7 mm is not necessary to determine laser classification. The value *a* could have been used only for the laser range equation Eq B7 (see B4.1).

Example 16: Classify a 632.8-nm visible laser (He-Ne) used as a remote-control switch. The laser is electronically pulsed with a 1-mW peak-power output, a pulse duration of 0.1 s (hence an energy of 10^{-4} J per pulse), and a beam diameter of 1 cm. The recycle time of the laser is 5 s (maximum prf = 0.2 Hz).

Since the device is pulsed with an exposure duration of 0.1 s, the applicable MPE for intrabeam viewing from Fig. 4 is 3.2×10^{-3} W • cm^{-2} or 3.2×10^{-4} J • cm^{-2}. Using Eq B6, the emergent beam radiant exposure per pulse is 1.27×10^{-4} J • cm^{-2}, which is less than half the MPE, indicating the laser is Class I—exempt if repeated exposure is not considered possible. In the absence of biologic data and MPEs for exposures repeated at prf's less than 1 Hz, we should consider the exposures linearly additive. Following this rule, at least two exposures are possible considering all three aspects in a hazard evaluation (see 3.1); the prudent approach would be to apply a caution label to the device with the words "Do Not Stare Continuously Into Laser Beam."

B4. Formulas and Examples Useful in Evaluation of Various Laser Applications*

B4.1 Beam irradiance, *E*, or radiant exposure, *H*, for a nondiverging beam at range, *r*, which is attenuated by the atmosphere is given by

$$E = E_o e^{-\mu r} \qquad \text{(Eq B7)}$$

or

$$H = H_o e^{-\mu r}$$

NOTE: The attenuation coefficient, μ, varies from 10^{-4} per centimeter in thick fog to 10^{-7} in air of very good visibility. The Rayleigh scattering coefficient at 0.6943 μm is 4.8×10^{-8} cm^{-1}, and 1.8×10^{-7} cm^{-1} at 0.500 μm. The effect of aerosols in even the cleanest atmospheres usually raises μ at 0.6943 μm to at least 10^{-7} cm^{-1}.

B4.2 Average irradiance at range *r* (for a direct circular beam) is the total power in the beam at that range divided by the area of the beam at that range; and likewise, the radiant exposure in a nonturbulent medium is the total energy in the beam at that range divided by the total area.

$$E = \frac{\Phi e^{-\mu r}}{\pi \left[\frac{a + r\phi}{2}\right]^2} = \frac{1.27\,\Phi e^{-\mu r}}{(a + r\phi)^2} \qquad \text{(Eq B8)}$$

$$H = \frac{Q e^{-\mu r}}{\pi \left[\frac{a + r\phi}{2}\right]^2} = \frac{1.27\,Q e^{-\mu r}}{(a + r\phi)^2} \qquad \text{(Eq B9)}$$

*Adopted from *Control of Hazards to Health from Laser Radiation,* U.S. Department of the Army Technical Bulletin TB-MED-279. See Appendix C, References for Control Measures.

NOTE 1: These formulas are accurate only for small ϕ values; that is, accuracy better than 1% for angles below 0.17 rad (10°) and better than 5% for angles less than 0.37 rad (21°).

NOTE 2: For a Gaussian beam the maximum irradiance at the beam axis is calculated using Eq B3, if a and ϕ are defined at the $1/e$ points of maximum irradiance.

Example 17: Find the radiant exposure at 1 km (10^5 cm) of a 0.1-joule ruby laser which has a beam divergence of 1 mrad (10^{-3} rad) and an emergent beam diameter of 0.7 cm. Using $\mu = 10^{-7}\ \text{cm}^{-1}$,

$$H = \frac{1.27(0.1)e^{-(0.01)}}{[(0.7 + (10^5)(10^{-3})]^2} = \frac{(1.27)(0.1)(0.99)}{(0.7 + 100)^2}$$

$$= 1.25 \times 10^{-5}\ \text{J} \cdot \text{cm}^{-2}$$

B4.3 The minimum beam diameter, for a small ϕ, at range r is given by Eq B4:

$$D_L = a + \phi r$$

Example 18: Find the diameter of a laser beam at 1 km where the emergent beam diameter is 10 cm and the beam divergence is 0.1 mrad.

$$D_L = 10\ \text{cm} + (10^{-4}\ \text{rad})(10^5\ \text{cm}) = 10 + 10 = 20\ \text{cm}$$

B4.4 The reflected irradiance or radiant exposure for a diffuse reflector (for $r_1 >> D_L$) is given by

$$E = \frac{\phi \rho_\lambda \cos\theta_v}{\pi r_1^2} \quad \text{or} \quad H = \frac{Q\rho_\lambda \cos\theta_v}{\pi r_1^2} \qquad \text{(Eq B10)}$$

Example 19: Find the maximum reflected radiant exposure from a diffuse matte of reflectance 0.6 which would return a distance of 10 meters to the operator of a 0.1-joule laser ($\cos\theta_v = 1$).

$$H = \frac{(0.1\ \text{J})(0.6)}{(3.14)(10^3\ \text{cm})^2} = 1.91 \times 10^{-8}\ \text{J} \cdot \text{cm}^{-2}$$

Example 20: Find the minimum safe viewing distance for looking at a diffuse target having reflectivity $\rho_\lambda = 0.9$ from a laboratory argon laser with $\Phi = 2$ W and $a = 2$ mm. Since the emergent beam irradiance is more than 6.3 W $\cdot$ cm^{-2} for a 10-second exposure, we conclude that r_1 (safe) must be greater than D_L/α_{min}. Hence (the 10-s MPE is 1×10^{-3} W $\cdot$ cm^{-2})

$$r_1\ (\text{safe}) = \sqrt{\frac{\phi \rho_\lambda}{\pi E}} = \sqrt{\frac{(2)(0.9)}{(\pi)(10^{-3})}} = 24\ \text{cm}$$

B4.5 The ratio, G, of the radiant exposure or irradiance at the retina, when viewing is aided by an optical system, to that received if the eye were unaided is defined by the equations given in B4.5.1 and B4.5.2.

B4.5.1 Intrabeam viewing and specular reflection (or diffuse spot unresolved by eye and optical system):

$$G = \frac{D_o^2}{d_e^2} \quad \text{for } d_e \geq D_e \qquad \text{(Eq B11)}$$

and

$$G = \frac{D_o^2}{D_e^2} = P^2 \text{ for } d_e \leq D_e \qquad \text{(Eq B12)}$$

B4.5.2 Indirect viewing of a diffuse reflection, extended objects only (that is, the object subtends an angle greater than 0.6 mrad when magnified):

$$G = \frac{D_o^2}{P^2 d_e^2} \leq 1 \text{ for } d_e \geq D_e \qquad \text{(Eq B13)}$$

and

$$G = \frac{D_o^2}{P^2 D_e^2} = 1 \text{ for } d_e \leq D_e \qquad \text{(Eq B14)}$$

NOTE: The ratio G is affected by the optical transmission of the instrument but this is normally insignificant. If this is not the case, then this factor should be in the numerations of Eqs B11 through B14.

Example 21: View the diffuse reflection of the laser flash of Example 19 through a pair of 10 X 50 binoculars (that is, $P = 10$ and $D_o = 50$ mm). For night viewing find the relative hazard to the viewer's eyes. Since the exit pupil is $D_o/P = 0.5$ cm, estimating d_e to be 0.7 cm and using Eq B13 gives

$$G = \frac{(5\ \text{cm})^2}{(10)^2 (0.7\ \text{cm})^2} = \frac{25}{(100)(0.49)} = 0.51$$

The hazard is equivalent to a corneal radiant exposure on the naked eye of: $(0.51)(1.91 \times 10^{-8}\ \text{J} \cdot \text{cm}^{-2}) = 9.7 \times 10^{-9}\ \text{J} \cdot \text{cm}^{-2}$.

Example 22: View a specularly reflected beam at a point where the beam radiant exposure measures 2×10^{-9} J $\cdot$ cm^{-2}. If an operator were to view the beam through a pair of 7 X 50 binoculars, what would be the relative hazard compared with unaided viewing? The magnificiation, P, of the binoculars is 7 and, if inserted in Eq B12, will provide the simplest solution:

$$G = P^2 = 7^2 = 49$$

Thus, the operator would be viewing a level 49 times greater than with the naked eye, or a corneal radiant exposure of nearly 10^{-7} J $\cdot$ cm^{-2}.

B4.6 The corneal radiant exposure for a single exposure from a scanning laser beam is given by Eqs B15 and B16. (Repetitive-pulsed exposures depend upon geometrical considerations, scan rate, and frame rate.)

$$H = \frac{1.27\ \Phi e^{-\mu r}}{(a + r\phi)^2} \cdot \frac{d_e}{rS\theta_s} \quad \text{for } d_e > (a + r\phi) \qquad \text{(Eq B15)}$$

or

$$H = \frac{1.27\ \Phi e^{-\mu r}}{(a + r\phi)\ (rS\theta_s)} \quad \text{for } d_e < (a + r\phi) \qquad \text{(Eq B16)}$$

The applicable MPEs depend upon the repetitive nature of the exposure and the exposure duration of a single pulse, where

$$T = \frac{(a + r\phi)}{rS\theta_s} \quad \text{for } d_e \leq \quad (a + r\phi) \qquad \text{(Eq B17)}$$

or

$$T = \frac{d_e}{rS\theta_s} \quad \text{for } d_e \geq (a + r\phi) \qquad \text{(Eq B18)}$$

and the prf is S if each scan passes over the eye.

Example 23: Find the exposure of a scanning He-Ne laser system having the following parameters: $a = 0.1$ cm, $\phi = 5 \times 10^{-3}$ rad, $\Phi = 5$ mW, $\theta_s = 0.1$ rad, and $S = 30\ s^{-1}$ for an intrabeam-viewing distance $r = 200$ cm.

Step 1. The beam diameter $D_L = (a + r\phi) = (0.1 + 1) = 1.1$ cm; hence we apply Eqs B16 and B17.

Step 2. The prf at the eye is 30 pulses per second and the exposure time is

$$T = \frac{0.1 + (200)\ (5 \times 10^{-3}) - 0.7}{(200)\ (30)\ (0.1)} = \frac{1.1}{600}$$

$$= 1.8 \times 10^{-3} \text{ second}$$

Step 3. The radiant exposure is

$$H = \frac{(1.27)\ (5 \times 10^{-3})\ (1)}{(0.1 + 1)\ (200)\ (30)\ (0.1)} = 9.6 \times 10^{-6}\ \text{J} \cdot \text{cm}^{-2}$$

Step 4. The average irradiance at the cornea is

$$E_{avg} = H \cdot S = (9.6 \times 10^{-6})\ (30) = 2.9 \times 10^{-4}\ \text{W} \cdot \text{cm}^{-}$$

Step 5. The applicable MPE for a 0.25-s exposure is determined by the cumulative exposure of almost eight pulses, or (8) $(9.6 \times 10^{-6}\ \text{J} \cdot \text{cm}^{-2}) = 7.7 \times 10^{-5}\ \text{J} \cdot \text{cm}^{-2}$. This radiant exposure must be compared with the MPE for a single pulse of duration $\tau = (8)\ (1.8 \times 10^{-3}\ \text{s}) = 1.5 \times 10^{-2}$ s, which from Fig. 4 or by calculation (Example 1) is $\sim 7.7 \times 10^{-5}\ \text{J} \cdot \text{cm}^{-2}$. Hence, the exposure is permissible for momentary (unintentional) viewing.

Appendix C

References for Control Measures

Control of Hazards to Health from Laser Radiation, Department of the Army Technical Bulletin TB-MED-279, Department of the Navy Publication NAVMED P-5052-35, 1969. Available from Clearing House for Federal Scientific and Technical Information, Springfield, Virginia.

Electronic Product Radiation and the Health Physicist, Proceedings of the 4th Annual Midyear Topical Symposium, U.S. Department of Health, Education and Welfare, Bureau of Radiological Health, Division of Electronic Products, 70-26. Washington, D.C.: Government Printing Office, 1970.

GOLDMAN, L., and ROCKWELL, R. J., Jr. *Lasers in Medicine.* New York: Gordon and Breach Science Publishers, 1971, chapter 11.

GOLDMAN, L.; ROCKWELL, R. J., Jr; and HORNBY, P. Laser laboratory design and personnel protection from high energy lasers. In *Handbook of Laboratory Safety,* ed. Steere, N. V. Cleveland: The Chemical Rubber Company, 1971, 2nd ed., pp 381-389.

Laser Health Hazards Control, U.S. Department of the Air Force Manual AFM-161-8(C1), 1969. Available from Clearing House for Federal Scientific and Technical Information, Springfield, Virginia.

LITWIN, M.; FINE, S.; KLEIN, E.; FINE, B.; and

KRAEMER, H. Hazards of laser radiation: mechanisms, control and management. *Journal of the American Industrial Hygiene Association,* vol 28, Jan/Feb 1967, pp 68-75.

SLINEY, D. H. Evaluating health hazards from military lasers. *Journal of the American Medical Association,* vol 214, no. 6, Nov 1970, pp 1047-1054.

SLINEY, D. H., and PALMISANO, W. A. The evaluation of laser hazards. *Journal of the American Industrial Hygiene Association,* vol 29, Sep/Oct 1968, pp 425-431.

SWOPE, C.H. The eye protection. *Archives of Environmental Health,* vol 18, Mar 1969, pp 428-433.

Appendix D
A Guide for Organization of a Laser Safety Program

Introduction

The extent to which the various parts of the following guide are applicable to a specific organization is dependent on the magnitude of the laser hazards within that organization.

D1. Laser Safety Committee

D1.1 Necessity. If warranted by the magnitude of the potential hazards of laser operations within the organization, a Laser Safety Committee is to be formed.

D1.2 Membership of Laser Safety Committee. The membership should include persons skilled in laser technology and those with expertise in laser hazards assessment, for instance, technical management personnel, Laser Safety Officer, medical officer, etc.

D1.3 Policies and Practices. The committee will establish and maintain adequate policies and regulations for the control of laser hazards.

D2. Responsibility and Authority of Laser Safety Officer

D2.1 General. The Laser Safety Officer, who may be designated from among such personnel as the radiation protection officer, the industrial hygiene officer, the safety officer, etc, is to have authority for supervision of the control of laser hazards. Such authority is to include, but is not necessarily limited to, the fields of operation specified in D2.2 through D2.12 of this Appendix.

D2.2 Consultative Services. He is to provide consultative services on laser hazard evaluation and control.

D2.3 Regulations. If a committee does not exist, the Laser Safety Officer will establish and maintain adequate regulations for the control of laser hazards.

D2.4 Authority. The Laser Safety Officer has the authority to suspend, restrict, or terminate the operation of a laser system if he deems that laser hazard controls are inadequate.

D2.5 Maintain Records. He will maintain the necessary records required by applicable governmental regulations.

D2.6 Protective Equipment. If protective equipment is necessary, only that protective equipment approved by the Laser Safety Officer is to be used for control of laser hazards.

D2.7 Personnel Records. He will submit to the appropriate medical officer the persons' names that are obtained in accordance with D4.3, D4.4, and D5.3 of this Appendix. He will maintain appropriate records that scheduled medical examinations have been performed.

D2.8 Surveys. He will survey all areas using laser equipment as frequently as he considers necessary.

D2.9 Review of Planned Installation or Modification. When submitted, he will review the planned installation or modification of laser equipment relative to laser hazards and their control, approving of the installation or modification only if he is satisifed that laser hazard controls are adequate.

D2.10 Accidents. On notification of a real or suspected accident resulting from laser operation, he is to investigate said accident and initiate appropriate action.

D2.11 Approval of Laser System Operation. He will approve a laser system for operation only if he is satisfied that laser hazard controls are adequate.

D2.12 Warning Systems and Signs. He is to make certain that adequate warning systems and signs are installed in the appropriate locations.

D3. Deputy Laser Safety Officer

If necessary, a deputy laser safety officer is to be appointed by management. The deputy laser safety officer will perform the functions of the Laser Safety Officer when the latter is not available.

D4. Responsibility of the Laser or Laser System Supervisor

D4.1 Indoctrination. The supervisor is to provide for issuance of appropriate instructions on laser hazards and their control to all personnel who may work with lasers that are operated under his supervision.

D4.2 Laser Hazard Control. He will not permit the operation of a laser unless there is adequate control of laser hazards to employees, visitors, and the general public.

D4.3 Persons Scheduled to Work with Lasers. Except for Class I lasers (see 3.3), he is to submit the names of persons scheduled to work with lasers to the Laser Safety Officer. In addition, he will submit information requested by the Laser Safety Officer for medical surveillance scheduling.

D4.4 Reporting of Real or Suspected Accidents. When a supervisor knows of or suspects an accident resulting from a laser operated under his supervision, he will notify immediately the Laser Safety Officer.

D4.5 Medical Attention. If necessary, he will assist in obtaining appropriate medical attention for the employee involved in a laser accident.

D4.6 Approval of Laser System Operation. He will not permit the initial operation of a laser system which is under his supervision without prior approval by the Laser Safety Officer.

If any modification of a laser system is made which may result in additional laser hazards, the system is not to be energized without prior approval by the Laser Safety Officer.

D4.7 Approval of Planned Installation or Modification. When appropriate, plans for a laser installation or modification thereof are to be submitted to the Laser Safety Officer for approval.

D5. Responsibility of Employees Working with or near Lasers

D5.1 Authorization. An employee is not to energize or work with or near a laser unless he has been authorized to do so by the supervisor of the laser.

D5.2 Compliance. An employee must comply with the safety rules and regulations prescribed by the supervisor of the laser.

D5.3 Accident Reporting. When an employee knows or suspects that an accident has occurred involving a laser operated by him or other employees responsible to his supervisor, he is to immediately inform the supervisor of the laser, or if the supervisor is not available, the employee is to notify the Laser Safety Officer.

Appendix E

Medical Surveillance

E1. Eye Examinations

E1.1 Recommended Examination Requirement

E1.1.1 Minimal-Risk Personnel. This group is to be subject to the eye examination protocol (see E1.3).

E1.1.2 Moderate-Risk Personnel. This group is to be subject to a modified protocol consisting of the ocular history, visual acuity, and ocular fundus examination requirements of the eye examination protocol (with mydriatic instead of cycloplegic). If best corrected visual acuity in either eye is less than 20/20, or if any pathology is seen, the potential worker is advised to have a complete eye examination done. If any eye pathology (not based on refractive error) is found, the protocol for high-risk personnel is to be employed.

E1.1.3 High-Risk Personnel. This group is to be subject to the eye examination protocol (see E1.3).

E1.2 The Frequency of Medical Examination

E1.2.1 Minimal-Risk Personnel. These personnel are to undergo examination previous to participation in laser work.

E1.2.2 Moderate-Risk Personnel. These personnel are to undergo examination at the following times:

(1) Previous to participation in laser work
(2) On discharge from a laser environment
(3) Immediately after suspected laser eye exposure

E1.2.3 High-Risk Personnel. These personnel are to undergo examination at the following times:

(1) Previous to participation in laser work
(2) After completion of laser work
(3) Immediately after suspected laser eye injury

E1.3 The Eye Examination Protocol. This protocol is to include the departments specified in E1.3.1 through E1.3.10.

E1.3.1 Ocular History. The patient's past eye history and family eye history are reviewed. Any current complaints which he now has about his eyes are noted. The patient's general health status should be inquired about with a special emphasis upon diseases which can give ocular problems. The patient's present lens prescription, if any, is recorded.

E1.3.2 Visual Acuity. Tested and recorded in Snellen figures for 20 feet with and without lenses to 20/15. The visual acuity at near is tested at 35 cm and recorded in Jaeger-test figures with and without lenses, if any.

E1.3.3 External Ocular Examination. This includes examination of brows, lids, lashes, conjuctiva, sclera, cornea, iris and pupillary size, and equality reactivity and regularity.

E1.3.4 Amsler Grid. The Amsler grid sheet is presented to each eye separately and any distortion of the grid is noted by the patient and drawn by him.

E1.3.5 Manifest Refraction. This is to measure the patient's refractive error and the new visual acuity of the patient must be noted if the visual acuity is improved over that achieved with the patient's old lens prescription, or if he has no lenses at the time of the examination. This examination shall be carried out in all personnel whose visual acuity in either eye is less than 20/20.

E1.3.6 Measurement of Intraocular Pressure. The pupils are then dilated by the instillation of a mydriatic drop in each eye. The remainder of the examination is carried out with the eye under this medication.

E1.3.7 Examination by Slit Lamp. The cornea, iris, and lens are examined with this biomicroscope and described.

E1.3.8 Examination of the Ocular Fundus with an Ophthalmoscope. In the recording of this portion of the examination the points to be covered are: the presence or absence of opacities in the media; the sharpness of outline of the optic nerve; the color of the optic nerve; the size of the physiological cup, if present; the ratio of the size of the retinal veins to that of the retinal arteries; the presence or absence of a well-defined macula and the presence or absence of a foveolar reflex; and any retinal pathology that can be seen with a direct ophthalmoscope. Even small deviations from normal should be described and carefully localized.

E1.3.9 Photograph of the Posterior Pole of the Fundus. This includes the area of the macula and head of the optic nerve and is to be taken in color film. A positive color photograph should be mounted in the patient's chart.

E1.3.10 Other Examinations. Further examinations are to be done as deemed necessary by the examiner.

E2. Medical Surveillance for Possible Skin Damage from Exposure to Lasers

E2.1 General. The skin surveillance is to include a history, examination, and recommendations; it applies only to personnel in the high-risk category.

E2.2 History. The history is to cover the following data:

(1) Atopy

(2) History of exposure to x-ray, nuclear radiation, ultraviolet sources, high-intensity visible or infrared sources, cutting oils, arsenic, and antibiotics

(3) Presence or absence of reactions to drugs and chemicals

(4) Photodermatitis

(5) Intensity of exposure to sun

(6) Skin cancers, including type, location, and treatment

(7) History of previous skin disorders

(8) History of extensive or severe thermal burns

E2.3 Examination. The skin examination is to cover the areas specified in E2.3.1 through E2.3.3.

E2.3.1 Skin color is to be noted as to:

(1) Caucasian – light, medium, or dark

(2) Negro – light, medium, or dark

(3) Oriental – light, medium, or dark

If special instrumentation is available for precise analysis of skin color, this should be used and the data recorded.

E2.3.2 The presence or absence of the following changes are to be observed and noted under normal light and Wood's light on exposed and unexposed skin (body charts may be used):

(1) General skin type: normal (N), seborrheic (S), or ichthyotic (I)

(2) Pigmentation: type, severity, distribution

(3) Depigmentation: type, severity, distribution

(4) Keratoses: actinic, seborrheic; type, distribution patterns

(5) Skin malignancies: type, size, distribution patterns

(6) Skin disorders: type, distribution patterns

E2.3.3 Photographs are to be made if indicated, including black and white, color, and under Wood's light.

E3. Recommendations

E3.1 Specific recommendations are to be made as regards skin protection, both with respect to area control and personnel control, covering the use of gloves, special clothing, protective creams, templates, etc. Specific recommendations such as the following may also be made:

(1) No laser exposure

(2) Limited laser exposure

(3) Laser exposure with ordinary precautions

E3.2 At the discretion of the Laser Safety Officer, an annual examination may be required.

E3.3 It may be recommended that all high-risk personnel be required to have a final examination upon leaving laser work.

Appendix F

Special Considerations

Introduction

The operation of lasers and laser systems, like any industrial or technological process, involves various possible related hazards. The following special considerations apply to such associated hazards of laser or laser systems operation.

F1. Compressed Gases

The following standards, regulations, and guides should be observed while using compressed gases:

(1) American National Standard Method for Marking Portable Compressed Gas Containers to Identify the Material Contained, Z48.1-1971 (R-1954).

(2) American National Standard Compressed Gas Cylinder Valve Outlet and Inlet Connections, B57.1-1965.

(3) *Handbook of Compressed Gases,* New York: Compressed Gas Association, 1966.

(4) Pertinent regulations of the U.S. Department of Transportation.

F2. Cryogenic Liquids

Cryogenic liquids should be stored and handled in accordance with the instructions listed in the *Handbook of Compressed Gases* (see F1(3)).

F3. Ventilation

Local exhaust or general ventilating systems must be provided to maintain the levels of toxic fumes, gases, or dusts in occupied areas below the maximum allowable concentration as set forth in Code of Federal Regulations, Title 29, 1910.93, Subpart G. Oxygen levels should be maintained to at least 18% by volume (equivalent to pO_2 of 135 mmHg).

F4. Ionizing Radiation

The source, quality, and intensity of x-rays emanating from laser power supplies and components therein should be investigated and controlled in accordance with the provisions listed in American National Standard Safety Standard for Non-Medical X-ray and Sealed Gamma-Ray Sources, Z54.1-1963 (NBS Handbook H93), and any applicable federal, state, or local codes and regulations.

F5. Toxic Materials

Exposure of personnel to toxic materials used in laser targets or laser system elements is to be eliminated or reduced to safe levels by adequate ventilation, personal protective equipment, and by enclosure, shielding, or other appropriate engineering control methods. Exposure criteria are to be based upon federal safety and health standards.

F6. References

F6.1 Associated Hazards

Electrical Safety Guides for Research, Safety and Fire Protection, U.S. Atomic Energy Commission Technical Bulletin 13 (0-292-830). Washington, D.C.: Government Printing Office, 1968

GAFAFER, W. M., ed. *Occupational Diseases – A Guide to Their Recognition,* U.S. Public Health Service Publication 1097 (0-218-695). Washington, D.C.: Government Printing Office, 1966.

Hygienic Guide Series. Detroit: American Industrial Hygiene Association.

Industrial Ventilation – A Manual of Recommended Practice. Cincinnati: American Conference of Governmental Industrial Hygienists.

KOLLER, L. R. *Ultraviolet Radiation.* New York: John Wiley & Sons, 1965, 2nd ed.

PATTY, F. A., ed. *Industrial Hygiene and Toxicology,* vol 1. New York: Interscience Publishers, 1958, 2nd ed.

POWELL, C. H., and HOSEY, A. D., eds. *The Industrial Environment – Its Evaluation and Control,* U.S. Public Health Service Publication (0-730-352). Washington, D.C.: Government Printing Office, 1965.

SLINEY, D. H., and FREASIER, B. C. The evaluation of optical radiation hazards. *Applied Optics,* vol 12, no. 1, 1973, pp 1-22.

STEERE, N. V., ed. *Handbook of Laboratory Safety.* Cleveland: The Chemical Rubber Company, 1971, 2nd ed.

ZABETAKIS, M. G. *Safety with Cryogenic Fluids.* New York: Plenum Press, 1967.

F6.2 Optical Propagation

BUCK, A. L. Effects of atmosphere on laser beam propagation. *Applied Optics*, vol 6, no. 4, 1967, pp 703-708.

BURCH, D. et al. Absorption of infrared radiation by CO_2 and H_2O – absorption by CO_2 between 8000 and 10,000 cm^{-1}. *Journal of the Optical Society of America,* vol 58, 1968, p 335.

DIETZ, P. H. Twelve eye safety homographs. *Applied Optics*, vol 8, no. 11, 1969.

HOGG, D. C. Millimeter wave communication through the atmosphere. *Science*, vol 159, 1968.

MIDDLETON, W. E. K. *Vision through the Atmosphere*. Toronto: University of Toronto Press, 1952.

SLINEY, D. H. Evaluating health hazards from military lasers. *Journal of the American Medical Association*, vol 214, no. 6, 1970, pp 1047-1054.

TATARSKI, V. I. *Wave Propagation in a Turbulent Medium*. New York: McGraw-Hill Company, 1960.

Appendix G

Biological Effects on the Eye and Skin

G1. Minimal Biological Effects of Laser Radiation on the Eye

G1.1 General. Among the different laboratories some differences in standardization and calibration probably exist. This had introduced a certain spread among the data. The majority of the work in arriving at the MPEs in Section 8 has been concerned with how to compare the data and which data to weigh the most heavily. Where regression lines were available, they indicated that a factor of 10 below the 50% damage level gave a negligible probability of damage. Whenever possible these regression lines formed the basis for the level selected for any particular MPE. Where the data indicated a steeper regression line, a factor less than 10 was used.

G1.2 Corneal Damage

G1.2.1 Infrared (Spectral Region 1.4 to 1000 μm). Excessive infrared exposure causes a loss of transparency or produces a surface irregularity in the cornea. The Maximum Permissible Exposure (MPE) is well below the energy or power required to produce a minimal lesion. For the purpose of this standard, a minimal corneal lesion is a small white area involving only the epithelium whose surface is not elevated or swollen. It appears within ten minutes after the exposure. Very little or no staining results from fluorescein application. A minimal lesion will heal within 48 hours without visible scarring. These observations are based on experiments with CO_2 lasers; extrapolation to wavelengths other than 10.6 μm must be made with care.

Damage results from the heating of the cornea by absorption of the incident energy by tears and tissue water in the cornea. The absorption is diffuse, and simple heat flow models appear to be valid. The identity of the sensitive material or protein in the cornea is not known. Although the critical temperature threshold is not known, it does not appear to be much above normal body temperature, and there are indications that it is a function of exposure time.

G1.2.2 Ultraviolet (Spectral Region 0.2 to 0.4 μm). Excessive ultraviolet exposure produces photophobia accompanied by redness, tearing, conjunctival discharge, surface exfoliation, and stromal haze. The Maximum Permissible Exposure (MPE) is well below the energy required to produce (within 24 hours following exposure) any of the following changes: redness, tearing, conjunctival discharge, surface exfoliation, stromal haze, and photophobia.

Damage to the epithelium by absorption of ultraviolet light probably results from photochemical denaturation of proteins or other molecules in the cells. Some of the most important molecules are the deoxyribonucleic acids (DNA) and ribonucleic acids (RNA). The absorption is probably by selective sensitive portions of single cells. Thus the action of the ultraviolet radiation is photochemical rather than thermal, since the temperature rise calculated for experimental exposure is negligible.

G1.3 Retinal Damage (Spectral Region 0.4 to 1.4 μm). The Maximum Permissible Exposure (MPE) is well below the exposure required to produce a minimal (or threshold) lesion. For the purpose of this standard a minimal retinal lesion is the smallest ophthalmoscopically visible change in the retina. This change is a small white patch (apparently coagulation) which occurs within 24 hours of the time of exposure. At threshold the lesion is probably the result of local heating of the retina subsequent to absorption of the light and its conversion to heat by the melanin granules in the pigment epithelium.

Extended exposure lasting several minutes for a retinal image that is very small is difficult to accomplish, except by stabilized image optics. Thus there exist no experimental data for long exposures and small spot sizes. However, accidental retinal exposures which combine long periods of time and small spot sizes are very unlikely.

G1.4 Other Ocular Damage. There are two transition zones between corneal hazard and retinal hazard spectral regions. These are located at wavelength bands separating ultraviolet and visible regions, and separating near-infrared and infrared. The exact wavelengths are not known. In these regions there may be both corneal and retinal damage. Also, damage to intermediate structures, such as the lens and iris, can occur.

G1.5 References. The most important references are cited in this section. They cover the major portion of the data used in deriving the standard. Several of the references are review articles; their bibliographies should be used as a source of additional references.

ADAMS, D. O.; BEATRICE, E. S.; and BEDELL, R. B. Retina: ultrastructural alterations produced by extremely low levels of coherent radiation. *Science*, vol 177, 1972, pp 58-60.

BRESNICK, G. H.; FRISCH, G. D.; POWELL, J. O.; LANDERS, M. B.; HOLST, G. E.; and DALLAS, A. G. Ocular effects of argon laser radiation. I. Retinal damage threshold studies. *Investigative Ophthalmology*, vol 9, 1970, pp 901-910.

CLARK, A. M. Ocular hazards from lasers and other optical sources. *Critical Reviews in Environmental Control*, vol 1, no. 3, 1970, pp 307-339.

COGAN, D. G. Lesions of the eye from radiant energy. *Journal of the American Medical Association*, vol 142, 1950, pp 145-151.

DESVINGNES, P.; AMAR, L.; BRUMA, M.; and VELGE, M. Sur la generation d'ondes ultrasonores et formation de bulles dans le vitre d'un oeil humain par irradiation d'impulsion laser. *Comptes Rendus Hebdomadaires des Sceances de l'Academie des Sciences*, vol 259, 1964, pp 1588-1591.

DUNSKY, I., and LAPPIN, P. Evaluation of retinal threshold for cw radiation. *Vision Research,* vol 11, 1971, pp 733-738.

HAM, W. T., Jr; GEERAETS, W. J.; MUELLER, H. A.; WILLIAMS, R. C.; CLARKE, A. M.; and CLEARY, S. F. Retinal burn thresholds for the helium-neon laser in the rhesus monkey. *Archives of Ophthalmology*, vol 84, 1970, pp 797-808.

HAM, W. T., Jr; WILLIAMS, R. C.; MUELLER, H. A.; GUERRY, D.; CLARKE, A. M.; and GEERAETS, W. J. Effects of laser radiation on the mammalian eye. *Transactions of the New York Academy of Sciences*, ser 2, vol 28, 1966, pp 517-526.

HANSEN, W. P., and FINE, S. Melanin granule models for pulsed laser induced retinal injury. *Applied Optics*, vol 7, 1968, pp 155-159.

HAYES, J. R., and WOLBARSHT, M. L. Thermal model for retinal damage induced by pulsed lasers. *Aerospace Medicine*, vol 39, no. 5, 1968, pp 474-480.

KUWABARA, T. Retinal recovery from exposure to light. *American Journal of Ophthalmology, vol 70, 1970, pp 187-198.*

Laser Health Hazards Control, U.S. Air Force Manual AF 161-8, 1969; and Change No. 1, 1971. Available from Clearing House for Federal Scientific and Technical Information, Springfield, Virginia.

PEABODY, R. R.; ZWENG, H. C.; ROSE, H. W.; PEPPERS, N. A.; and VASSILIADIS, A. Threshold damage from CO_2 lasers. *Archives of Ophthalmology,* vol 82, 1969, pp 105-107.

PEPPERS, N. A.; VASSILIADIS, A.; DEDRICK, K. G.; CHANG, H.; PEABODY, R. R.; ROSE, H.; and ZWENG, H. C. Corneal damage thresholds for CO_2 laser radiation. *Applied Optics*, vol 8, 1969, p 377.

PITTS, D. G.; BRUCE, W. R.; and TREDICI, T. J. *A Comparative Study of the Effects of Ultraviolet Radiation on the Eye*, U.S. Air Force School of Aerospace Medicine Report SAM-TR-70-28, Aerospace Medical Division (AFSC), Brooks Air Force Base, 1970.

SLINEY, D. H. The development of laser safety criteria. In *Laser Applications in Medicine and Biology*, vol 1, ed. WOLBARSHT, M. L. New York: Plenum Press, 1971, pp 163-238.

VASSILIADIS, A. Ocular damage from laser radiation. In *Laser Applications in Medicine and Biology*, vol 1, ed. WOLBARSHT, M. L. New York: Plenum Press, 1971, pp 125-162.

VOS, J. J. A theory of retinal burns. *Bulletin of Mathematical Biophysics*, vol 24, 1962, pp 115-128.

WHITE, T. J.; MAINSTER, M. A.; TIPS, J. A.; and WILSON, P. W. Chorioretinal thermal behavior. *Bulletin of Mathematical Biophysics*, vol 32, 1970, pp 315-322.

G2. Biological Effects of Laser Radiation on the Skin

G2.1 General. The large skin surface makes this body tissue readily available to accidental and repeated exposures to laser radiation. The biological significance of irradiation of the skin by lasers operating in the visible and infrared regions is considerably less than exposure of the eye, as skin damage is usually reparable or reversible. Effects may vary from a mild reddening (erythema) to blisters and charring. Depigmentation, ulceration, and scarring of the skin, and damage to underlying organs may occur from extremely high-powered laser radiation.

Latent and cumulative effects of laser radiation are not known at this time. The possibility of such effects occurring, however, should not be ignored in planning for personnel safety in laser installations.

Little or no data is available describing the reaction of skin exposed to laser radiation in the 0.2–0.4-micrometer spectral region, but chronic exposure to ultraviolet wavelengths in this range can have a carcinogenic action on skin as well as eliciting an erythematous response.

On the basis of studies with noncoherent ultraviolet radiation, exposure to wavelengths in the 0.25–0.32-μm spectral region is most injurious to skin. Exposure to the shorter (0.20–0.25 μm) and longer (0.32–0.40 μm) ultraviolet wavelengths is considered less harmful to normal human skin. The shorter wavelengths are absorbed in the outer dead layer of the epidermis (stratum corneum), and exposure to the longer wavelengths has a pigment-darkening effect. However, the sensitivity of skin to the longer wavelengths may be increased by known or inadvertent usage of photosensitizers.

G2.2 References

American Conference of Governmental Industrial Hygienists. *Threshold Limit Values of Airborne Contaminants and Physical Agents,* with changes adopted by ACGIH for 1971. Cincinnati: American Conference of Governmental Industrial Hygienists, pp 62-70.

BUCHANAN, A. R.; HEIM, H. G.; and STILSON, D. W. Biomedical effects of exposure to electromagnetic radiation. Part I. Ultraviolet. *Wadd Technical Report,* 60-376, DDC-AD No. 244-786, May 1960.

GOLDMAN, L., and ROCKWELL, R. J., Jr. *Lasers in Medicine.* New York: Gordon and Breach Science Publishers, 1971.

LAOR, Y.; SIMPSON, C.; KLEIN, E.; and FINE, S. Pathology of internal viscera following laser irradiation. *American Journal of the Medical Sciences,* vo. 257, no. 4, 1969.

PARR, W. H. *Skin Lesion Threshold Values for Laser Radiation as Compared with Safety Standards,* U.S. Army Medical Research Laboratory Report 813 (AD 688 871), 1969.

URBACH, F., ed. *The Biological Effects of Ultraviolet Radiation (with Emphasis on the Skin),* Proceedings of the First International Conference, 1966. New York: Pergamon Press, 1969.

Appendix H

Measurements

H1. Radiometric Measurements

It is important to note that certain types of radiometrically calibrated instruments can exhibit substantial errors when used to measure coherent radiation. These errors are often associated with interference phenomena (constructive and destructive) or with spatial inhomogeneities, and instrument readings can change greatly with small changes in angle-of-incidence, wavelength, or distribution of radiation over the surface. It is, therefore, important to select instruments that minimize such effects or are designed specifically for use with lasers. As an example, current National Bureau of Standards laser power and energy standards are based on calorimetric, rather than radiometric techniques (see West et al in Section H4 of this Appendix).

Because of these problems, there is currently some question about the use of radiometric instruments and terminology when applied to coherent radiation – questions that hopefully will be resolved in the next few years.

H2. Radiant Exposure Profiles and Radiant Exposure Estimates for Pulsed Lasers

Several types of photosensitive or thermally sensitive papers or emulsions may be used as aides in the determination of laser beam radiant exposure and for estimating beam profiles of high-energy lasers.

H3. Beam Divergence

The divergence of the laser beam may require determination when evaluating exposures at far distances from the laser. It is recommended that beam divergence be determined by using either maximum beam irradiance or radiant exposure at two different distances from the laser, by calculating with Eq B4 in Appendix B3.2.3. (See also 3.4.2.)

H4. References

Applied Optics, vol 8, no. 12, 1969.

BIRNBAUM, G., and BIRNBAUM, M. Measurement of laser energy and power. *Proceedings of the Institute of Electrical and Electronics Engineers,* vol 55, no. 6, 1967, pp 1026-1031.

GEIST, J. *Fundamental Principles of Absolute Radiometry and the Philosophy of this NBS Program (1968 to 1971),* National Bureau of Standards Technical Note 594-1. Washington, D.C.: Government Printing Office, 1972.

HEARD, H. G. *Laser Parameter Measurement Handbook.* New York: J. Wiley & Sons, Inc, 1968.

HONEY, R. C. Laser instrumentation and dosimetry. *Electronic Product Radiation and the Health Physicist,* Bureau of Radiological Health, Division of Electronic Products, 70-26, 1970, pp 263-272. Available from Clearing House for Federal Scientific and Technical Information, Springfield, Virginia.

JENNINGS, D. A. et al. *Laser Power and Energy Measurements,* National Bureau of Standards Technical Note 382. Washington, D.C.: Government Printing Office, 1969.

KILLICK, D. E.; BATEMAN, D. A.; BROWN, D. R.; MOSS, T. S.; and de la PERRELLE, E. T. Power and energy measurement techniques for solid state lasers. *Infrared Physics,* vol 6, 1966, pp 85-109.

LANDSBERG, H. E., and VAN MIEGHEM, J. *Advances in Geophysics,* vol 14. New York: Academic Press, 1970.

NICODEMUS. F. E. Optical resource letter on radiometry. *Journal of the Optical Society of America,* vol 59, no. 3, 1969, pp 243-248.

Optical Design, Military Standardization Handbook MIL-HDBK-141. Washington, D.C.: Defense Supply Agency, 1962.

ROSS, M. *Laser Receivers.* New York: J. Wiley & Sons, Inc, 1966.

SLINEY, D. H. Evaluating hazards – and controlling them. *Laser Focus,* vol 5, no. 15, 1969, pp 39-42.

SLINEY, D. H.; BASON, F. C.; and FREASIER, B. C. Instrumentation and measurement of ultraviolet, visible and infrared radiation. *Journal of the American Industrial Hygiene Association,* vol 32, no. 7, 1971, pp 415-431.

SMITH, R. A. Detectors for ultraviolet, visible and infrared radiation. *Applied Optics,* vol 4, no. 6, 1965, pp 633-638.

WEST, E. E.; CASE, W. E.; RASMUSSEN, A. L.; and SCHMIDT, L. B. *NBS Journal of Research,* 76A, 13, 1972.

ZIMMERER, R. W. Power and energy measurement. *Laser Focus,* vol 6, no. 9, 1970, p. 39.

ZIMMERER, R. W. Update on power measurement. *Laser Focus,* vol 8, no. 3, 1972, p 25.

In addition, periodically updated listings of commercially available laser measuring instruments are included in the following publications:

Electro-Optical Master Catalog. Available from Milton S. Kiver Publications, Inc, 222 West Adams Street, Chicago, IL 60606.

Instrumentation for Environmental Monitoring – Radiation. Available from Environmental Instrument Group, Lawrence Berkeley Laboratory, University of California, Berkeley, CA 94720.

Laser Focus Buyers' Guide. Available from Advanced Technology Publications, Inc, 246 Walnut Street, Newtonville, MA 02160.

The Optical Industry and Systems Directory. Available from Optical Publishing Company, 7 North Street, Pittsfield, MA 01201.

AUTHOR INDEX

E

F

G

H

M

N

O

P

R

S

T

U

V

W

Y

Z

SUBJECT INDEX

I

J

K

L

M

N

O

P

Q

R

S

T